# GEORGE & ROBERT
# STEPHENSON

*For Alexander David Miguel Bueno Ross*

A man shall be the Temple of Fame. He shall walk, as the poets have described that goddess, in a robe painted all over with wonderful events and experiences; – his own form and features by their exalted intelligence shall be that variegated vest. I shall find in him the Foreworld; in his childhood the Age of Gold; the Apples of Knowledge; the Argonautic Expedition; the calling of Abraham; the building of the Temple; the Advent of Christ; Dark Ages; the Revival of Letters; the Reformation; the discovery of new lands; the opening of new sciences, and new regions in man. He shall be the priest of Pan, and bring with him into humble cottages the blessing of the morning stars and all the recorded benefits of heaven and earth.

Ralph Waldo Emerson, *History*

Great industrial inventions are never the fruit of a sudden and complete conception; they never emerge in a perfected state, or from the genius of a single inventor.

Marc Seguin, *Des Chemins de Fer*

Tyneside's the place where i' glory shine
The stars o' the canny toon;
Industry and genius byeth combine
To presarve wor greet renoon,
George Stephenson here first showed them
Te improve upon the past,
An' the Tyneside Collier gain'd the day,
Wiv his wundrus wark at last.

From Joe Wilson, 'George Stephenson', in *Tyneside Songs and Droleries*

The Stephensons made the railways, and the railways made the Stephensons.

Anonymous saying

# GEORGE & ROBERT
# STEPHENSON

## *A Passion for Success*

### DAVID ROSS

First published 2010

The History Press
The Mill, Brimscombe Port
Stroud, Gloucestershire, GL5 2QG
www.thehistorypress.co.uk

British Library Cataloguing in Publication Data.
A catalogue record for this book is available from the British Library.

ISBN 978 0 7524 5277 7

Typesetting and origination by The History Press
Printed in India by Nutech Print Services

# CONTENTS

# ACKNOWLEDGEMENTS

Research for this book was done in many different places, and I am grateful to all for their help: Birmingham Reference Library (Boulton & Watt Archive); Bodleian Library and Radcliffe Science Library, Oxford; Bristol University Library, Special Collections; Chesterfield Library, Local Studies Centre; Durham Records Office; Gateshead Reference Library; Institution of Civil Engineers Library and Archives; Institution of Mechanical Engineers Library and Archives; John Rylands Library, University of Manchester; Lilly Library, University of Indiana; Literary & Philosophical Institution, Newcastle; London Library; Montrose Public Library; The National Archives, Kew; National Railway Museum, York; Northumberland Records Office; Robinson Library, University of Newcastle; Royal Society Library; Royal Yacht Squadron Archives, Cowes; St Andrews University Library, Special Collections; Science Museum Library, London and Swindon; South Shields Local Studies Centre; Stephenson Locomotive Society Library; Sunderland Public Library; Tyne & Wear Archives Service, Newcastle; Whitby Library; York Reference Library.

I am indebted to Professor Maggie Snowling of the University of York and Professor P.G. Aaron of Indiana State University, for assessments of George Stephenson's orthography. Thanks are also due to A.G. Miller, MICE. My account of Robert Stephenson's involvement with the Isthmus of Suez was published in the *Journal of the Railway & Canal Historical Society*, November 2007, and I am grateful to the Editor for permission to reuse it here.

For help in sourcing illustrations I am grateful to Carol Morgan of the Institution of Civil Engineers (ICE), Liz Bentley of the Royal Meteorological Society, Mike Claxton of the IMechE, the Chesterfield and Darlington Local Studies Centres, the Northumberland Archives and the Victoria & Albert Museum.

# INTRODUCTION

George Stephenson was not a modest, self-effacing sort of man, and if he had been, many things might have happened differently. In him, will-power, tenacity, technical ability and self-confidence were fused into intense and strongly focused ambitiousness. To these qualities can be added courage, bloody-mindedness and gross insensitivity to the feelings of other people. Perhaps if he had found it easy to pursue his natural gifts, he would have achieved less than he did. A formidable personality was made even more so by having to overcome the obstacles set before an illiterate working boy with aspirations. When George was born, railways were horse-drawn wagon-ways. Twenty-two years later, at Robert's birth, that was still the case. By the time of George's death, the steam railway had transformed industry and commerce, reshaped and expanded the economies of several countries, hugely influenced everyday life and set in motion an ever-expanding series of further developments. He was not the only begetter, but at a crucial time, he was its main driving force. Success made him a phenomenon, a walking, talking legend whose great subject, in the end, was himself. For his son, having George Stephenson as a father was to be handed a ticket to life printed with very clear directions. It was no free ride and he was offered no optional routes. Before Robert reached his teens, father and son were working together at mechanical contrivances. Perhaps inevitably, the boy became an engineer. In that role he faced and framed for himself some breathless challenges, but none greater than the private one, of defining his own character and approach to life against the hugely compelling yet unrepeatable example of his father.

To write a coherent life of either Stephenson without full reference to the other is really impossible, and so this is a double-headed biography. Yet their personalities were very different, as were the ways in which they went about things. Often there is a sense of the son looking at his father's methods and doing the opposite. And, having jerked a slow-turning planet on to a new, faster-spinning and unpredictable course, each reacted in his own way to the changed order of things which, with no particular intention, they had done so much to bring about.

Considering their impact on the world, there have been few biographies of the Stephensons. First in the field was Samuel Smiles, who persuaded Robert Stephenson in 1849 that he was the right person to produce a biography of George. Robert told him that:

> ... there had been some talk of writing the life of his father, but nothing had been done. Indeed, he had given up hope of seeing it undertaken. Besides, he doubted whether the subject possessed much interest; and he did not think the theme likely to attract the attention of literary men of eminence.
>
> 'If people get a railroad,' he said, 'it is all that they want; they do not care how or by whom it is done.'[1]

Having warned Smiles against wasting his time, labour and money, Stephenson agreed to help and took a close interest in the project.[2] Smiles was able to gather first-hand information from many people who had known George, even as a young man. His book, published in 1857, became a bestseller and established its author's name and fame. An important source, it is not wholly reliable, glossing over some aspects of George's behaviour and character, and playing down the role of other people in his achievements. Later it became clear that Smiles had shaped his portrait of George as the prime exemplar of his personal idea of 'self-improvement', spelled out in his *Self Help: Or Illustrations of Conduct and Perseverance*, written before the Stephenson biography but only finding a publisher after it, when Smiles had become a big name. Revised versions of the biography were published in 1862 as *Lives of the Engineers*, Vol. 3, and separately in 1864 and again in 1873. Also in 1864, John Cordy Jeaffreson published a two-volume *Life of Robert Stephenson*, with chapters on Stephenson's engineering works contributed by Professor William Pole. Jeaffreson, a professional author, was commissioned to write it in 1860 by George Parker Bidder, a close friend of Robert's and the most active executor of his will.[3] He was not Bidder's first choice, and makes no claim to have known Robert Stephenson personally, but he too visited Northumberland to gather oral memories and to look at parish and colliery records. He refers to 'letters submitted to my perusal by a great number of the engineer's friends' and to papers provided by his executors. In the practice of the time, Smiles and Jeaffreson provide no references for their quotations and statements. Thomas Summerside's short anecdotal memoir of George appeared in 1878, and Francis Grundy's *Pictures of the Past* (1879) has a chapter, 'Memories of Tapton House', based on recollections from John Hart, who was George's secretary from 1840 to 1843. No other attempt at a researched biography was made for almost a century, though some very useful and important books and journal articles dealt with aspects of the Stephensons' work, most notably J.G.H. Warren's centenary history of the Robert Stephenson locomotive works in 1923. In 1960 L.T.C. Rolt published *George and Robert Stephenson: The Railway Revolution*, which, though a long overdue corrective to Smiles, is seriously marred by factual errors, confusions and hasty-seeming judgements. Hunter Davies' *George Stephenson: Father of Railways*, published in 1975 at the time of the 150th anniversary of the Stockton & Darlington Railway, includes some new material. In 2003 Michael Bailey, author of numerous scholarly articles on Robert Stephenson in particular, edited *Robert Stephenson: The Eminent Engineer*, in which he and other specialists examine most aspects of Robert's work. This is a valuable and generally authoritative book, to whose scope and detail any writer attempting a more general account of Robert must be grateful. Not a systematic biography, its focus is very much on his engineering works. *Robert Stephenson: Railway Engineer* (2005), by John Addyman and Victoria Haworth, is another well-researched book with an engineering emphasis. The present book is the first to set out the interlocked careers of George and Robert in chronological order, and to consider their activities within what Emerson called 'the idea of the age'. Only in this way can the pattern of their lives and their involvement with the intellectual, political, commercial and social currents of the time, be clearly seen, and an assessment made of what sort of men they were and of the scope – and limitations – of their achievement. The evidence is here for the reader to judge, but, like the glimmer of a miner's lamp advancing through the dark, one thing becomes steadily more apparent. By comparison with most of their peers and rivals, to the many qualities possessed by the Stephensons must be added an extra one. Luck plays some part in it, but far more important is the urge to strive and survive, to compete and come out on top – a passion for success.

Note: Quotations from original documents occasionally have odd spellings. Rather than interrupt their flow with a [sic] I have left them as they are and made elucidatory comments only.

# PART ONE: GEORGE

## 1

## A WORKING BOY

In the summer of 1854 Robert Stephenson made a nostalgic visit to Tyneside, accompanied by Samuel Smiles who was writing the biography of George Stephenson. At the village of Wylam they met a few ancient men who remembered George as a youth, and recalled his parents. One 'old Wylam collier' is quoted as saying of 'Old Bob' Stephenson that: 'Geordie's fayther war like a peer o' deals nailed thegither, an' a bit o' flesh i' th' inside; he war queer as Dick's hatband – went thrice aboot, an' wudn't tie. His wife Mabel war a delicat boddie, an' varry flighty. They war an honest family, but sair hadden doon i' th' world.'[1]

Wylam had changed less than many places, but seventy-three years on from George Stephenson's birth, and six from his death, it was already hard to recreate the nature and spirit of the community in which he grew up during the 1780s and '90s. Viewed from the mid-1850s, the later years of George III's reign seemed in many ways as remote as the times of King Henry. Familiar landmarks were shrunk into a new context of encroaching streets and sky-darkening chimneys. Even the recent past was diminished in perspective by the huge rapidity with which population, commerce and industry had expanded, and by the impact of scientific and technological advance.

But every generation is modern in its time, and Wylam in the last years of the eighteenth century was not an old-fashioned place. Coal-mining had gone on in this region for centuries, but during George's boyhood many new pits were being sunk to cater for the growing demands of local industry as well as the London market. Shafts were beginning to go deep, with workings leading off at different levels, making a pattern of intersections and connections which every underground worker had to imprint in his brain. Above those pitch-dark labyrinths, steam engines of the sort built by Boulton & Watt of Soho, Birmingham, great sighing monsters, operating at low pressure, powered the winding gear and pumps. When one section was worked out, a new pit was sunk and as much machinery as possible transferred to it, along with the workforce.

Wylam from the south

New buildings to house machines and people were put up, using the local rubble-stone. Pit-villages were quite small, their population numbered in hundreds. To an outsider, the inhabitants might seem all of a kind, but each community was as structured as a tribe, with individuals' status relating to tasks, skills and wages. At the bottom in every sense, the men and boys[2] who hacked at the coal face with picks had the most exhausting and dangerous work, suffering a high toll of injuries and deaths. Broken or crushed heads and limbs from roof collapses and rock-falls, or death by asphyxiation or burning, were a daily hazard. Gas seeped constantly into the Northumberland mines, both lighter-than-air methane and dense low-lying 'choke-damp'. Surface workers had a relatively easier life. Most were coal-shovellers or wagon-pushers, but some had been trained to operate machines or to record output and payments. All worked hard for small wages, twelve-hour shifts for six days a week. In the 1790s and early 1800s most of the miners were Northumbrians and had family connections in the farms and villages round about, but a sense of kinship between the new pit communities and the old agricultural settlements was thinning out. The rhythms of life were too different. Existence in farm and pit cottages was at an equally basic level in terms of space, furnishing and sanitation, but the colliers lived in a cash-based society. Ready money, plus hard labour and a shift-based work pattern that kept men and youths in groups, encouraged pastimes such as fist-fights, dog-fights and wrestling, made even more exciting by cash bets on the results, and all fuelled by ale. It was a rough-and-ready lifestyle, viewed with apprehension by outsiders at the time, and with horror by later writers. Here is the respected French commentator on the condition of the English people around 1815, Elie Halévy:

> The miners lived like utter savages absolutely cut off not merely from the middle class, but also from the other sections of the labouring classes ... it required constant effort to overcome the obstinate carelessness of the miners. Savages are always careless, and the miners lived, as we said above, like absolute savages both in the dirty and ruined villages in which they spent the night and in the subterranean galleries where of necessity there was less supervision than in the workshops of a factory.[3]

Distaste for, even loathing of, a sub-human Morlock species oozes through every phrase. Brutish conditions encourage the production of brutish people, and in the mining villages drunkenness and violence were commonplace. Most of the inhabitants could not read or write. Violent sport, gambling, drinking and casual sexual liaisons were elements of their social life. But that is to shine the light only on the grubbier side of the seam, while decent living in almost intolerable conditions, a strong sense of community, pride in skills and a long-established self-reliance are all ignored. Illiteracy encouraged self-entertainment through an oral tradition of folk songs, ballad-making, verbal wit and good-natured (mostly) mockery of themselves and their 'betters'. A Londoner visiting the Northumbrian coalfield wrote in 1816: 'I find them a very industrious quiet people, perfectly subservient in every respect to their viewers or other masters; and amongst them are many who, though illiterate characters, are possessed of much scientific practical knowledge and wonderful natural ability.'[4] The writer had met George Stephenson, but evidently did not consider him unduly remarkable. These communities also harboured men like William Locke, who from 1795 was a leading banksman at the Water Row colliery pit-head, his job being to note the amount of coal as it came up by the corf-ful and apportion it among the miners, who were paid accordingly. He carried in his pocket a copy of Alexander Pope's *Essay on Man*, rich in such phrases as, 'Order is Heaven's first law', and the by-then almost proverbial, 'An honest man's the noblest work of God'. One of his grandchildren would be Alfred Austin, Tennyson's successor as poet laureate, who wrote of him: '... he was that extraordinary thing, a Roman Catholic Puritan. Of scrupulous honour and severe integrity, he applied to others the standard of speech and conduct he imposed on himself.'[5] William Locke, clearly not a savage, was not local to Tyneside, but he was a working man, not a 'gentleman apprentice'.

Nor were the Stephensons savages, although the family was illiterate. Old Bob maintained a young family on 12s a week. But their life seems to have been happy. Bob's duties, if ill-paid, were not excessively onerous, even for such a slightly-built man. He enjoyed the company of children and had a reputation as teller

of tales like 'Sinbad the Sailor' and 'Robinson Crusoe' (neither is a traditional
English folk-tale, though both had by then found their way into an English oral
tradition), and also invented yarns of his own. He had a strong feeling for the
natural world, and an affinity for animals and birds. In winter, robins came to his
engine-house for crumbs, and at home he liked to tame blackbirds. In summer
he took his young sons on country rambles looking for birds' nests, and on nut-
ting expeditions in autumn. Mabel Stephenson's father, George Carr, worked as
a bleacher and dyer at Ovingham, a few miles up the river. His wife was Eleanor
Wilson, a farmer's daughter, who had been strong-minded enough to defy her
father in marrying a man considered to be her social inferior. Her name would be
passed on to the eldest daughter of Bob and Mabel. Until 1789 the family lived in
the cottage known as High Street House, which still exists, now as a Stephenson
memorial site. It looks quite substantial with its upper storey, but four families
inhabited it, with the Stephensons in a single room on the ground floor. Between
1779 and 1792 Mabel bore six children, all of whom survived into adult life. The
eldest, James, had his father's easy-going and unambitious nature. George was
next, born on 9 June 1781, followed by Eleanor in 1784, Robert in 1788, John
in 1789 and Ann in 1792. Nothing suggests that Bob and Mabel had any aspirations for their children other than to be decent and respectable working folk
like themselves. School was out of the question – it could not be afforded, even
for the two older boys. Perhaps, however, the father's feeling for nature and gift
for story-telling brought a kind of masculine gentleness into home life that was
not typical of all families. The unavoidable proximity of old and young, male and
female, could turn a single-room dwelling into a crucible of violence or a pit of
physical and moral squalor. To make it clean, trim and homelike, to feed everyone,
if only on oatmeal 'crowdie', and keep them clothed and healthy, was a constant
struggle. It was a rough life, but later, when George Stephenson would be happy
to relate how his boyhood was hard and impoverished, he never referred to it
with distaste or loathing.

A region of about five square miles, with the River Tyne flowing between
Ovingham and Newburn as its southern boundary, formed the early world of
young George: a rural countryside, of farms and woodland, sloping gradually up
from the river towards the blue rim of the moor lands. Only three miles away were
Hadrian's Wall and the remains of the Roman fort at Vindolanda. An enquiring
child might have wondered what buildings these grass-grown foundations had
once supported, but this boy responded to a more modern music. A few yards
from the front door, a set of wooden tracks ran parallel to the river and, on these
rails, horses pulled trains of chaldrons – coal wagons – downstream to where the
coal would be transferred on to flat-bottomed keels (the Tyne becomes tidal just
below Wylam); they would then be floated down to the estuary to be reloaded in
brigs for shipment as the 'sea-coal' that kept the myriad fires of London burning.
Far beneath the ground lay the coal seams, and a local expertise had developed in

assessing the best places to sink pits. Among the woods and green fields, and the occasional parklands set around country houses, new structures rose: the winding gear, engine-houses and chimneys of collieries.

In 1789 the pit at Wylam closed down, and the Stephensons left High Street House for another one-roomed home, away from the river at Dewley Burn colliery, where Bob was taken on as a fireman. With a playmate, Bill Thirlwall, George paddled and puddled in the little clayey streams, made sluices for water-mills and model engines out of clay. They had a model winding-machine set up on a bench outside Bill's home, with twine for a rope and hollowed-out corks for the corves or baskets in which coal was brought to the surface.[6] Once a boy was 7 or so, it was desirable to find him a job, for a penny or two a day, to help the family budget and get him out of the way, but also to find him something to occupy his time. George began to earn money, at the rate of 2*d* a day, to look after the cows of Grace Ainsley who lived at Dewley Farm. The cows grazed around the unfenced colliery wagon-way, and the boy had to keep the track clear for coal wagons and make sure the animals stayed on their own territory. As he grew older and stronger, George moved on to leading plough-horses and hoeing turnips, doubling his wages to 4*d* a day, but his older brother Jemmy was already employed at the pit, and George was keen to follow him. Barely in his teens, he was taken on as 'a "corf-batter" (knocking the dirt from the corves employed in drawing the coals)' and pick-carrier, taking miners' picks to the smithy for sharpening.[7] Shortly afterwards his farm experience stood him in good stead when he was employed to drive the gin-horse on its unending trudge round the windlass, at a wage of 8*d* a day. He moved or was transferred to Black Callerton pit, about 2 miles away, to drive the gin-horse there. From this time he was recalled by one of Smiles' informants as 'a grit growing lad with bare legs and feet … very quick-witted and full of fun and tricks'.[8] George would have worn breeches without stockings and no shoes, at least outside winter time. Despite a 2-mile walk at each end of his shift, he had time to catch and tame blackbirds, as his father did, and apparently to let them fly about inside the cottage. His mother's thoughts on this are not recorded.

George was about 14 when he was promoted from the gin-horse to be an assistant fireman to his father. Soon after this, Dewley Burn pit was worked out and the family had to move again, first to a short-lived pit, 'the Duke's Winning', owned by the Duke of Northumberland. Still in a one-roomed home, even more uncomfortably full with the three older children by now adolescent, and three beds taking up most of the space, one for the parents, one for the boys and one for the girls, they lived at Jolly's Close behind the village of Newburn. At 15, George was taken on as a fireman at the 'Mid Mill Winning' pit. He was growing tall and rangy, developing a sinewy strength. When this pit closed two years later, he went to fire a pumping engine at Throckley Bridge close by, where his wages were raised to 12*s* a week: 'I am now a made man for life,' he announced.[9]

His father was transferred from the failed pit to a new one at Water Row, close to the river about half a mile upstream from Newburn. The engineer in charge was Robert Hawthorn, himself once an engineman, now superintendent of all machinery at the Duke's pits, and the first to spot George's talents. While Old Bob was, as ever, the fireman, young George was given the job of plug-man, or engineman – a post of greater responsibility and better paid. At 17 he was earning more than his father. The plug-man had to ensure that the pumping engine was working properly and drawing efficiently. Plugging came into it when the shaft was temporarily dried out and he had to go down and plug the suction tube so that the pump should not draw. How George had attracted Hawthorn's attention is not known, but it can be reasonably supposed that he had a more than usual interest in the workings of the machinery and showed it in his remarks or questions when Hawthorn came round to make his inspections.

Seen in his own perspective, this period, just at the turn of the eighteenth century, is the crucially defining one of George Stephenson's life. Until now he had been just another collier boy, noticeably brighter than most, but not seen as significantly different to his own brothers or workmates. He had a responsible job and a wage that few working-class youths could match. The rest of his life could have been spent as an engineman. Something, internal or external, or both, prompted Stephenson not to stop there. Perhaps Hawthorn encouraged him, but far more important was the inner drive, or daemon, that turned interest into ambition and wish into action.

As he grew into young adulthood, this drive took full possession of George, making him examine himself and recognise his gravest deficiency, his inability to read, write or calculate. More than in any of his later great tests of will and determination, he had to confront this alone and decide what to do about it. It was not just the effort involved or the time to be put in after a long shift. A convention was being breached – unwritten but powerful – that said in effect: 'You are a pit-head workman and the son of a pit-head workman. Reading and writing are not necessary for the likes of you. You can do your job and live your life without that.' Among his own kindred, it might be taken as a sign of repudiation of his family and their way of life. In any case, it was putting his head above the parapet: the ambition to be more than an engine-minder could not be concealed, since the understood reason for acquiring these skills was to 'get on'. Other able men saw opportunities for themselves and did the same thing; Stephenson was exceptional not in tackling his illiteracy, but in what he did with the chances he made for himself. If he had an urge to be one of the bosses, riding around giving instructions, as Robert Hawthorn (apparently rather imperiously) did, it was not the prime one: the nostril-tickling whiff of steam was his intoxicant. Above all, he wanted to understand machines; to know how and why they worked, what natural laws they obeyed, how their dimensions could be calculated and their workload defined. Already he had an intuitive grasp of how a steam engine functioned, as though his brain was ready-wired for the purpose. Now he was hungry to know more, partly because other things interested him too, ideas which he had picked up because they were in the air, being discussed, however vaguely, by technically-minded people everywhere. One of these was the perpetual motion machine. Theoretical knowledge of physics and dynamics had not yet reached the point at which 'perpetual motion', which requires a machine to give out more energy than it consumes, would have to be ruled out. George was intrigued by the idea. Perhaps he could devise a perpetual motion engine.

With such thoughts in mind, he subjected himself to the necessary discipline, attending 'night schools' – the only way of learning for working boys and youths. Bringing one or two friends, he first attended the class of Robert Cowens in Walbottle, a colliery hamlet behind Newburn, where he studied three times a week for a penny a session; then transferred in the winter of 1799 to a more convenient one run by a 'Scotch dominie', Andrew Robertson, in Newburn, where arithmetic was also on the curriculum (Scotland, with a much more extensive system of cheap education and four universities, exported many teachers to England). L.T.C. Rolt asserts astonishingly of Stephenson that 'all reliable evidence indicates that his brain totally lacked the capacity to store theoretical knowledge, even of the simplest kind'.[10] There is certainly plenty of evidence that George mistrusted theoretical knowledge, sometimes with good reason, as most sciences were still in their infancy, and also because some of the most vocal theoreticians among his later critics had scant grasp of practicalities. But even if he could not express it in formal terms, much of his own empirical approach would have ended up as floundering in the dark if he had lacked a basic grasp of theory. George's mathematical education did not extend to trigonometry, and his delegation of such tasks as taking surveys and measurements would lead to problems. Thomas Summerside, a boyhood friend of Robert's who knew George well in later years, wrote that George did not get much beyond the rule of five in his arithmetical education,[11] but Summerside was somewhat carried away by the idea of the elder Stephenson as a natural, untaught genius. Robert Gray, who studied with him for a while, was impressed by how George 'took to figures so wonderful'.[12]

What may have misled Rolt, and others, is the fact that George never mastered conventional spelling and grammar. Undoubtedly this was an embarrassment to him and perhaps heightened his sense of being set apart from 'educated' men. Later, he would write only to family members and close friends, dictating his business correspondence to secretaries. Scrutiny by experts in 2009 suggests strongly that he was dyslexic[13] – a condition still not fully understood, and wholly unknown in his day, when its indications would be put down to poor teaching or low intelligence. Dyslexia is also identified as causing difficulties in reading, and George, on such evidence as there is, was no great reader and preferred to be read to. This undetectable hindrance to learning must have laid additional strain on his capacities for perseverance and for compensatory adaptation. In financial and business matters he always showed a very clear head. He was a man of immense intelligence, with a brain both quick to absorb knowledge and capable of using it.

If the picture of an earnestly drudging, late-teen monomaniac threatens to emerge, the open-air George blows it away. Still interested in birds, breeding rabbits in cages behind the house, owning a dog which he trained to carry his dinner from house to pit in a tin can slung round its neck, he can seem a naturalist manqué. He did not have time, in any sense, for the colliers' favourite entertainments of cock-fighting and dog-fighting, nor for the boozing of pay-night, when the workmen's wages were handed over, conveniently, in a pub. But he was said to have been an ace hammer thrower and weightlifter. On one occasion he lifted a weight of 60 stones[14] (381kg; the current world record is 472.5kg). He was a wrestler and always relished his prowess in this sport. There can be no doubt that he was respected for physical and mental qualities, and that his presence was an intense and compelling one. In a Tyneside mining community though, to be a striking personality could also invite a bit of taking down, and as a tall poppy, the young George Stephenson was a target for those who reacted negatively to his positive drive.

One of the most responsible surface jobs was that of the brakesman, in charge of the winding gear or 'whim' that hoisted and lowered not only the corves, but also the underground workers. Jerky operation could result in serious injuries or death of those being transported – just how easily can be understood since: 'The method of raising the men and boys, being that only two out of four or five were slung in the noose at the end of the rope, the other two or three clinging to the rope by the mere grasp of their hands and knees.'[15] At Water Row, one of the brakesmen let George try his hand on the winding engine, but this venture by the plug-man was not well received by some other workers. With William Locke as their spokesman, they downed tools when Stephenson was working the brake. In Locke's view, 'young Stephenson couldn't brake, and what was more, never would learn, because he was so clumsy'.[16] But Charles Nixon, the pit manager, took a different view and allowed Stephenson to continue his practising. This led to another minor crisis, and a fist fight, when one Ned Nelson accused him of clumsy operation of the whim and threatened to give him a kicking. Nelson was reputed to be a rough fighter, but when the bare-knuckle bout took place, Stephenson thumped him into submission.[17] Another fight followed George's protest against the ill-treatment of a dog; as he recalled later: 'They said I was beat … but I never knew it myself, and just as I was feeling faintish like, Billy Bull gave in … But I wasn't just myself either for a week or two.'[18] Despite the phase of antagonism, William Locke and George became good friends. Locke left the Tyne in 1802 for a job as a colliery viewer in Yorkshire, but the two did not lose touch, and William's son, Joseph, would later be one of George's 'young men' before making his own career as a great railway engineer. After a couple of years at Water Row, George was given the job of brakesman at the Dolly Pit, located at Black Callerton, out in the open country between Newcastle and Ponteland. It was beyond daily walking distance from the family home outside Newburn, so he took lodgings at the house of Thomas Thompson, a local farmer.

Working mostly on night shifts, he found time to do other things as well as operate the whim. Soon he was trying to devise an improved brake. By now he had mastered reading and writing, but was still doing arithmetic with Andrew Robertson, who had moved his own lodging to Black Callerton. George also took up cobbling and became a proficient shoe-mender and even shoemaker, supplementing his income by repairing the shoes and boots of his workmates. Whether it was for the extra income or from a need to keep himself occupied, it was typical of him that he should turn the urge to use his hands to good financial account. He was as canny and thrifty a character as the banks of Tyne ever produced. And, now aged 20, he was thinking about marriage. Thompson employed two sisters as domestic servants, Hannah and Frances Henderson. George first of all fixed his eye on their younger sister Ann, also in domestic service locally. But, although he made her a pair of shoes, Ann did not favour his advances. The Henderson girls' father was a farmer in a small way, and Ann may have been influenced by the farming community's feelings of resentment towards the intensified coal-mining, which was transforming the appearance of the landscape and hugely enriching the landowners, but doing nothing for the tenant-farmer. George does not seem to have been deeply affected by the rebuff. The shoes, and his affections, were presented instead to Frances, who lived in the same house as he did, and whose respectability, domestic abilities, personal virtues and other charms he was in a good position to assess.

Fanny Henderson had worked for Thompson since 1791, when he had acquired her services from the previous tenant of the farm, along with a testimonial which stated her to be 'a girl of a sober disposition, an honest servant, and of a good family'. Already 12 years old when George was born, she had been engaged to the village schoolmaster, but he had died in 1794. Marriage between a man of 21 and a woman of 32 was unusual in 1802. George Stephenson, handsome, high-spirited, hard-working, evidently a man of promise, might have seemed a desirable *parti* to many a damsel between Wylam and Black Callerton, and his courting of a woman already considered an 'old maid' has seemed strange. Perhaps he was too full of ideas and restless energy to appeal to a conventionally-minded girl of his own age, or to find such a girl appealing. Jeaffreson, the biographer of his son Robert, says of Robert's mother that 'it was not for her to object to the disparity of their ages, since he was willing to marry a woman so much older than himself',[19] and also suggests that Ann was vexed by George's swift change of tack, though she acted as bridesmaid to her sister. William Fairbairn, who got to know George and Frances in the early days of their marriage, described her as 'a very comely woman'; and Smiles' informants seem to have been unanimous in praising her sweetness, kindness and good sense. George was marrying a wife of his own social class, and there was no question of Mr Henderson providing his eldest daughter with a dowry, although Fanny had accumulated her own savings. They were married at Newburn Church on 28 November 1802, and Mr Thompson laid on a wedding breakfast for his former servant. The signature on his marriage certificate is the first example we have of George's handwriting; he added Fanny's maiden name beneath it.

The parish church was the only place where a couple could be legally married. Among the colliers, Methodism and other forms of non-conformity were more prevalent than conventional Anglicanism, and indifference more frequent still. Always more interested in the mechanical than the metaphysical, George Stephenson never gave religion much thought. The newly-weds lodged in a cottage at Black Callerton, but only for a short time. Robert Hawthorn reappeared to offer George a new job, as brakesman on the ballast lift being erected at Willington Quay, on the north bank of the Tyne just east of the village of Wallsend. Seagoing ships came as far as here and unloaded their ballast of stone and rubble in order to load up with coal. Alongside the river a 'ballast mountain' was growing, and a powerful winding machine was needed to hoist newly dumped material to the top – the converse of the colliery whim in action. George took the job. With his wife riding behind him on a borrowed horse, they made the journey through Newcastle and Wallsend to Willington Quay, where they had arranged to rent an upper room in a cottage close to the engine. They bought new furniture, using some of her savings. By May 1803, if not before, they would have known that Fanny was pregnant.

In his time at Willington, industrious as ever, George continued to work on the side as a cobbler, and also taught himself clock-cleaning and repairing. According to Smiles, this came about as a result of his own clock being stopped by dust and soot when the cottage chimney caught fire.[20] Families accustomed to shift work liked to have a clock, despite the attentions of the 'knocker-up' who summoned each night-time shift, and a clock-mender could earn useful extra money. Pendulums and cogwheels were also potential parts of a perpetual motion machine, though the model with which George was busy at this time was a large wooden wheel, to whose perimeter glass tubes partially filled with mercury were fixed. As the wheel went round, the downwards motion of the mercury was intended to impart enough momentum to maintain the turning movement.[21] George also made friends with a mechanically-minded boy of 15, William Fairbairn, born in Kelso, Scotland, and an engineering apprentice at Percy Main Colliery, 2 miles away, where his father was steward of a farm belonging to the colliery owners. Later Fairbairn, as the country's leading iron founder, would be closely involved with Robert Stephenson's first tubular bridges; Samuel Smiles makes a little word-picture of him and George at this time, shovelling ballast of an evening: 'It is pleasant to think of the future President of the British Association thus helping the future Railway Engineer to earn a few extra shillings by over-work in the evenings, at a time when both occupied the rank of humble working men.'[22] In fact, Fairbairn minded George's engine while George earned extra cash by shovelling.[23] They discussed mechanics and, as Fairbairn would teasingly remind George many years later, engaged in wrestling bouts.

On 16 November 1803[24] Fanny Stephenson gave birth to a boy child. The baby was small and delicate, and, at a time when infant mortality was high, his chances of survival were poorly rated by some. Undaunted, the proud father arranged a family party for his son's christening, which took place in the Wallsend village schoolhouse, as the parish church at the time was in danger of collapse into the mine-workings underneath. The boy was named Robert, and his sponsors were his aunt Ann and Robert Gray, George's fellow night-schooler, who had also been best man at the wedding. Gray did not rise to eminence, and when Smiles met him as an old man, he was living on a pension provided by George Stephenson. Bob and Mabel Stephenson came across for the party, as did George's brothers and sisters: the youngest, Ann, being only eleven years older than her baby nephew. Robert Stephenson was toasted by his relatives in Newcastle ale and Scotch whisky, but the guests' unspoken opinion was that here was 'a wee sickly bairn not made for long on this earth'.[25]

Soon after the birth of her son, it became apparent that Fanny was not at all well. A terrible, unrelenting cough and other symptoms revealed that she was suffering from consumption, as tuberculosis was then called. It was incurable and sufferers could only hope for arbitrary periods of remission. The move to Willington Quay had seemed a step-up in life for George, but at the end of 1804 the little family moved to the township of Killingworth in Longbenton parish, north of Newcastle. George had taken the post of brakesman at West Moor colliery there. Smiles says that it was not without considerable persuasion that George was induced to leave the Quay, because of the loss of earnings, but does not shed any light on what form it took or who made it. West Moor, one of many pits owned by a consortium of wealthy aristocrats known as 'the Grand Allies', was very deep, 720ft (219m) and needed a skilled brakesman. Various reasons for the move are possible. The inland moor, though scarcely 50ft above sea level, may have appeared a healthier place for the mother and child than the misty riverside. Fanny was again pregnant. And George Stephenson's relationship with Robert Hawthorn may have become uncomfortable or unfriendly. During the Willington period, 'his intercourse with Robert Hawthorn first took the form of personal intimacy',[26] but Hawthorn did not last as a friend. Stephenson's growing conviction that he understood engines better than anyone else may have caused strains. Another not implausible reason for the move is that the focus of inventive development had settled on the collieries. Although the event had made no impression on the nation as a whole, the enginemen of the north-east knew that the Cornish engineer, Richard Trevithick, had successfully built a steam-powered travelling engine, which had pulled wagons along cast-iron tram rails at Penydarren in South Wales, in February 1804. For men who lived and worked among fixed steam engines and horse-hauled colliery tramways, this was portentous news.

The most dynamic reaction came from George's old home and workplace at Wylam, where the local land- and coal-owner, Christopher Blackett, followed up Trevithick's success by placing an order for a locomotive to work on the Wylam tramway.[27] Built in Gateshead in the first months of 1805, to drawings supplied by the inventor, and supervised by a Tynesider, John Steel, who had worked on the Penydarren locomotive, this machine, the world's second steam locomotive, suffered from the same problem as its precursor: the tracks broke under its weight. Though it made demonstration runs in the yard of Whinfield's foundry in Gateshead, it was not taken to Wylam and was used only as a stationary engine. Blackett let his interest lapse for several years, during which the practicability of the 'travelling engine' on rails remained a talking point only.

At Killingworth, the Stephensons had a two-roomed house to themselves, in a terrace known as Paradise Row. Not far away, among its trees, was Gosforth Hall, the home of Charles Brandling, a member of a wealthy land- and coal-owning family, who was keenly interested in technical matters. They had a little front garden and the place was undoubtedly airy, but Fanny's health, after a brief improvement, deteriorated again. She gave birth to a second child, a girl, in July 1805. The baby was baptised as Frances Stephenson but died when only three weeks old. After that, Fanny was more or less permanently an invalid, and George could only watch as her condition grew steadily worse. She died on 14 May 1806, and was buried in Longbenton churchyard two days later.

# 2

# TO SCOTLAND,
# AND BACK AGAIN

'... in his distress and restlessness,' wrote W.O. Skeat, George 'engaged a house-keeper to look after Robert and took himself to Montrose, Scotland, where he supervised the working of some Boulton & Watt machines in a textile factory.'[1] According to Smiles, he was invited by the factory proprietors,[2] which on the face of it seems improbable, as he was a totally obscure figure. A middle-man was almost certainly involved. James Watt Jr himself had visited the Newcastle area in October 1801, calling on customers and contacts: he dined with the engineer William Chapman on 8 October, 'breakfasted with Hackworth'[3] on the 12th, and on the 13th 'engaged William Walker to go to Falmouth'. Clearly the engine builders went fishing in the small pools of skilled men, and it is likely that one of Boulton & Watt's representatives made the necessary arrangements for George. Montrose's flax-spinning and linen-making industry had expanded dramatically in the first years of the nineteenth century, and the 'stupendous five-storey build-ing known as Richards' Mill'[4], erected in 1804–05, where Stephenson was to work for a year, was among the first flax mills in Scotland to be driven by steam instead of water power.

Few, if any local men would have had the necessary experience to manage steam engines. It is still surprising that George Stephenson should have left his job, his son and his family to tramp some 200 miles to take up another place, having arranged with a local girl, Anne Snaith, to act as housekeeper and minder of the child. Distress and restlessness have been generally taken as the reasons for his departure. Certainly he was a man of strong feelings and emotions, but the Scottish interlude is a unique episode in his life and nothing else in his career sug-gests that he was likely to run away from difficulty or to take refuge in an escapist fugue. One might read into it the action of a romantic hero, striding off into the unknown to explore the mystery of his destiny. The possibility of some kind of psychological crisis cannot be dismissed, but it may also be that mere ambition prompted his decision.

In Montrose he had two engines, of 12 and 25 horsepower respectively, to work and maintain, whilst at Killingworth he was a brakesman with no immediate prospect of promotion. His pay in Scotland was enough for him to save £28 in a year, which implies a wage in excess of £1 a week. Since he kept his Killingworth house and on returning was promptly reinstated in his job, it is also possible he was 'lent' for a year to his Montrose employers. Had he felt inclined to settle at Montrose, he could in due course have brought his son to live with him. Fond as he became of looking back on his early struggles, George never made any public reference to his time there, but he evidently talked to Robert about it, though the only story Robert passed on is, typically, about a technical achievement. One of the mill engines was employed in pumping up water, which was needed both for steeping the raw flax and for making steam. Deep sand deposits held the water reserves, and sand, sucked in by the pump, rapidly abraded the leather flaps of the valves which admitted water to the pipe, clogging the pump. George devised a 12ft-long wooden 'boot', which was placed at the foot of the well and allowed water to flow in, but kept out the sand, greatly increasing the pump's efficiency.[5] Close to the mill, rows of single-storey cottages were being put up for the work-ers to rent, and it is likely that George rented or shared a room here in a street where the dust or 'mill-pob' from the spinning machines was blown about by the east wind. Like colliers, the mill-workers were a segregated community, con-sidered to be a rough lot and eyed askance by the burghers of Montrose. Affable though he was, and from a similar background, the English Stephenson may have found himself something of an outsider with both the factory workers and the townsfolk. He may also have missed the opportunities for discussing machines which Tyneside provided. After a year, probably his initial contractual period, George left Montrose and walked back to Killingworth.

These were uneasy times. Trafalgar had been fought and won in October 1805, but still rumours arose of French invasion and French spies, and officialdom was on the lookout for possible Jacobin or radical agitators. A solitary walker might be eyed with suspicion. But a working man out on the open road was far from an uncommon sight. Whatever their reason for being on the move, walking was the only option for most people. Even on roads which had a stage coach service, such a means of travel was only for the well-off. George, physically robust and long-legged, would have easily managed 25 miles a day. At night, cheap inns in towns and villages offered food and accommodation for a few pence. The luxury of a single room was not even thought of: travelling men were expected to share a bed. A Smiles anecdote relates that on one occasion, caught by darkness, Stephenson asked for shelter at a farm cottage. Somewhat dubiously he was admitted, but of course his genial personality quickly won over his hosts, and they sent him on his way next day refusing all offers of payment. Years later, Stephenson revisited them, 'and when he left the aged couple, they may have been reminded of the old saying that we sometimes "entertain angels unawares"'.[6]

During his absence, Stephenson probably knew nothing of what was happening at Killingworth. His brother Robert was also acquiring a night-school education, but letter writing was a preserve of the well-to-do. On George's return to West Moor he found his house empty, but Anne Snaith had not absconded. The delicate little boy had been taken to her own home. In March 1808 she married Robert Stephenson[7] and became one of several doting aunts whose ministrations helped to compensate the small Robert for the lack of a mother whom the child would hardly remember. For that matter, his father must have been a stranger to the 4-year-old at first. George's job as brakesman was restored to him, he re-established his home, took Robert into his own care and employed a resident housekeeper. This woman did not remain long and was succeeded by a much more satisfactory person, from every point of view. George's sister Eleanor, aged 23 in 1807, had come back from domestic service in London to marry, only to find that her 'intended' had married someone else. The Stephensons always looked out for one another, and Eleanor's humiliating predicament ended with her becoming housekeeper to her brother and a virtual mother to his son. 'Aunt Nelly' ran the household. At this time, there was a surge of interest in Methodism, with its dynamic preaching, powerful hymns and ethos of social care. George was impervious, but Eleanor became a keen Methodist, attending services and prayer meetings, and involving herself in charitable work among the poor and sick. His aunt's enthusiasm grounded Robert in religious feelings, which lasted through-

Stephenson's Cottage, West Moor

out his life, and though not a Methodist, his sympathies were always closer to the chapel than to the emerging High Church trend in Anglicanism.

At Jolly's Close the situation was not happy. Old Bob Stephenson had been caught in a blast of escaping steam in a boiler house accident, which left him blind and severely scalded. Unable to work and with no pension, he was in debt and needed medical care. George used almost half his savings to pay off his father's debts, which amounted to £15, and found a new home for his parents in a cottage at West Moor, ensuring from then on that the old couple had enough to live on. Effectively, he was the head of the family, its biggest earner and strongest personality, but from his own point of view, the responsibility must have been more of a burden than anything else. If he reviewed his situation – as he undoubtedly did – it appeared to have more minuses than pluses. After eight years of work and study to learn what was known of the principles of mechanics, and his own experiments with brake mechanisms and perpetual motion, he was still a colliery brakesman. It was intensely frustrating – his ideas were running far beyond this role: 'as the engineer of the colliery lived in the adjoining cottage to him, they frequently, as he often told me, discussed matters relating to machinery and the steam engine, and as he always thought he knew more of the principles of the steam engine than his superior … these discussions were, no doubt, very animated.'[8]

George resumed his shoemaking and clock-mending to supplement his wages. Britain was at war with France, as had been the case for most of his adult life, and though industry was thriving as a result, taxes were high and able-bodied young men were in demand for the army. Normal recruiting was supplemented by a ballot conscription system in which the names of eligible men were drawn at random. George's name came up, but he had no taste for soldiering and used up almost all the rest of his savings to pay the necessary sum to have someone else substituted in his place. This was a blow – his cash reserve would have been enough to pay for emigration. Memories of his unhappiness were deeply etched, and much later he said to Thomas Summerside: '"Summerside, you know that road which passes from my house and past the fiery heap?" I replied, "Yes, sir." "Well," he said, "when I see it I am reminded of my early struggles and difficulties, for I knew not where I should have found my destiny, and wept …"' Stephenson went on to say that he was thinking very much of emigration at that time.[9]

The house at Paradise Row was the centre for a discussion group on mechanical matters. Among the regulars was a farmer, Captain Robson, who had forsaken the sea for the land; Anthony Wigham, also a farmer of a rather struggling sort; and his son John, men of some education who enjoyed talking mechanics and physics. Another visitor was John Steel,[10] who had supervised the building of the Trevithick locomotive at Gateshead. For Steel to be a friend suggests that George already had a local reputation as someone with an uncommon interest in technical matters. Prospects did not seem good, however. John Wigham joined with George in considering emigration to the United States; either with George's sister Ann, who had recently married, or following her and her husband to the fast-growing coal and iron metropolis of Pittsburgh, Pennsylvania. Looking back at this period, Stephenson recalled that: 'he applied to a Manufactory of Steam Engines at Newcastle to take him into their employ, but no notice was taken of his application; he then intended to go to America where he imagined he might have had a chance of getting into some Manufactory of Steam Engines.' Two friends were to accompany him: a farmer's son and 'an able-bodied labouring man'. If Stephenson's ambitions for a factory job succeeded, 'he was to be their Master but if he failed in that they were to try farming, and if they got on in that business the young farmer was to be master ... the arrangement got broken up by his two friends getting married and their wives prevented their going'.[11] Frustration nagged at him. Inside the brakesman, the engineer was bursting to emerge: he yearned to put his ability, his knowledge and his ambition to the test and build his own engines.

In the mining industry, the hierarchy was clear. At the top were the proprietors – in Stephenson's case the Grand Allies: Lord Strathmore, Sir Thomas Liddell (later Lord Ravensworth) and Mr J. Stuart Wortley Mackenzie (later Lord Wharncliffe). Such men did not usually concern themselves with the technicalities of their industry. Among the less grand, though still wealthy owners like the Brandlings and Blackett, a real interest in the nuts and bolts was more frequent. Each colliery had its viewer, or overseer, who in turn had assistants and apprentices. Apprentices were usually the sons of gentlemen, or at least of families who could pay the premium for their on-site education. Like university fees, this expense would be compensated for many times over by the young man's earnings once he had completed his period of indenture. Between these managerial figures and the workforce, even its upper echelons of foremen and specialist workers, a barrier existed, not impassable but nonetheless daunting. On the far side were affluent families, school education, comfortable homes, social skills, well-placed friends and relations, confidence in authority, all the advantages of being in the middle-to-higher strata of a class-based society. On the near side were child labour, lack of education – and of any capital other than tiny savings from a small wage – one-room homes, uncouth speech and no experience of drawing rooms, no well-placed relations; and instead of a confidence in authority, an awareness of its other side – the requirement that the labouring classes should know their place and stay in it. All Englishmen might be freeborn, but George Stephenson was part of the great majority whose life, if it allowed him aspiration and ambition, also required him to work twelve hours a day, six days a week, to earn his living.

To pass through the barrier it was not enough to be merely literate or clever. Thousands of people more intelligent than many on the privileged side would never cross it. Something else was needed, a talisman whose effect in individual cases made the obstacle melt away. It took the form of a special skill in dealing with some aspect of the new industrial era: a talent for invention or management of complex processes or the handling of finances. Of the most successful, it was written, 'Never had men passed with steps so sure and swift from poverty to wealth, from obscurity to renown. To recite the names on this new roll of fame, from Brindley to Stephenson, from Davy to Arkwright, from Telford to Peel, is like reciting the names of Napoleon's field marshals ...',[12] but the analogy merely emphasises the exceptional nature of the achievement. To George Stephenson in the 1800s there would have seemed nothing sure, and certainly nothing swift, about the process; but he did possess the talisman, which in his case was his understanding of the steam engine. Although it had been around for more than a century, there was a hugely increased focus on steam power between 1805 and 1815, caused in part by Trevithick's development of a high-pressure, movable engine, and also by the commercial pressures of a major war, pushing the price of horses, their feed and equipment to unheard-of levels, and placing heavy demands on manufacturing capacity and manpower. Efficient steam engines were a way of sidestepping these costs and speeding up production. But men who understood the steam engine well enough to adapt and improve it were extremely rare; and, on the privileged side of the barrier, no tradition or educational system existed to produce such people. It was all much too new. On the unprivileged side they were more numerous, because from their ranks came the engine-minders and among those who minded the engines were also some who thought about how they worked. Mastery of the new technology and the prestige associated with it would make them men to be wooed and respected. George Stephenson was too much in the middle of events to make such an analysis of his chances, and 1807–10 was a difficult and depressing period for him, though brightened by the liveliness and obvious intelligence of his young son.

Robert Stephenson's earliest recollections were of sitting on his father's knee, 'watching his brows knit over the difficult points of a page, or marking the deftness and precision with which his right hand plied its craft ... His seat was always on George's left knee, his body encircled by his father's left arm.'[13] This, Jeaffreson believed, affected the development of Robert's left arm, which 'gradually developed into a permanent defect'. The sickly-seeming baby was growing into an active small boy who worried his relatives only by his susceptibility to colds and sweats, a reminder of his mother's fatal illness. Aunt Nelly kept up with Fanny Stephenson's sisters and took the child to visit them: Hannah, the wife of an innkeeper, and Ann, who had indeed married a farmer. Robert later recalled these outings with pleasure, though it was the Stephenson, rather than the Henderson side of the family to which he stayed closer during his adult life.

Meanwhile, George worked and worked. In a speech made at Newcastle on 18 June 1844, on the opening of the Newcastle & Darlington Railway, he reminisced:

Perhaps for more than twenty years he had to rise at one or two in the morning and work till late in the evening ... when Robert was a little boy, he saw how deficient he was in education, and made up his mind that his son should not labour under the same defect, but that he would put him to a good school, and give him a liberal training. He was, however, a poor man, and how do you think he managed? He betook himself to mending his neighbours' clocks and watches at nights, after his daily labour was done, and thus he procured the means of educating his son.[14]

Though Smiles paraphrases this with a tear in his eye,[15] attendance at the Longbenton village school cost at the most 6d a week, which George could readily afford, and by the time that Robert was a pupil at the more expensive Mr Bruce's academy in Newcastle, George was no longer in any need of supplementing his income by clock-mending or shoemaking. The conventional upbringing of a boy in a mining village did not involve school: two or three years of juvenile freedom separated infancy and employment; and his way of life was formed within the powerful shaping forces of local custom and the labour market. For young Robert, his father had a different intention.

Theories on child rearing were not lacking in the 1800s. Among cultivated and literary people, the ultra-liberal ideas of Rousseau were well known, though not necessarily well received. William Wordsworth published his ode on 'Intimations of Immortality in Early Childhood' in 1807, celebrating the response to natural beauty and including a phrase which became proverbial: 'The Child is father of the Man.' Whether or not the Lakeland poet, now 37 but still a controversial literary figure, featured in the table-talk at Killingworth, he was expressing ideas whose time had come. Country children had always had a degree of freedom to roam, but for the Wordsworthian child there was a philosophy behind this freedom, in which nature stimulated and formed sensitivities which the intellect would later explore and codify. At the same time, a diametrically opposite approach to upbringing was being applied by another father–son combination.

James Mill, eight years older than George, a Scotsman of high intellect who had settled in London, had a son, John Stuart Mill, who was three years younger than Robert. James and John would be as prominent in the fields of political economy and social philosophy as George and Robert in the sphere of engineering. James Mill was the chief apostle of Jeremy Bentham, who developed the philosophical idea of Utilitarianism. As a very young boy, John Stuart Mill was 'chosen for the onerous honour of continuing his [James'] work', and subjected to 'perhaps the most severe and painful discipline which any great mind has ever survived ... By the age of twelve his classical reading covered more than most of those who take a first in the classical school of a university can boast of ... No holidays were shared lest the habit of work should be broken, or a taste for idleness acquired.' John later recalled: 'I was never a boy.'[16] He suffered a nervous breakdown at the age of 20, from which he salvaged himself by, among other things, reading the poetry of Wordsworth. Robert Stephenson would also experience a crisis at that age, and find his own solution.

The most obvious comparison to make is with Isambard Kingdom Brunel, born in Portsea in 1806. As a boy in France, Marc Brunel had had to struggle against his father's wishes in order to achieve his ambition of becoming an engineer. Remembering this, he did not try to constrain his own young son, but was delighted by the boy's obvious aptitude for mechanics and his interest in engineering matters. Isambard was taught arithmetic, scale drawing and geometry at home by his father. Following the already established pattern for the well-off in England, Marc sent his son to a boarding school in Hove; then, less conventionally, to his own native France in 1820, where the 14-year-old studied at the College of Caen in Normandy and the Lycée Henri IV in Paris. He also worked briefly with Abraham Louis Breguet, a master craftsman of watches and scientific instruments.

Of this trio of brilliant fathers and sons, Robert Stephenson had much the nearest thing to a Wordsworthian childhood. Barefoot, stockingless, he could be as his father had been, a bird's-nesting, apple-scrumping, clay-puddling youngster secure in a semi-rural community where everyone knew his name. In the recollection of one elder, as a little boy he was 'a sad hempy'[17] (a cheeky fellow). George's views on child rearing may have been discussed with his friends, but came primarily from his own experience, observation and intention for his son, with more shrewdness and pragmatism than theory. Most vitally, Robert was to have the early education that George had been denied: reading, writing and numbers. 'Mind the buiks' was the constant paternal injunction. But there was another strand: although the Killingworth household was a very different one to the old home at Wylam and there was no economic need for Robert to work, his father set him the task of carrying miners' pick-axes to the blacksmith at Longbenton, for sharpening or re-forging. This he did on the way to school, collecting the finished article on his way home. By having his son taste manual labour in this fashion, George showed him that physical work was a necessity that underlay every aspect of human life; and, more immediately, drew the boy into the community and the continuity of his own family. It would be surprising if Robert did not often hear about his father's cow-minding and pick-carrying as a child. But as long as he did his allotted tasks, he was also free to run about and play under, or away from, the indulgent eye of Aunt Nelly.

His father's and grandfather's fondness for birds, animals and plants extended the range of interest beyond daily tasks (though in adult life Robert was no countryman). George was still a workman and Robert was a workman's son, and other aspects of Robert's nurturing reflected the home background and the still-limited horizons of his father. It was not in George's nature to assume the manners of a social class to which he did not (yet) belong; and young Stephenson was not to

look down on pit villages and their ways. But 'Mind the buiks' always came first
– and fortunately Robert was a clever boy. His first teacher was Tommy Rutter,
the Longbenton parish clerk, who made his income by acting as book-keeper to
local farmers and businessmen, as well as teaching, so he knew the importance of
numeracy. The school was in a one-up, one-down house on the main street, and
while Tommy taught the basics to eight or ten boys on the ground floor, his wife
taught sewing and domestic subjects to girls upstairs. Robert's age when he first
went to school is not known, but he was 'quite a little fellow, when he [perhaps
only figuratively] first felt his master's cane'.[18] The schoolhouse was about a mile
away, across a stretch of empty moor.

An anecdote attested by Robert himself again shows George as a shrewd
parent. At gleaning time, when the women and children went to the fields to help
with the last of the harvest and fill their own bags of gleanings, Robert wanted to
go with Aunt Nelly instead of going to school. His father at first refused permission, then agreed, on conditions: 'Weel, gan; but thou maun be oot a' day. Nae
skulking, and nae shirking. And thou maun gan through fra the first t' th' end o'
the gleaning.' At the end of a long, tiring day in the field he was greeted by his
father with: 'Weel, Bobby, hoo did thee come on?'

'Vara weel, father,' said Bobby stoutly.

Next day, the gleaning continued. At the end, the same question was asked, to
which the answer was: 'Middlin', father.' At the end of the third day's toil, coming
home with his small bag of gleanings, and asked the same question again, Robert
burst into tears.

'Oh, father, warse and warse, warse and warse; let me gan to school agyen.'[19]

Whether or not it ever crossed George's mind that Robert might wish to
become something other than an engineer, like Marc Brunel he could savour
the paternal pleasure of having a son who spontaneously, from the earliest, liked
nothing better than to be involved with what his father was doing. The house in
which Robert was growing up reflected George's passions both inside and out; its
curiosities reaching a Dickensian level of English eccentricity, as bizarre and individual as the 'castle' of the Wemmick father and son in *Great Expectations*. Giant
leeks and cabbages were cultivated in the little garden and the elder Stephenson
prided himself on their being bigger than anyone else's. A tame blackbird flew
inside the cottage, but outside a wind-powered scarecrow flailed its arms to keep
birds off the garden beds. Mechanical contraptions in various states of development were everywhere, starting at the garden gate, which could only be opened
by someone who knew its secret mechanism. Inside were models of engines,
of self-acting inclined planes for wagons to run on and further efforts at perpetual motion machines; along with clocks and watches under repair – and, in the
autumn of 1815, parts and prototypes of a miner's safety lamp.

As that safety lamp would show, there was a great deal more than dabbling and
superficial scientific curiosity going on in the cottage at West Moor. In 1811 the

moment came when opportunity and ambition coincided. A new shaft had been sunk by the Grand Allies at Killingworth in the previous year; known as the High Pit and intended to produce a high yield. To pump water from the workings, a steam engine was installed of a type designed by the engineer John Smeaton, who had died in 1794: a modified Newcomen engine and so of a kind more than a century old, using atmospheric rather than steam pressure to operate its piston. Though like all such engines it guzzled coal, it completely failed to keep the shaft from flooding and the High Pit remained unusable. For a year its ineffective wheezing continued, during which time George Stephenson went more than once to look it over, and was heard to claim that he could improve it to the point where it would dry out the pit within a week. When he was finally given a chance to prove his boast, it was because the colliery's head viewer, Ralph Dodds, was 'quite in despair', and promised that if the brakesman succeeded he would see him 'a made man for life'.[20]

George recalled that he was afraid to turn up when summoned by Dodds, thinking he was going to be reprimanded for talking out of turn.[21] When given the job, knowing that the men already employed on the engine would not take kindly to an outsider, not even a trained engineer, telling them what to do, he got permission to pick his own team, chiefly of blacksmiths. In three days of intense activity, the tank holding water to be fed into the boiler was raised by 10ft to give more force to the feed, the injection valve was enlarged and the steam pressure was doubled from 5 pounds per square inch to 10. Now the engine could make more steam more quickly and use steam pressure as well as atmospheric pressure to drive the piston down. At the end of the stroke, condensation of the steam in the cylinder formed a vacuum, which 'sucked' up the piston and the water behind it. At the top of the stroke, a fresh blast of steam into the cylinder set the process off again. Thus invigorated, the engine caused alarm when it was fired up: the massive iron beam swung wildly between its stops and violently shook its supports, but Stephenson understood what was happening. The pit was so full of water that very little of the available power was needed at first. As the level fell, and water had to be drawn up from greater depths, the motion settled to a steady rhythm, and in the space of three days and nights the pit was pumped dry. Later, the shaft was sunk 100ft deeper and the engine continued to keep it dry. Stephenson was intensely conscious of the fact that he had 'broken Smeaton's rules' and made a better engine.

He was rewarded with £10, but for some months it appeared that nothing much else was going to happen, and any hope for quick relief of the frustrations that had been hemming him in had to be dropped. Looking back more than thirty years later, George stated that when Dodds finally made good his promise, it was not because of the engine work, but for another reason. A coal seam caught fire and the miners fled from the pit. It seemed that the whole colliery might be lost, and George, with one other man, went underground to assess the situation as well as they could. He decided that if a wall was built across the tunnel, denying access of air, the fire would go out. Using bricks and lime that were to hand, they blocked the tunnel; the fire did go out and the pit was saved. As George recalled: '... the person who went with Stephenson was a working man but soon after that he got an overlooker's situation, and he, Stephenson, got a horse to ride on and his salary increased to £100 per year and two additional rooms added to his cottage.' Smiles, however, from information given to him directly by one of the participants, dates this fire incident at 1814, two years after Stephenson's promotion. According to this account, six men went down with Stephenson, and people had been killed in the fire. In either version, George's mettle is well attested. Dodds' delay was probably simply due to waiting for a suitable opportunity.

In 1812 the engineman at Killingworth, George McCree, was killed in an accident, and Stephenson was not only appointed to his place, but given responsibility for all the machinery at the collieries worked by the Grand Allies in the Newcastle area. To do this job he had to travel, hence the horse. No longer did he have to offer his views or suggestions with the diffidence of a man conscious of venturing beyond his 'place'; he was in a position to give orders and with the authority to ensure that they were carried out.[22] Dodds and he became friendly, though when the head viewer invited him to drink ale in the morning, George politely refused after a couple of occasions.[23] Never a teetotaller, he was nevertheless generally a temperate drinker, and at this time in particular he had good reason to keep a clear head, both for his new responsibilities and the heightened sense of purpose which they lent to his home studies and experiments.

# 3

# STEAM
# AND FIREDAMP

Ralph Dodds may not have suspected quite what a pent-up torrent of constructive energy he was releasing. Stephenson was not the first to set up steam engines underground, but he was the first, or among the first, to apply them to haulage, adapting an underground pumping engine to 'drag the coals up a sloping plane from the dip side of a downcast dyke of twenty fathoms; and this was one of the earliest engines so employed underground'.[1] In 1812–14 the focus of his attention was very much on stationary engines. Killingworth's underground system eventually had three, providing rope haulage for coal trams along the mine galleries; they were called 'Geordie', 'Jemmy' and 'Bobby' after George and two of his brothers. One was as much as 1,500yds from the shaft, with flues to carry the smoke all the way to the open air. It was 'the most extensive system of engine planes underground in the trade at that time'[2] – the new engine-wright was determined to make it a 'state of the art' pit. Supplementing the slower pace of pony-hauling or the drudging toil of boy-hauling (two pulled, one pushed), the engines speeded up the movement of coal from seam to surface, enabling galleries to be extended and improving the productivity of the mine. Keen to keep abreast of the latest developments, having heard of 'a float of a superior kind, used in a boiler about four miles off',[3] George and a brakesman, George Dodds, went there one midnight and, in order to discover the nature of the device, removed the man-hole cover from the boiler. Steam rushed out and they narrowly escaped a severe scalding. Summerside narrates it as a prank, but evidently Stephenson at this time was not above a little industrial espionage. Above ground, he supervised the laying of new tracks to speed up the movement of coal and slag, matching the improvements below. He had an enthusiastic youthful colleague, orphan son of a Durham tenant farmer and protégé of Sir Thomas Liddell, named Nicholas Wood, who had come to Killingworth in 1811 aged 16 to learn the profession of coal viewer under Dodds.

Describing his career to the House of Commons Committee which was considering the Liverpool & Manchester Railway proposal in 1825, George stated that he left the employment of the Grand Allies in 1813 to work in a partnership formed with George Dodds and Robert Weatherburn, who was a Killingworth brakesman. The trio worked by contract as engine-wrights rather than as waged employees. The partnership was not of long duration. Dodds, also able and ambitious, became a railway engineer in Scotland and in later life was not on friendly terms with Stephenson. Weatherburn, sixteen years later, was a locomotive driver on the Liverpool & Manchester Railway. Dodds was not an uncommon name, and George and Ralph should not be confused with each other. Ralph's contacts and influence were helpful to Stephenson in obtaining work in collieries outside the 'Grand Alliance', for which he also continued to work on a semi-detached basis for his £100 a year. He was also now in a position to learn something of colliery management. Apart from engine work, he designed and supervised the layout of a self-acting inclined plane running down to the coal-loading staithes at Willington Quay, arranged so that the weight of the downwards-travelling coal trams was used to pull the upwards-bound empties to the top. There was plenty to do, but at this time the most intriguing activity was going on very close by, though he was not involved. Christopher Blackett, at Wylam, had resumed his interest in travelling engines.

1812 was a year of portentous events. Britain, already at war with Napoleonic France, also became engaged in war with the United States of America. The Prime Minister, Spencer Perceval, was shot dead in the House of Commons by an infuriated bankrupt. The demands of war on a large scale, by land and sea, were beneficial for most industries, but also created great economic distortions. The price of horses was at an all-time high because of the demands of the army. Financial pressure, as well as latent scientific curiosity, prompted Blackett to try mechanical haulage again. Richard Trevithick had switched his attention to road vehicles, but others now were taking an interest in rail steam traction. In 1812 the iron founders and engine builders Fenton, Murray and Wood of Leeds, built a steam locomotive, the first to have two cylinders. Designed by Matthew Murray, it ran by means of a cogwheel acting on a toothed track on one side of the rails on the nearby Middleton Railway, a colliery line owned by Charles Brandling, Stephenson's neighbour at Gosforth Hall. In the same year, William Chapman, a Durham civil engineer, and his brother Edward, a Wallsend rope-maker, applied for a patent for a steam locomotive designed to move itself by winding a chain cable laid between the rails.

From 2 September 1813, a cog-rail locomotive was put to work at the Kenton and Coxlodge colliery, about 3 miles from West Moor and owned by the Brandlings, where George's brother Robert was engineman.[4] Its first run was attended by a large crowd and afterwards 'a large party of gentlemen connected with coal mining partook of an excellent dinner provided at the Grand Stand for the occasion, when the afternoon was spent in the most agreeable and convivial manner'[5] – doubtless George Stephenson was in the crowd, though probably not at the dinner: in 1813 he was not a gentleman. A second locomotive of Chapman's was tried at Heaton colliery, equally close by, in October and November. Like his first, it was built with the support of John Buddle, part-owner of the colliery and a leading figure in the coal industry. Murray, in particular, made valuable contributions to locomotive development, but crucial experiments had also been carried out in October 1812 by Blackett's viewer, William Hedley, who had a hand-crank trolley built to scientifically measure the tractive power of 'smooth' wheels on a smooth rail. Satisfied that there was no need for racks or chains, Blackett had a self-propelling locomotive built by Thomas Waters of Gateshead. Delivered in spring 1813, it was not a very effective machine, and later that year a team at Wylam, with Hedley in charge and including his foreman blacksmith Timothy Hackworth, set out to build their own. The locomotive, later known as 'Puffing Billy', first took to the rails in March 1814,[6] and then they started on another, 'Wylam Dilly'. Tyneside was a cockpit of activity in the new technology, with pride in display vying with possessiveness and anxiety to obtain patent protection. Obtaining a patent was not a simple matter; at that time only a 'gentleman' could apply and seven different government offices were involved, with payments required at each stage.

In September 1813 George Stephenson was given the longed-for instruction by Sir Thomas Liddell to arrange for construction of a steam travelling engine in the West Moor colliery workshops, to run on the Killingworth wagon-way. Familiar with the Wylam experiments, George chose wheel-on-rail traction, with flanged wheels on flat-edged rails; and having observed both Murray's and Hedley's locomotives, the latter being single-cylinder as Trevithick's first one had been, he opted for two cylinders. A locomotive driven by only one cylinder needed a heavy flywheel to maintain the motion when the piston reached the end of its lap; two cylinders gave better balance and a smoother drive, though they also required more steam and added to the building cost. The first Killingworth locomotive had four wheels on the rails of 3ft diameter, their axles turned by a set of geared cogwheels. Originally, a chain linked the rear axle to the front axle of the four-wheeled 'convoy cart', carrying coal and water, to assist with traction, but this (though patented by Stephenson and Dodds in 1815) was found to be unnecessary. George later said the engine was first named 'My Lord' after Liddell, but this is unlikely as Liddell was not made Baron Ravensworth until 1821. The name given was *Blücher*, a sign of the esteem felt for the Prussian marshal who was Britain's military ally.[7] Having taken ten months to build, it made its first run on 27 July 1814: '... upon a piece of road with the edge rail, ascending about one yard in four hundred and fifty, and was found to drag after it, exclusive of its own weight, eight loaded carriages, weighing altogether about thirty tons, at the rate of four miles an hour; and after that time continued regularly at work.'[8]

Nicholas Wood's level tone gives no hint of the nervous excitement and glee they must have felt when the steam valves were first opened and their new beast, with successive ear-bashing blasts of steam and vibrant meshing of its gear-wheels as the connecting rods jiggled them into movement, started forward; nor of the fact that 'Hundreds if not thousands' of people were there to watch and cheer.[9] At his first attempt, Stephenson had built a locomotive which worked at least as well as the other pioneer machines, but shortcomings were not lacking. The cog drive was not very effective in transmitting power. Instead of Trevithick's turned-back 'double-flue' boiler, *Blücher* had only a single large fire-tube inside its boiler, and consequently about half as much heating surface to generate steam. This, and subsequent Killingworth locomotives, would be chronically short of steam, until Stephenson adopted the return-flue. His reasons for clinging so long to the inadequate single-flue boiler are not known. Perhaps he simply wished to avoid paying a patent fee (Murray and his colleague Blenkinsop paid £30 to the owner of Trevithick's patent).

In his second, or possibly third engine, he resolved the drive problem, becoming the first engine-builder to use crank pins fitted to the driving wheels, now of 4ft diameter, which were turned by the direct action of hinged rods connecting the cranks to the pistons. First attempted on 6 March 1815, it was 'found to work remarkably well'.[10] Virtually all steam locomotives afterwards were to use this drive method. Stephenson did not invent the crank drive, which had been used to turn fly-wheels on stationary engines, but he and Dodds patented it in February 1815 for use on locomotives. In his private diary, though not in his *Treatise*, Wood noted that a similar drive, with a chain link between the axles, had been employed on a locomotive tried at Newbottle colliery 'before the patent was dated'.[11] How similar is not known, nor is there any other information on this engine.[12]

Demands beyond the ability of current technology were made by the locomotives of 1813 and onwards. Apart from the inadequacy of the tracks, no foundry could produce leaf springs of a strength and elasticity to support a 5-ton body, or forge crank axles of sufficient strength to withstand constant use. As a result, such features, though possible in theory, were not fitted to Stephenson's earliest locomotives. The success of his first attempt ensured that he was asked to continue, and apart from its normal activities, the smithy at West Moor became a locomotive workshop. In the next seven years he built at least four locomotives for the Killingworth colliery lines, and twelve more for other users. From now on he could consider himself a builder of travelling engines.

Busy as he was with locomotive construction added to his other commitments, in his spare-time, George had moved on from the dream of perpetual motion to a more immediate concern. Miners could not work in total darkness and the only lights available to them were either candles or lanterns with naked flames, which could not safely be used in the presence of gas; or the steel-mill, in which a small wheel and flint produced a stream of sparks, unlikely to cause an explosion but giving very poor light. Such desperate expedients as 'the phosphorescence of decayed fish-skins' were tried[13] to provide at least some safe light. Methane gas, known as firedamp and often assumed to be hydrogen, issued from cracks or 'blowers' in the seams often with an audible hiss, but it could not be seen or smelled. It was not poisonous, but ignited on contact with a flame, and explosions and fires killed hundreds of miners every year. As the pace of the industry quickened, and galleries reached deeper and further, and water seeped into worked-out seams, forcing gas upwards, the number and violence of blasts increased. In 1812 ninety underground workers died in an explosion in the Felling pit, on the south side of the Tyne near Gateshead, owned by the Brandling family, and another explosion in the same pit in December 1813 killed twenty-three men and boys and twelve horses.[14]

The lack of safe lighting in mines was a general problem, but it particularly concentrated minds in north-east England. Much coal was left uncut because the miners either could not see it, or could not safely work it. Coal owners have generally had a bad press as uncaring exploiters of their workers, but there is no reason to doubt that humanitarian feelings, as well as their own business interests, moved Northumbrian and Durham proprietors, headed by John Buddle of Wallsend, the region's leading mining engineer, to form a committee to consider the problem.[15] Already one successful attempt at a safety lamp had been made, by Dr William Clanny of Sunderland: a rather cumbersome piece of equipment in which air had to be pumped through a water container, by means of a small hand-bellows, into the thick glass jar which enclosed the flame. Clanny displayed his lamp to the Royal Society in May 1813,[16] but his solution had no appeal to coal viewers because it required an extra 'unproductive' boy to work its bellows, and not one example was used underground. On 1 October 1813 the Society for the Prevention of Accidents in Mines was formed in Sunderland,[17] but had little effect at first, and it was not until mid-1815 that, through the Rev. Robert Gray of Bishopwearmouth, it invited Sir Humphry Davy, the most famous scientist in the country, to visit the coalfield and advise on whether a convenient solution could be found. Davy came to Sunderland in late August and stayed in the area for some three weeks,[18] meeting Clanny, who showed him a specimen of his lamp. From the middle of October, back in London, he got to work, using samples of gas sent down by another member of the Society, the Rev. John Hodgson, vicar of Jarrow.

Davy worked speedily and in the last two weeks of October produced three successive lamp designs in the Royal Institution's laboratory. He ascertained that if two vessels connected by a narrow tube were both filled with explosive gas, the gas in one chamber could be burned without any effect on the gas in the linked vessel. From this he concluded that a flame inside a lamp, fed with air through small orifices, would not ignite a surrounding air-gas mixture within the mine. On 19 October he sent notes on it to Gray and Hodgson,[19] and by the 30th he had made a lamp in which the air entered and left through narrow tubes. His prediction of success prompted the president of the Royal Society, Sir Joseph Banks, to arrange for him to deliver a paper on the subject on 9 November.

Meanwhile, George Stephenson was also working hard on a lamp design. Their speed might suggest a competitive element, but Davy at least had no notion of Stephenson's activity, and even if he had, would scarcely have given it attention. More questionable is George's sudden decision to pursue the matter just at the time of Davy's visit. In sworn evidence given on 1 November 1817 to a committee examining his own achievement, Stephenson denied having had any 'knowledge of Sir H. Davy's discoveries or experiments, nor of any communications made to the coal trade thereupon',[20] but this is not to say he did not know that Davy was at work on the problem. Nicholas Wood, however, 'can vouch that from the period when he first directed his attention to the subject, up to the time when his lamp was tried in the mine, he had never been informed that Sir Humphry Davy was engaged in a similar enterprise'.[21] Smiles claims that Stephenson had been conducting experiments with firedamp 'in his own rude way' for several years, and he may have long intended to try to do something about the gas hazard. In any case, now he went into very rapid action, as of course he was perfectly entitled to do. He had direct experience of underground fire, and on that occasion had conquered it. In his 'I can do that' mode, he would have a go at anything, but this was very different to clocks, shoemaking or even steam engines. Knowledge of the properties and behaviour of flames, gas and air was needed, and he gained this at first hand. Most of his experiments were made in 'Bob's Pit', where Thomas Summerside's father was overman,[22] and the pitmen were startled to see him approach gas blowers with a lighted candle in his hand. From this he 'observed that a flaming "blower" of firedamp was extinguished by the burnt gases from lighted candles held upstream from it'.[23]

Both experimenters followed similar 'evolutionary' methods.[24] Stephenson, watching the action of flames in and out of a firedamp-rich atmosphere, saw that the gas burned at the base of the flame and not at the top, and decided that if he could build a lamp with a strong up-draught, the presence of what he called 'burnt air' (carbon dioxide) at the top would prevent the flame from coming into contact with any surrounding gas, while the draught would prevent ignition from reaching downwards. As Davy did, he also 'showed by controlled experiment that the firedamp flame would not flashback along a fine tube'.[25] He had already decided that a stout outer case was necessary to avoid breakage of the glass lamp-jar; his first model had a movable shutter but subsequent ones had perforated cases. During October and November, he designed and arranged for the making of three successive oil lamps, testing each prototype both down the mine and in Nicholas Wood's house, with Wood, who shared half the costs,[26] and Robert as helpers. Home-made equipment was not used: apparatus was borrowed from the Literary & Philosophical Society in Newcastle; the Northumberland Glass House made glass funnels to specification; and a tinsmith, Hogg of Newcastle, made the first case from Wood's drawings. It had a single air tube and a sliding panel in the base to control air supply.

On 21 October George, Robert, Wood and the under-viewer, John Moody, took this lamp down the mine to try it out, in the only way which would ever convince the miners. At a point where gas was hissing from a fissure, George had had wooden screening erected in order to concentrate its presence. Moody, sent forward to assess the condition of the air, returned to say that an explosion would be inevitable if any flame was present, and neither he nor Wood would follow George, who went forward alone with his lamp lit. Fearing an explosion at any second, they were relieved to hear him call to them – the lamp had behaved as he had expected. Rather than ignite the gas, it had gone out through lack of oxygen. Relighting the lamp, he went back into the gas zone. As it encountered the gas, the flame increased in size, changed to a bluish colour, then again went out. This was not yet a proper safety lamp; the point was to have light.

Risks were run by the experimenters above ground as well as down the mine; speaking before the Committee on Accidents in Coal Mines in 1835, George recalled an occasion when they were using a supply of water to keep the burning gas at a certain level in the glass tube. Nicholas Wood was operating the water-cock on the apparatus: 'As I saw the flame descend in the tube, I called for more water, and Wood unfortunately turned the cock the wrong way; the current ceased, the flame went down the tube, and all our implements were blown to pieces, which at the time we were not very able to replace.'[27]

At a meeting of coal owners in the Newcastle Assembly Rooms on 3 November, a letter from Davy was read out, 'and to the same company Mr R. Lambert, chief agent for Killingworth and other mines on this river', announced Stephenson's discovery.[28] Next day, George tested his second lamp, with three narrower air tubes in the base,[29] and on the 9th Davy presented a paper at the Royal Society, *On the Fire-Damp of Coal Mines and on Methods of lighting the Mine so as to prevent its explosion*, with a demonstration of his lamp, as it then was: a glass jar with no outer covering. Neither it nor Stephenson's were yet practical for use down a mine. A third Stephenson prototype, tried out successfully on 30 November, replaced the air tubes by a 'perforated plate';[30] it was completed just in time for a discussion which had been arranged at the Literary & Philosophical Society of Newcastle for 5 December 1815.

At this meeting, a paper from Dr John Murray of Edinburgh on his lamp design was read out; Dr Clanny gave a demonstration of his bellows lamp; and another safety lamp (not of a satisfactory nature), designed by one of the Brandling brothers, was also shown. Some eighty interested persons attended, and George, struck by diffidence, let Wood make the presentation – a full practical demonstration with firedamp provided in pigs' bladders. Such was the interest and the flow of questions that George was unable to resist amplifying his assistant's answers and came forward to give his own explanations. J.H.H. Holmes, a fan of Clanny's, was present and not wholly impressed, partly because 'Mr Stephenson appeared totally ignorant of the manner in which the air and gases operated upon the light', and partly because the lamp looked too pristine to have been tried out down a coal-mine six weeks before;[31] he evidently did not appreciate that he was looking at the newly-made Mark III model. But he felt that Stephenson 'undoubtedly claims great merit, if the invention produced was from his own genius'.[32]

By 18 December[33], if not before, Davy knew that a Northumbrian colliery worker, of no scientific background or technical qualification, had designed a lamp that employed the same principle as his own. At this time he engaged in a 'fundamental rethink', abandoning narrow tubes for what he referred to at first as 'fire sieves', in their simplest form no more than a finely perforated stopper that could be fitted on a conventional lamp-glass. In the space of a few days he developed this into an inlet system and then, by 29 December, had a complete wire-gauze cylinder, dispensing with the glass lamp-jar altogether: 'Quite how he discovered he could use gauze in this way is not known, but he soon began to exploit it.'[34] On that date he wrote to Hodgson: 'when a candle or lamp is enclosed in a wire gauze cylinder & introduced into an explosive mixture, the flame of the wick is extinguished but the mixed gas burns steadily within the wire gauze vessel ... I can confine this destructive element like a bird in a cage.'[35] On 11 January 1816 he read a short paper to the Royal Society, displaying the gauze case, and in the 'Advertisement' to a revised version of his 'Firedamp' paper, wrote of it as 'a discovery which appears the most important in the whole of these researches'.[36]

For a mechanically-minded boy, the cottage at Killingworth was a sorcerer's chamber. The sorcerer's son, 12 years old in 1815, was an active and excited participant in the safety lamp tests and in the explosive aftermath. In this year he left Tommy Rutter's to attend Mr Bruce's School in Newcastle. His father's intention for him was clear. George wrote in 1823 that 'I have onely one son who I have brought up in my own profeshion'.[37] Robert's future was to be in engineering, if his father had anything to do with it. But Rolt goes too far in saying that: ' ... *recognising his lack of knowledge and feeling himself cut off from the accumulated fund of experience and learning which others had set down in print, he determined to realise his ambition by means of a working partnership with his son in which Robert would supply what he lacked.*'[38] By the time Robert was sent to school at Longbenton, his father had been able to read and write for almost ten years. The corpus of technical and mechanical knowledge so far set down in printed form was not vast, and George had been discussing the details of steam power and basic mechanics with others since before his son was born. With John Wigham he read through Ferguson's published *Lectures on Mechanics*[39] and they tried out the experiments for themselves. Everything suggests that he had ambitions for Robert, but the idea that he would wait a decade or more for his son to 'supply what he lacked' is negated not only by what he accomplished while Robert was still a schoolboy, but also by his own ability, personal confidence and drive. If Robert Stephenson had died as a child, or shown no interest in mechanical things and decided to become a parson or a lawyer, would George still have been a pioneering railway engineer? The answer, surely, is yes. Would there have been a *Rocket* in 1829? The answer is less certain – but there would certainly have been something, and most probably something not very different from the actuality.[40] Undoubtedly, Stephenson saw his son's future as an engineer and steered him in that direction, but it was a long-established social convention that a son, particularly an elder or only son, would follow the same trade or profession as his father; it was also by far the most obvious choice of professional career for a boy in Robert's situation.

Newcastle had two schools for older boys, the Grammar School and Mr Bruce's, so George had to make a choice. John Bruce was a scholar and writer; friend and biographer of Charles Hutton, the celebrated Newcastle-born mathematician who had measured the density of the earth. His academy in Percy Street had a higher reputation at that time than Newcastle's famous Grammar School[41] and its fees were relatively modest at £10 a year. What would have appealed to George Stephenson about the establishment was its 'modern' approach to learning. It offered, and found takers for, a useful education. Greek and Latin were available, but not obligatory, and more than half of the hundred or so pupils did not study the classics, concentrating on mathematics and what was known as natural philosophy, embracing most of the emergent sciences. French was also on the curriculum, and George had Robert learn it, not for Racine or Corneille, but for its commercial usefulness (the Legion of Honour might almost have been foreseen).

From West Moor into the town was a 5-mile walk. Ten miles a day, on top of all the learning, was a heavy stint, but George bought his son a donkey to ride to school. Intensely interested in what Robert was studying, he was avid to keep up, and school did not end when the boy got home, since he had to show his father what he had been taught, as well as do whatever homework he was set. To his schoolmates, sons of town merchants, ship-owners, officials and gentry, Robert Stephenson was something of an oddity, coming in on his cuddy, in the blue jacket and working boy's corduroy breeches made by the Killingworth tailor: 'A thin-faced, thin-framed delicate boy, with his face covered in freckles', was how one remembered him.[42] In that era of rich regional accents, they all spoke with the Northumbrian burr and long vowels, but Robert's rustic diction seemed outlandish in the town. Here was a ripe set-up for bullying, but a boy from a pit village was not to be easily intimidated. Robert's good-natured manner seems to have stood him in better stead than truculence, however. Having to ride home made it difficult to become friends with boys who lived near one another in town, but he made one or two lasting friendships from his schooldays. At West Moor, his father waited eagerly to hear of continuing progress with 'the buiks', but home also offered experiences unknown to his classmates. His grandfather was still alive in his first years at Bruce's, and Robert remembered riding his cuddy right into the cottage, where the blind old man would pat the animal, 'after which he would dilate upon the shape of his ears, fetlocks and quarters, and usually end by pronouncing him to be a "real blood". I was a great favourite with the old man, who continued very fond of animals, and cheerful to the last; and I believe that nothing gave him greater pleasure than a visit from me and my cuddy.'[43] Old Bob died in June 1817, his age given as 79, and Mabel died in May 1818.[44]

More in his father's interest than his own, Robert enrolled as a 'reading member' of the Newcastle Literary & Philosophical Society, which had been founded in 1793. Here, as in other provincial towns, the 'Lit and Phil' was a centre for the acquiring, discussion and exchange of ideas among those citizens who cared about such things – an alternative or supplementary focus to the Corn Exchange, the churches and the taverns. Monthly talks on contemporary issues were given by visiting speakers and by members, and of course the safety lamp had been demonstrated there in 1815; but most importantly, for the Stephensons and many others, the combined subscriptions (a guinea a year) of the members helped to stock the library with books which, as individuals, they could not have afforded. Many books could only be read on the premises, and Robert had to do much transcribing and memorising. Borrowed volumes, mostly encyclopaedias and instruction texts, were read and discussed round the table with visitors like Wood and Wigham. As Secretary, the Unitarian minister, William Turner, kept up a progressive and liberal ethos. Robert was appreciative of his consideration to a village boy: '… always ready to assist me with books, with instruments, and with council, gratuitously and cheerfully.'[45]

Ideas, of course, can be as explosive as methane in their way, and in some places the 'Lit and Phil' was a seed-bed of political radicalism, not confined to those who felt themselves denied rights that were worth arguing or fighting for. Across the country in Liverpool, young Henry Booth, an enthusiastic member of that city's 'Lit and Phil', and later to be a collaborator, friend and ally to George, was a radical in politics despite his comfortable background as the son of a well-established corn merchant. But arguments for reform or revolution did not prevail at Paradise Row. In the political world, the victory at Waterloo in June 1815 underwrote fifteen years of reactionary governments which would resist demands for social reform. George Stephenson would have approved, though what was being hatched in his front room would galvanise social change in ways that the most radical thinkers had not imagined. His lack of a vote in parliamentary elections troubled him not at all, and he had no political or social vision of what his work would achieve.

Thomas Carlyle, fourteen years younger and a more profound observer, saw in those times 'a deep-lying struggle in the whole fabric of society; a boundless grinding collision of the New with the Old'.[46] He could have had the elder

Stephenson in mind when he wrote: 'Men are grown mechanical in head and in heart, as well as in hand ... Not for internal perfection, but for external combinations and arrangements, for institutions, constitutions – for Mechanism of one sort or another, do they hope and struggle. Their whole efforts, attachments, opinions, turn on mechanism, and are of a mechanical character.' Carlyle had mental processes and moral states in mind, not pistons and connecting-rods. But if George Stephenson exemplifies perfectly the modern type which Carlyle (in 1829) rather mournfully defined, he also stands full in an older English tradition. Immortalised in folk tales from that of Dick Whittington, if not before, and proved by many examples including, in the preceding century, Sir Joseph Arkwright of the 'spinning jenny', it is the story of the individual Englishman who makes his own luck and his own fortune. At first indistinguishable from a mass of other handy chaps, he is endowed with the resolution and determination that can break down obstacles and cope with rebuffs. Indeed, it is one of the great virtues of England that it produces such men, testing them severely, but ultimately giving them acceptance and honour, however humble their beginnings. Such was the national myth, and yet not wholly a myth, because it was known to have worked in real lives. It helped to keep the social fabric in shape and allowed the dominance of the small proportion that lived above the barriers, while not wholly denying opportunity to the very large proportion that lived below. George Stephenson, with his superabundant energy, his genius for machinery, his determination to 'get on' and his powerful sense of individuality, had all the necessary characteristics. He had no wish to change the social system because he could use it to his benefit.

It is intriguing to see how in the Stephensons' lives at this time, much of the 'Utilitarianism' expounded by Jeremy Bentham and to be adapted by John Stuart Mill was already in action, though driven by pragmatic reasoning rather than any ideal conception of the greater good. In a sense, what the Utilitarian philosophers did was to place their gloss upon a process that was already unstoppably under way. The key doctrine of Mill's was the principle of utility as a standard for ethics and politics, both personal and public. Decisions should be made and actions taken solely on the basis of which had the most beneficial effect, summed up in the phrase 'the greatest good of the greatest number'. Like all such phrases, this over-simplifies a closely reasoned set of arguments, but utilitarianism had a harsh side: it was a work ethic and its emphasis on 'moral arithmetic' tended to reduce human beings to the equivalence of mechanical parts in a vast social machine. Few people appreciate being dictated to for their own good. George Stephenson and his friends no doubt discussed Benthamism, just as they also argued about the slave trade, parliamentary reform and the Corn Laws, but to them these were the background issues of the day. More immediately compelling were the problems they could do something about next morning, with the help of a file or a spanner. There was no textbook on locomotives. Earnestness, energy and argument went into creating the know-how and the what-for of the new technology just as into ethical and political issues.

While at Bruce's school, Robert 'did not exhibit any marked enthusiasm for the subjects in which his father was most warmly interested', and Jeaffreson suggests that the teenager balked at his father's 'goading', adding that some of the villagers considered George 'an o'er strict father' because Robert was made to read so many books.[47] Smiles does not mention this father-son friction, but it would not suit his 'self-help' theme to do so. Given George's forcefulness, single-mindedness and his huge appetite for absorption of information on mechanical matters, it seems very likely that he might often have pressed Robert to a point which the boy found oppressive. To help in exciting experiments with bangs and flashes was one thing; to read column after column of technical prose, aloud, was not at all the same. As far as the neighbours' comments are concerned, though, to have books in the house at all was unusual. Robert, on the evidence of contemporaries, was still a lively boy, with time to join in the pastimes and tricks of the village lads, including stone-putting and hammer-throwing. At 16 he accidentally knocked himself out by failing to let go of a 28lb hammer quickly enough. In a more scientific adolescent jape, prompted by reading about Benjamin Franklin's experiments, he 'proceeded to expend his store of Saturday pennies in purchasing about half a mile of copper wire at a brazier's shop in Newcastle'.[48] A fraction of that length seems more likely. Attaching the wire, Franklin-style, to a large kite on a thundery day, and controlling it with a length of non-conductive silk cord, he brought the end of the wire to touch the tail-ends of Farmer Wigham's cows as they grazed in the field by the front gate, and sent them skipping. George watched from the window with a paternal smile, until the wire was transferred to the rump of his own horse 'Squire', tethered to the gate post. When it reared up at the shock, he issued from the front door with his whip and the words: 'Ah! Thou mischeevous scoondrel – aal paa thee.'[49] But the youthful experimenter fled.

4

# EMERGENCE OF
# AN ENGINEER

In the course of 1815 and 1816, George Stephenson's work on travelling engines and on his safety lamp – soon known as the 'Geordie lamp' – made him something of a celebrity in the Newcastle area and widened his range of friends and acquaintances. Among them was William Losh, a member of a prominent family in the town, senior partner of Losh, Wilson & Bell, an iron-founding business at Walker, on the Tyne east of Newcastle. George's eagerness for the ironworks to improve their technology and supply new products in order to keep apace with what he, as a locomotive builder, needed, made Losh a valuable contact. This applied not only to the forging of mechanical parts, but also to the rails on which the locomotives ran. From the Trevithick locomotive's first outing, the impact of unsprung wheels on light and brittle cast-iron rails had been a serious problem. Rail lengths were short, only 3ft (1m) or so, and if not well secured they rose up when the weight came on one end and frequently cracked or broke. Stephenson appreciated that locomotive and track formed an operational unity, and in Losh he found a ready willingness to respond to the challenge. His 'partnership' with Dodds did not prevent him from arranging to spend two days a week at Losh's works, adding a salary of £100 a year to his £100 from the Grand Allies, and Losh's name joined his on a patent they took out in 1816 for a new kind of overlapping rail joint, intended to avoid the problem of the rails springing up (it also included 'steam springs'[1] and a wheel design of malleable iron). For George this was a busy, stimulating, satisfying time. Aged 35 in 1816, he was at last able to concentrate on what he wanted to do, and explore the potential that had so long seethed within him. The support and encouragement of men like Liddell, Losh and the Brandling brothers was vital, but he was grasping his own opportunities and taking a place in the select band among creative geniuses who make sure that invention goes hand in hand with earnings.

In 1816 he and his son collaborated on a new domestic project, the design, calibration and placing of a sundial above the front door at Paradise Row, dated

The sundial

'August 11th, MDCCCXVI'. The 13-year-old Robert had at first protested that he did not know enough astronomy and mathematics to make an accurate sundial, but his father '... would have no denial. "The thing is to be done," said he; "so just set about it at once." Well, we got a "Ferguson's Astronomy," and studied the subject together. Many a sore head I had while making the calculations to adapt the dial to the latitude of Killingworth. But at length it was fairly drawn out on paper, and then my father got a stone, and we hewed, and carved, and polished it, until we made a very respectable dial of it.'[2] Robert's drawing for it was presented to William Losh. The sundial is still in place.

Dim as their light might be, the coal industry now had two safety lamps which had brought a new confidence to underground work. Neither of the inventors attempted to patent his design, allowing its free use in the interest of life-saving. Both made further modifications: Stephenson — as his critics did not fail to note

– adopted Davy's trimming screw and the wire gauze, which gave more light than his perforated tin casing, and Dr Clanny[3] also brought out a new design. Through the busy years of 1816 and 1817, controversy about who first invented the safety lamp flashed and rumbled, in a lengthy saga which taught both Stephensons much about the realities of public life. The issue cut in various ways, including political antagonism between Whig (on the Davy side) and Tory (on the Stephenson side) coal-owners.[4]

Sir Humphry came to Newcastle to be formally thanked on 25 March 1816, and the *Philosophical Magazine* recorded that 'the coal trade has liberally presented 100 guineas to Mr Stevenson of Killingworth Colliery, for his ingenious lamp … which, though superseded by Sir Humphry Davy's more perfect invention, not only evinced great ingenuity, but offered much comparative safety to the miners'.[5] Clanny received a silver medal on 31 May from the Society of Arts. Stephenson's friends considered the 100 guineas to be a wholly inadequate recognition of his achievement and the dispute became entrenched and bitter. On 31 August 1816, the committee, which had approached Davy met again and a subscription was raised 'for the purpose of presenting him with a reward for his invention of the safety lamp'. This form of words drew criticism from others, led by R. W. Brandling, who considered Stephenson to have the prior claim. Buddle's committee was embarrassed rather than gratified that a local man should also have solved the problem, and hostile remarks of Davy's were repeated by some members in the local press. Brandling's motives were assumed to be personal and political, and acrimony bubbled in a confidential letter from Buddle to the Rev. John Hodgson: 'The fool Stephenson is made a catspaw of the knave Brandling – I burn for shame at his audacious calumny of Sir Humphry.'[6] Quite reasonably, his committee was honouring its own man: they had not asked George Stephenson to involve himself (incidentally, as Chapman's patron and collaborator, Buddle was finding Stephenson a rival in locomotive design also). It was not only the disproportion of the two rewards, but also the knowledge that Davy and his friends had been belittling Stephenson's achievement, even accusing him of pirating Davy's work, that prompted George's supporters to mount a campaign for proper recognition of his invention. Captain Robson, the Killingworth farmer, made a counter-accusation that Buddle had got to know of George's ideas and passed them on to Davy.

At a mine-owners' meeting on 11 October 1816, R. W. Brandling unsuccessfully proposed an investigation into whether Davy or Stephenson was the inventor of the safety lamp, and George himself entered the arena by sending a letter to local newspapers, enclosing another by Brandling endorsing him as the first discoverer of the safety lamp principle.[7] Encouraged by the Brandlings and William Losh, with the Grand Allies also in support, he published a sixteen-page pamphlet, *A Description of the safety lamp, invented by George Stephenson, and now in use in the Killingworth Colliery*, in January 1817. Explaining that '… my habits, as a practical

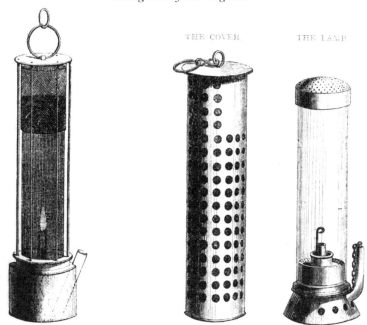

THE COVER    THE LAMP

Stephenson's third safety lamp, with Davy's final version

mechanic, make me afraid of publishing theories, and I am by no means satisfied that my own reasons, or any of these that I have seen published, why hydrogen gas will not explode through small apertures, are the true ones',[8] he did not set out his reasoning, but described and illustrated his successive lamps. Referring to the Davy lamp, he said: '... the use of the wire gauze is certainly a happy application of a beautiful manufacture to a very useful purpose, but I confess I cannot consider it in any light than as a variation in construction.'[9] Another pro-Stephenson pamphlet followed, quoting letters favourable to Stephenson from the north-east newspapers (*A collection of all the letters which have appeared in the Newcastle papers, with other documents relating to the safety lamps, London, 1817*).

By now, the *Philosophical Magazine* had swung to Davy's side and claimed Stephenson had 'borrowed from what he had heard of Sir Humphry Davy's researches'.[10] On 11 October 1817 an array of silver plate to the value of £2,500 was presented to Davy at the Queen's Head Hotel in Newcastle; and on 1 November a subscription to reward Stephenson was launched, citing him as '... having discovered the fact that Explosion of Hydrogen gas will not pass through Tubes and Apertures of small dimensions, and having been the first to apply that principle in the construction of a Safety Lamp'. Davy, by now referring privately to Stephenson as a thief and 'illiterate pirate' who deserved the pillory[11], wrote in protest to the named organisers, who returned frosty responses. Lord Strathmore replied:

I beg leave to inform you that George Stephenson is, and has been for many years, employed at Killingworth and other Collieries in which I am concerned, and that no other Safety Lamp but that of his Invention ever has been used in any of them ... Is it to be wondered at that I should be anxious to reward a very deserving, unassuming Man, who has to his employers always proved himself a faithful servant and whose abilities, if they had been aided by the advantage of Education, would probably have rendered him conspicuous in the annals of Science.[12]

Under the rubric of the Royal Society, on 20 November Sir Joseph Banks issued a statement awarding 'sole merit' to Sir Humphry Davy in the creation of the safety lamp, and six days later Buddle's group held a meeting to issue a statement – evidently Davy-inspired – claiming that Stephenson's first lamps did not embody any principle of security and that his later lamp was 'a coarse imitation of one of Sir Humphry Davy's earlier lamps'.[13] At the same time, a new pamphlet was issued by the Brandling camp, *Report Upon the Claims of Mr George Stephenson ...* with sworn evidence from Stephenson, a verified timetable that showed he could not possibly have made use of any of Davy's findings and first-hand accounts from John Moody and Nicholas Wood, as well as corroborative evidence from the tinsmiths and others who could testify that Stephenson had worked independently.[14] Even-handedly, the Newcastle 'Lit & Phil' bestowed honorary membership on both inventors on 2 December. On 12 January 1818, a public dinner was held in the Newcastle Assembly Rooms with Charles Brandling in the chair and George Stephenson as guest of honour. A thousand pounds had been raised and he received this, less the value of an inscribed silver tankard, as well as a silver watch paid for by contributions from individual coal miners. In the first public speech of his life, written out by himself and learned by heart, he thanked the subscribers, not only for their gifts:

> I still feal more by being honour'd by such and highley respectable meeting the Gentlemen of which not only having rewarded me beyond any hopes of mine for my endeavours in constructing a safity lamp but has supported me in my claims as to priority in my invention to that of that distinguished Pholosipher S H Davy. For when I consider the manner that I have been brought up and liv'd the manner of which is known to many of the Gentlemen present and when I consider the high station of S H Davy his high Character that he holds among society and his influence on scientific men and scientific bodys, all of which Sir lays me under a Debt of Gratitude to the Gentlemen of this meeting which Gratitude shall remain with me so long as ever I shall live.[15]

From this first occasion, the theme of contrast between Stephenson's humble origins and present distinction is brought in, to be repeated in many speeches, from ever-greater eminence. He emerges from the long controversy with more credit

than his distinguished and involuntary rival. His perforated plate predated and perhaps prompted Davy's 'twilled gauze', though the gauze covering set the pattern
for the final safety lamp. The episode justified his confidence in his own abilities
and showed him that he had influential friends who would back him publicly, even
against big local figures like Buddle and Hodgson, and a national authority like Sir
Humphry Davy. In mining communities, he was now a respected household name.
He also fully absorbed the lesson that it was not enough to make an invention
– the world did not automatically award credit unless it was vigorously claimed.
This was useful and necessary knowledge, but in two ways the effect was less beneficial. Having matched the arch-practitioner of the theoretical approach with his
own practical method, he felt able to turn his deficiencies in technical education
into a virtue. And the personal contempt expressed by Davy and his supporters
contributed to a lasting resentment against 'London' savants and their snobbish,
blinkered, dismissive attitude to the achievement of someone who was far outside
their charmed circle. Lord Strathmore (more than a trifle condescending himself)
had got something wrong: his employee might indeed be deserving, but George
Stephenson was not the humble, 'unassuming' figure suggested by the earl's letter.

Peace in Europe had brought many changes, including an economic slump as the
war economy wound down and huge numbers of soldiers and sailors returned to
their homes. Horse prices fell towards pre-war levels and interest in the potential
of locomotives sank accordingly. It was once supposed that only the West Moor
workshop continued to build travelling engines after the little rush of activity in
1812–15, though engines built by Murray and Blenkinsop, Hedley, and Chapman,
continued to work, but recent research[16] shows that John Buddle was involved
with colliery locomotive projects into the 1820s. Nothing disturbs the view that
from his base at Killingworth, George Stephenson was the only person making a
serious and consistent effort to develop a steam-powered railway, though the kind
of railway he envisaged at this time was a slow-speed, coal-hauling one. In 1817
he supervised the building of a six-wheeled locomotive ordered by the Duke
of Portland for the Kilmarnock & Troon Railway in Ayrshire – this was the first
locomotive to run in Scotland. Almost immediately it was withdrawn because it
broke the cast-iron rails; the problem of getting a viable track was still unresolved.

 With work in hand, and the opportunity to develop his abiding interest,
Stephenson had little reason to feel concern. He had a good income and his own
savings apart from the subscription fund. Smiles records that he lent £1,300 to a
gentleman near Durham, with land as security; and in December 1820 he invested
£700 as a partner in a 21-year lease of Willowbridge colliery, close to Bedlington,
about 8 miles north of Killingworth, and laid a wagon track there with new
rails supplied by the Bedlington Ironworks.[17] On the home front, too, there were
satisfactions. In 1818 Robert, now 15, had left Bruce's academy to begin a three-
year apprenticeship with Nicholas Wood at Killingworth. Although he had won

no prizes at school, he had been well grounded in mathematics and would later write to Bruce's son to praise the teaching he had been given.[18] Wood, eight years his senior, was his friend as well as his master and the benevolent if piercing eye of his father was never very far away. This was a crucial step on Robert's part. Until now, his useful education could have led in various directions, even if only one had a large arrow pointing down it and a large hand propelling the lad that way. Robert did not object. If fate and his father, two forces he might well see as one, intended him to be an engineer, he was not unwilling. No other vocation made a stronger claim on him. And no engineering apprentice has ever been more fortunately placed. He was at one of the prime locations and moments in the world where human technology was opening a new perspective into the future. His teachers were his father and Wood, who, despite his aptitude for turning handles the wrong way, would prove to be an excellent recorder and describer of the mechanical innovations being made.

Killingworth's activities attracted numerous visitors, one of whom was Robert Stevenson, famous for lighthouses but an all-round civil engineer. Having taken an interest in railway developments, he came down from Edinburgh in 1818 and wrote: '... some of the most striking improvements in the system of railways are the patent inventions of Mr Stephenson of Newcastle, particularly his Locomotive Engines.'[19] His surname would have been of special interest. Part of Stephenson family lore was that Old Bob's father had come from Scotland and had connections with a landowning family, the Stevensons of Mount Grenan in Ayrshire.[20] The Edinburgh Stevensons had their origins in the same Carrick district of Ayrshire. But Robert Stevenson was a brusque, go-getting man of business, and no gossip about remote cousinship appears in the occasional letters he afterwards exchanged with George. A Frenchman, M. de Gallois, chief engineer of the *Corps Royal des Mines*, came to Killingworth on a tour of English railways and wrote one of the first accounts of the subject, *Des Chemins de Fer en Angleterre*, published in 1818. No foreign orders followed, however, though George built a locomotive for Scott's colliery at Llansamlet in South Wales in 1819.[21]

Just over a year after Robert's apprenticeship began, a hazardous incident might have ended it, and his father's career, permanently. Late in December 1819, fire broke out underground. George wrote to a former workmate, Joseph Cabry: 'The soot in the Geordy Flues caught fire beside the damper and set the Coal on fire nearly all the way to the shaft before we got it under.'[22] Robert added a P.S. to Joseph's son Thomas: 'My father has almost wrought me to Death in the Flues but he himself has been two or three times dropt with the choak damp but I took care not to go so far as that – But where I was I think I would have not been long in making a joint of Meat Ready ...' Above ground, threats of a different sort were perceived. In this year of the Peterloo Massacre, the possibility of civil disorder from radicals, reformers and disaffected workers had prompted local magistrates to form volunteer troop squads, and George went on drill:

I am sorry to tell you that I have become a soldier – we send a Dozen every day to Mr Brandling's to learn exercise – & I do assure you we handle the sword pretty well – Mr Wood makes an excellent soldier – But I hope we shall never be called to action as I think if any of us be wounded it will be on the *Back* ...[23]

George's local reputation was sufficient for Arthur Mowbray, manager of the Vane Tempest collieries at Rainton in County Durham, to commission him to survey a new coal railway in 1819. When it was decided not to go ahead, Stephenson offered to build the line himself and work it for a year, charging the old rate and expecting to make good his costs within the time.[24] Mowbray was replaced by John Buddle, no friend to Stephenson, but he became manager of a new mine, intended for high-volume production, being sunk in 1820 by the Hetton Coal Company in County Durham. Included in the project was an 8-mile railway, from the pit-head near present-day Hetton-le-Hole to staithes on the River Wear at Sunderland. George Stephenson obtained the job of planning and building this line. For him it was another hugely important step: this time, out of the mechanic's shop and into the wider world of a railway contractor. It was his own initiative, putting himself forward as the right man in the right place at the right time. Here was the chance to produce a complete 'package', embracing track, fixed engines, locomotives, wagons, everything. He had laid tracks around Killingworth and had set out at least one inclined plane before, but Hetton demanded a very different level of project management and responsibility. Here he worked out the methods of estimating for materials, equipment and labour, the calculation of costs and the finding and managing of sub-contractors, which he would apply in later contracts. George's only surviving estimate refers to horse, not locomotive traction,[25] perhaps because this was the original specification. Locomotives were certainly to feature in his plan, though only on 3.5 miles of the line. Concern about their impact on the tracks was one reason; another was Stephenson's own concern about their ability to mount a gradient with a load. The single-track line climbed to a maximum height of about 250ft (80m) before gradually dropping, over undulating terrain, to the river. To supervise construction, George, with his other commitments, managed to install his younger brother Robert – the only other Stephenson brother to become a professional engineer – as resident engineer. Two sections were worked by locomotives, three by fixed cable engines, and there were five self-acting inclined planes. Ten separate stages were thus taken for wagons to go down and up again, but the capacity of the line was much greater than horse haulage would have provided. Up to 1822 George built four locomotives for Hetton. With single-flue boilers, their steam-raising capacity and consequent power were modest. In most ways, they were hardly an advance on the engines he had built after *Blücher* in 1814–15.

All in all, the outlook for the 'travelling engine' in the years around 1820 did not seem overly bright, and when the promoters of a railway between

Darlington, on the Auckland coalfield, and Stockton-on-Tees agreed in August 1818 to have a line surveyed, it was horse haulage they had in mind. Still, in the hinterland of Newcastle, steam locomotives could be seen at work daily, and though the horse-breeders and dealers and the canal proprietors felt under no threat, some far-sighted visitors came – from great distances in some cases – to see these machines with their own eyes and gather as much useful information about them as possible.

On the home front an important change happened at this time. After fourteen years as a widower, George remarried. Aunt Nelly left Paradise Row and in her place came Elizabeth Hindmarsh to be the second Mrs George Stephenson. Such matters are not always so well managed, but from their wedding on 29 March 1820, in Newburn parish church where her husband had also married his first wife, Elizabeth seems to have glided into family life with ease. In the first edition of his biography, Smiles wrote that she had actually been George's first love, before either of the Henderson girls, but that her father, a farmer at Black Callerton, had forbidden the pair to meet. George had then looked elsewhere, and she had never married. In later editions he withdrew this story, having been informed by Elizabeth's brother Thomas that she and Stephenson had never met in those early days. Thomas Hindmarsh stated categorically that he had introduced Stephenson to his sister, at Stephenson's request, in 1818 or 1819. Subsequent biographers have repeated the more romantic version, without any new evidence. It seems unlikely that if an old love had been available, George would have waited for so long to resume the relationship. He was 39 and his new bride was 43. The marriage was a happy one and Elizabeth, always known in the family as Betty, appears to have accepted the absences and separations entailed by George's relentless pursuit of gainful work. It remained a household dominated by male voices and masculine activities.

During the first half of 1821, William James came to look at railways on Tyneside. A man of consequence, ten years older than George Stephenson, he had built up a large land-agency business, and was a surveyor, land developer and colliery-owner in his own right. In the course of his nationwide activities, he had become enthused by the idea of a national railway system, to ease and speed up the transport of goods and passengers, and for two decades had been a tireless publicist for railways. Much more than an airy theorist, James had been in touch with potential backers of a Liverpool–Manchester railway as long ago as 1803, and made a preliminary survey of its line. Having first anticipated horse traction, he became interested in what locomotives might achieve. His visit to Killingworth made him a complete convert, envisioning locomotives puffing about the country on a railway network planned by himself. George Stephenson was away on James' first visit, but the railway promoter returned in July, with William Losh. James' expansive manner and gift for imaginative projection were beguiling. George had not met anyone quite like him before, and the enthusiastic approach and compliments of a businessman who was passionate about railway development, were powerful prompts to raise his own sights beyond the horizon of the colliery railway. With entrepreneurial foresight, James made an arrangement with Losh and Stephenson whereby in consideration of 'giving his recommendation and best associates for the rising and employing the locomotive engines for which we … have obtained two letters patent …', they granted him '… one fourth part or share of our rights and patent in the exclusive use of the locomotive engine for working on railroads … and of the profits arising'. The arrangement, sealed on 1 September 1821, related only to engines made, used or sold south of a line between Liverpool and Hull.[26] James' selection of Stephenson as a collaborator helps to confirm George's pre-eminence in locomotive building at the time. From now on, Stephenson embraced the idea of railway development on a nationwide scale. In an era when canals were considered the only possible means of bulk inland transport, he foresaw that railways would do it better. James inspired that vision, and a few others shared it, but Stephenson alone combined it with the ability and willingness to tackle the nuts-and-bolts labour which would make the dream a reality.

After many arguments, several surveys and one rejection of their scheme by Parliament, the promoters of the Darlington–Stockton Railway had produced a new proposal in 1820, based on a new survey, and submitted it to Parliament. Delayed by the death of King George IV, the Bill was approved by the Parliamentary Committee at the end of March 1821. On the Committee, the leading figure was Edward Pease, head of a prominent family in Darlington, where they were involved both in woollen manufacturing and in banking. The Peases were Quakers and Edward Pease held fast to that introspective and sometimes deeply expressive form of religious practice. Despite the numerous surveys already done, he appears to have lacked confidence in their quality, and also in George Overton, the line's latest surveyor and prospective engineer. On 19 April 1821, by coincidence the day on which his railway's Bill received the royal assent, Pease met George Stephenson for the first time at his own house in Darlington.[27] Accompanied by Nicholas Wood, Stephenson rode to Newcastle, took a coach to Stockton, walked the length of the proposed line and knocked at Pease's door. Then aged 53, the Quaker was a man of stately appearance with narrowed, penetrating eyes and a judicious expression, but it did not take him long to assess his visitor. 'There was such an honest, sensible look about him, and he seemed so modest and unpretending … He described himself as "only the engine-wright at Killingworth; that's what he was".'[28] The engine-wright, setting his diffidence aside, assured Pease that: 'I think, sir, I have some knowledge of craniology, and from what I see of your head, I feel sure that if you will fairly *buckle* to this railway, you are the man to successfully carry it through.'[29] The meeting was a fruitful one, and Pease wrote to George on the 20th to confirm a verbal agreement which had been struck during the visit. Stephenson's reply, sent on 28 April, makes it clear that they had discussed his taking over entire responsibility for the project:

> From the nature of my engagements here and in the neighbourhood I could not devote the whole of my time to your Rail Way, but I am willing to undertake to survey and mark out the best line of way within the limits prescribed by the Act of Parliament and also to assist the Committee with plans and estimates and in letting to the different contractors such work as they might judge it advisable to do by Contract, and also to superintend the execution of the work.[30]

He might have crossed his fingers as he dictated this letter, for it marked another vital step forward, right into the province of the civil engineer. This was to be a public railway, not a colliery line, and any established engineer would have leapt at the chance to be its constructor. In previous years the S & D promoters had already used the services of such leaders of the profession as Robert Stevenson, whom they still retained as a consultant, and John Rennie the elder. Planned as a 12-mile tram road, for horse traction, with flanged rails, it offered an ideal opportunity for further use and development of edged rails and the steam locomotive – if the promoters could be persuaded. Pease's dominance of his Committee was sufficient for its next meeting, on 23 July, to decide to make it as a proper railway and ratify the request to Stephenson to set out his charges and cost estimates. On 27 September his terms were accepted at 2 guineas a day plus expenses. Overton was dropped, protesting about having been 'superseded not in a handsome manner at all'.[31]

On 18 January 1822 George presented his survey plans, cost estimates and general report to the S & D board, and within four days all were approved and he was appointed Engineer at a fee of £600 per annum, this to also cover his expenses and the salaries of his assistants, though it was accepted that it was only a part-time position for himself. The first of these assistants was John Dixon, who was to be one of the longest serving of George's 'young men'. George also prevailed upon Nicholas Wood to release Robert from his apprenticeship so that he too could help with the survey. Robert was glad to get away from underground work, which was affecting his health: 'His lungs were weak and manifested symptoms of tubercular disease.'[32] Robert had never been more than a day's journey away from Killingworth, but before starting work on the survey, he went by himself on a visit to London with a purse of money from his father. He intended to keep a diary but after a few entries, recording visits to the Tower, St Paul's Cathedral, the Custom House, the Water Works (he had been well briefed by someone beforehand), 'Summersite House' and an exhibition of a facsimile Egyptian tomb, the record stops.[33] Though younger boys worked on the coal brigs to London, it was quite a momentous thing for a provincial youth of 18 to make a trip of this sort.

George had not been to London; in this, as in applied mathematics, Robert was ahead of him.

William James, still pursuing the idea of a Liverpool–Manchester railway, and in touch with potential investors in Liverpool, was keen to bring Stephenson into the scheme. George, despite his commitments in the north-east, was not loth. A frequent visitor now to Pease, he saw the daughters of the house one day engaged in embroidery, and in typical style began to tell them how to do it: 'I know all about it'[34] – yet another accomplishment he had acquired while engine-watching by night was stitching buttonholes in coats. Out on the moors in fine autumn weather, Robert's health improved rapidly. One of the numerous Pease sons, Joseph, saw George and his son at work one day, Robert 'a slight, spare, bronzed boy', both arguing at the tops of their voices. Visitors to Paradise Row had also noticed that a vigorous dialogue went on between the two Stephensons on the details of locomotive design; to a degree George was perhaps deliberately leading Robert on, but clearly Robert was developing his own ideas and was not inhibited about expressing them. Though George had persuaded Pease and his board that locomotives should be employed, it was only to be on the flatter, western section of their line. Horses, inclined planes and winding engines were also to be used.

During the summer of 1821, George Stephenson went to examine an important new technical development. John Birkinshaw, an engineer at the Bedlington Ironworks, had produced and patented a new kind of running rail, of an L-shape, made of wrought or 'malleable' iron in 15ft lengths. These were a great improvement on any previous rail and George, always conscious of the shortcomings of the existing track, was generous in his praise, writing to Robert Stevenson: 'These rails are so much liked in this neighbourhood, that I think in a short time they will do away with the cast iron railways. They make a fine line for our engines, as there are so few joints compared with the other.'[35] He recommended the new rails to the S & D board, despite their cost being almost double that of the old type, at £12 a ton from Bedlington compared to £5 10s.[36] William Losh, who might confidently have expected to supply cast-iron rails with the special joint which he and Stephenson had patented, was angered by this recommendation of a competitor's product, even if it was better. A letter from Losh to Pease, suggesting that Stephenson and Michael Longridge, owner of the Bedlington works, were in collusion,[37] was followed by a furious confrontation between Losh and Stephenson, which marked the end of their partnership and friendship. While George must have regretted the breach, the cessation of his retainer and the earnings from the patent rails, he could reflect on the credit his disinterested action gave him with the shrewd and observant Edward Pease -- but far more important was getting the best road for his engines. That was building into the future, and that was what mattered.

# 5

# MUSIC
# AT MIDNIGHT

The break with Losh meant that Stephenson had to find another iron founder to supply cast- or wrought-iron parts. Michael Longridge was already a friend, but he could not provide the range of items that Losh, Wilson & Bell could make to order. George made contact with the brothers Isaac and John Burrell, who had a foundry at Orchard Street and South Street in Newcastle. Smiles implies that he was already friendly with them and states that he had even discussed the possibility of emigrating with one of the Burrells to set up a steamship-building business on the Great Lakes.[1] Though this may have been speculative rather than serious talk, it suggests that he might have been concerned about business prospects in Britain. What he did now shows how much his *modus operandi* had expanded: he formed a partnership with the Burrells. Their foundry made the girders for the S & D bridge, designed by George Stephenson and built over the River Gaunless in 1823–24. Although it has been described as the world's first iron railway bridge, this was not the case,[2] but it was a completely novel form – four short wrought-iron spans in double-bow formation, supported on splayed cast-iron pillars and with the track on top. Invited by William James, and with his father's approval, Robert joined the small team making a new route survey for the long-projected Liverpool–Manchester Railway and spent a few weeks in September and October 1822 on this task. Working with other young men, it was an enjoyable time, though they met hostility from some landowners. George, finding himself increasingly in demand for railway projects in other parts of the country, became impatient and wrote to James on 17 September: 'I am very much in want of Robert, you will send him off as soon as possible …'[3] – but Robert did not come back until early in October, and then for only a brief interlude at home before it was time to set off on another new venture. He was going to university.

As a baptised member of the Church of England, he was eligible to enter a college of Oxford or Cambridge, but the chosen institution was the University of Edinburgh. It was the closest of the three to Newcastle, but this is unlikely to have influenced the Stephensons' choice. Robert Bald, an Edinburgh technical writer who had visited Tyneside and become a friend of the Stephensons, was the main advocate. His article on 'Mines' in the *Edinburgh Encyclopaedia* of 1820 refers to a meeting with George. In the previous century, the Scottish universities had reformed and modernised their teaching methods and subjects to a much greater extent than the English ones, and Edinburgh, even if by the 1820s it was beginning to coast on the momentum of earlier renown, was the place where geology was being elaborated into a science and where other sciences, though still grouped under 'Natural Philosophy', were treated as modern topics by teachers who kept up with new developments and in some cases still led the field. The Scottish universities were also more democratic institutions than Oxford or Cambridge; open to the full range of the (male) population so long as they fulfilled the professors' academic criteria and could afford the modest fees. Oxford and Cambridge catered for the socially elite groups with which George Stephenson felt no affinity. 'Robert must wark, wark, wark, as I ha'e warked afore him,' he remarked to more than one acquaintance, and Jeaffreson also quotes him, in 'his own words', that it was not his wish to 'make his son a gentleman'. It was apparently by the suggestion of 'several gentlemen who came in contact with him [Robert] during the survey for the line' that George agreed to send his son to university.[4] If so, it was a very rapid decision and it is reasonable to suppose that Robert himself was keen to go.

'If a man is poor, plain, and indifferently connected, he may have excellent opportunities of study at Edinburgh,' wrote Sir Walter Scott in April 1824;[5] 'otherwise he should beware of it'. Robert avoided the 'speculating mammas and flirting misses' of the city. At 19 he was older than many of his fellow students, some of whom were only 14, and found that much of the information in some lectures was already familiar. He attended classes on Natural Philosophy, Natural History and Chemistry, taking assiduous notes which he later preserved in bound volumes. His cicerone was Robert Bald, from whom he had introductions to Dr Hope (Chemistry), Professor Robert Jameson (Natural History) and Professor John Leslie (Natural Philosophy), who were to be his lecturers. Knowledge of his father's work on the safety lamp (rather than on locomotives) made Robert someone of interest to the teachers, particularly the chemists. He found Jameson on Natural History tedious; writing to Michael Longridge he observed that: 'Natural historians spend a great deal of time in enquiring whether Adam was a black or white man. Now I really cannot see what better we should be, if we could even determine this with satisfaction; but our limited knowledge will always place this question in the shade of darkness.'[6] Jameson on geology was much more to his taste, and after the end of the academic year he joined a group of students who accompanied the professor on a botanical and geological field trip through the Highlands as far as Inverness. This involved a passage along Thomas Telford's recently opened Caledonian Canal through the Great Glen, but Robert made no comment on that feat of engineering. Late in life, he called the excursion the 'first and brightest tour of my life'.

His sojourn at university seems short, but a year's academic work was packed into the October–April period, with a lengthy summer break. It was not intended that he should do a full degree course; as was not unusual, he attended the classes of certain professors and received 'class tickets' attesting what he had studied. Professor Leslie presented him with a prize book, appropriately Charles Hutton's *Tracts on Mathematical and Philosophical Subjects*, inscribed in Latin: *Iuveni meritissimo Roberto Stephenson ... Aequo Certamine Optime Partum* [best result in an even contest]. One of his fellow students was George Parker Bidder, a Devon-born child prodigy known as 'The Calculating Boy' for his ability to work out vast calculations instantaneously in his head. Aged 16 in 1822, he had already been at the university for two years. A memoir by Bidder's friend, James Elliot, confirms that Bidder knew Robert, a 'tall young man, with high shoulders', but they were not particular friends in their university days.[7] He does not appear to have made any links with the 'lighthouse Stevensons', though Robert's sons were very much of his own age. His own son's six months at Edinburgh cost George Stephenson £80. Even if Robert had not learned a great deal that was specific to civil and mechanical engineering in that time – rather surprisingly he did not attend the advanced mathematics course under Professor William Wallace – he had unquestionably benefited from the association with scholarly men, and geology would remain a lifelong interest.

While Robert was at Edinburgh, on 31 March 1823 his father wrote one of his forceful and expressive letters from Killingworth to his former workmate at Water Row, William Locke, now the manager of a coal pit at Barnsley in South Yorkshire. Locke and he had been in touch some years previously when Locke had considered a move back to the north-east:

Dear Sir, – From the great elapse of time since I seed you, you will hardley know that such a man is in the land of the living ... This will be handed to you by Mr Wilson a friend of mine who is by profeshion an Atorney at law and intends to settle in your neighbourhood. You will greatly oblidge me by throughing any Business in his way you conveniently can ... If, I have the pleasure of seeing you I shall give you a long list of occurences since you and I worked together at Newburn. Hawthorn is still at Walbottle I darsay you will well remember he was a great enamy to me but much more so after you left. I left Walbattle Colliery soon also after you and has been very prosperous in my concerns ever since. I am now far above Hawthorn's reach. I am now concerned as Civil Engineer in different parts of the Kingdom. I have onely one son who I have brought up in my own profeshion he is now near 20 years of age. I have had him educated in the first Schools and is now at Colledge in

Edinbro' I have found a great want of education myself but fortune has made a mends for that want.

I am dear sir yours truly,
Geo. Stephenson.[8]

As a guide to how George saw himself in 1823, this letter is illuminating; this preening to his friend is quite different to his modesty, 'only the engine-wright at Killingworth', in front of Pease two years before. But even allowing for the fact that we can all behave differently in particular situations, Stephenson's self-perception had changed in the course of these first years of the 1820s; his belief in himself expanding in line with his business commitments and also hardening in the face of criticism. A hostile report on some unspecified work on which he and his brothers were engaged referred to: '… such a mechanic as Mr Stephenson, who can neither calculate, nor lay his design on paper, or distinguish the effect from the cause, may do well for repairing engines when they are once constructed, but for building new ones, he must be a great loss to his employers …'; and he was compared to a fly going round and round on a crank axle while shouting, 'What a dust I am kicking up!'[9] Finding that William Locke's 17-year-old son, Joseph, was at a loose end, he took the boy on as a pupil in the early summer of 1823, on a three-year unsalaried apprenticeship. The Hetton colliery railway had opened on 18 November 1822, and the Stockton & Darlington works were in progress. There was no need to advertise; anyone seriously considering the possibility of building a railway would soon meet the name of Mr Stephenson of Newcastle. He had been to London in 1822 to organise the purchase of wooden blocks on which to lay the rails. Stone blocks were his preference, but there was no suitable quarry close to the Darlington–Stockton line. These blocks were not transverse sleepers, as later used; this was initially because of the need to provide an unimpeded path for horses, but Stephenson would be slow to follow the example of full-width sleepers. Another trip had been to the Neath Abbey Ironworks in South Wales, to order cast-iron rails. Despite his advocacy of malleable iron, the directors opted for cast-iron rails on the 'loop' sections where trains would pass each other.

Construction of the Stockton and Darlington Railway was entrusted to a large number of small contractors, each responsible for the preparation of a short section. James and John Stephenson were involved either as foremen or sub-contractors. The first rails were laid on 23 May 1822, but some sections could not yet be started. Stephenson's survey diverged so far from the original line approved by Parliament that a revised Bill had to be prepared, and a new Act was approved in May 1823, receiving the royal assent on the 23rd. In the Bill, the opportunity was taken to allow the company to use steam locomotives, and a new word entered the official vocabulary:

… it shall be lawful for the proprietors to make and erect such and so many loco-motive or moveable Engines as the said Company shall from time to time think proper and expedient, and to use and employ the same in or upon the said Railways or Tramroads or any of them, for the purpose of facilitating the trans-port conveyance and carriage of Goods, Merchandise and other articles and things upon and along the same Roads, and for the conveyance of Passengers upon and along the same Roads.

Although the Bill was not opposed, it still had to be explained to Parliamentary Committees of both Lords and Commons, and George Stephenson went to London with the rest of the team to explain what a locomotive was. It would seem that Lord Shaftesbury, chairman of the Lords Committee, had some trouble with the concept, and George, having gone back to the north-east and then asked to return to London, wrote crossly to Francis Mewburn, the S & D's company solici-tor: 'Your letter by this day post has cut me most sadly … Lord Shaftesbury must be an old fool. I always said he had been a spoilt child but he is a great deal worse than I expected.'[10] This was the 6th earl, not a Parliamentarian of great distinction, aged 55 in 1823. His family's traditional Whiggism would not appeal to George.

The S & D was the first public railway in the world to be authorised to use locomotives, though they would often still be referred to as 'travelling-engines'. George Stephenson had confided to William James in a letter of 21 December 1821 that, 'I fully expect to get the engine introduced on the Darlington railway';[11] and in the summer of 1822 Edward Pease and his cousin Thomas Richardson had gone to Killingworth to see the colliery locomotives at work. Richardson, also a Quaker, was an influential London-based banker, partner to Samuel Gurney with whom he founded the great banking concern of Overend, Gurney & Co. The visitors were impressed and Pease convinced his board that the employment of locomotives would be beneficial to the company.

But where would these locomotives be built? West Moor was only a colliery workshop, the property of Lord Ravensworth, who had allowed Stephenson to build locomotives for collieries with which he was connected; but this was a different proposition. On 23 June 1823, very soon after the second S & D Act, George Stephenson, Edward Pease, Thomas Richardson and Michael Longridge agreed to form a new company, to be known as Robert Stephenson & Co., its business being 'Engine builders, Millwrights, etc.' Usually referred to as the world's first purpose-built locomotive works, it was in fact a general engine-builder from the beginning, similar to other works, though unique in that locomotive con-struction was the prime intention. Robert, not yet 20 years old, and newly back from Scotland, was to be managing partner, at a salary of £200 a year. His role in the discussions, which must have begun while he was still at Edinburgh, is not clear. The initial capital of the business was £4,000, and of the ten shares, Pease held four, and Longridge and the Stephensons held two each. Richardson did not become a partner at this time. Robert's £400 was lent to him by Pease, presumably because his father's capital was already tied up. Despite this thrusting of the youthful Robert into prominence, the partners ensured that his appointment was 'upon condition that his Father George Stephenson furnish the Plans etc., which may be required, and take the general charge of the Manufactory as long as required by the Partners',[12] and provision was also made for acquisition of patents held by George and sharing in any new ones.[13] While the new business could confidently expect to build locomotives for the Stockton & Darlington Railway, it was in every other way an act of faith in the future – an investment in two unknown quantities, the demand for steam locomotives and the abilities of Robert Stephenson.

The obvious thing would have been to call it George Stephenson & Co.: after all, it was George's name that was being mentioned in every new account of railways, from Robert Stevenson's *Report on Railways* to William James' pamphlets. Only George is likely to have been the proposer of his son's name, and its acceptance by his fellow investors shows them as ready to defer to him. Robert himself had clearly impressed them sufficiently to accept his nomination, despite his inexperience. No time was lost in acquiring a site adjacent to the Burrells' works in Newcastle, in the angle of South Street and Forth Street, which was leased, in anticipation of purchase, from 26 July.[14] George Stephenson, incidentally, was still a partner of the Burrells and remained so until the end of 1824.[15] Robert Hawthorn, eldest son of George's one-time supervisor, also had an engineering workshop in Forth Street, which opened in 1817. Meanwhile, George had other work for his son, delegating to him the responsibility for surveying, preparing plans and estimating costs for a proposed branch to be added to the S & D line 4.75 miles (7.6km) long, at the Darlington end, linking Lord Strathmore's colliery at St Helen's Auckland to the railway. Robert was designated the engineer, and John Dixon assisted him as surveyor. Dixon later recalled an utterance of George's from that time which shows how much the elder Stephenson was in tune with the idea of the age: 'The time is coming when it will be cheaper for a working man to travel upon a railway than to walk on foot.'[16] 'Time is money' was already a proverbial phrase, but it was to be one of the fundamental precepts underlying the industrialisation of both processes and people.

In his new job, once again, Robert was being given invaluable experience, but his father, one feels, was forcing the pace. By comparison, Robert Stevenson in Edinburgh expected an aspiring engineer to spend three years at college as well as twelve to eighteen months in a foundry, before taking on active responsibility.[17] Having completed his survey work, Robert went with his father to London in August 1823 to promote the engine works.[18] George found an excellent opportunity to publicise his own steam designs by demonstrating the ineffectiveness of a much-touted 'high-pressure' engine designed by the American-born inventor Jacob Perkins, showing he could stop its motion with only his hand. They went

John Dixon

on to Ireland in September – visiting Dublin and Cork. Ireland had no railways, but some cloth and paper-mills had steam engines, and George was pursuing an engine order for the Dripsey Paper Works at Cork. In 1823 Ireland was no tourist venue, and a letter from Robert to Longridge reflects this. As the Cork mail coach left Dublin, '… we were not a little alarmed, when it stopped at the post-office, to see four large cavalry pistols and two blunderbusses handed up to the guard, who also had a sword slung by his side. I can assure you, my father's courage was daunted, though I don't suppose he'll confess with it.'[19] Robert noted the distress and poverty of the people, 'indeed, many of them appeared to be literally starving', but with the equanimity of a detached observer, and his account shows no interest in the causes of this state of affairs.

On their way home, father and son spent a few days in Shifnal, Shropshire, near Coalbrookdale, where the earliest ironworks in England had been established. In the course of 1823, the family moved from the cottage at West Moor to a new development of handsome townhouses in Eldon Street, Newcastle: a further indication of George's changed status. At this time, if not before, his arrangement with the Grand Allies came to an end. He was too busy to spend time checking on the condition and action of colliery engines; he was a consulting engineer, a partner in the engine-building business, sharer of a lease in a colliery

– his plate was full, and his fortunes were on the upward path. His friend William James was experiencing the exact opposite. In 1823 he was made bankrupt, and the energetic part he had played in the proposing, planning and surveying of a Liverpool–Manchester railway came to a halt, just as the project itself was gaining momentum among a group of wealthy Liverpool merchants. James struggled on with two smaller railway projects he had sponsored, the Canterbury & Whitstable and the Stratford & Moreton. His personal disaster prompted the sympathy of Robert, who wrote to him on 29 August 1823, just before leaving for Ireland with his father:

> It gives rise to feelings of true regret when I reflect on your situation; but yet a consolation springs up when I consider your persevering spirit will for ever bear you up in the arms of triumph, instances of which I have witnessed of too forceful a character to be easily effaced from my memory. It is these thoughts, and these alone, that could banish from my thoughts feelings of despair ... Can I ever forget the advice you have afforded me in your last letters? and what a heavenly inducement you pointed before me at the close, when you said that attention and obedience to my dear father would afford me music at midnight? Ah, and so it has already ...[20]

The interesting thing about this letter is not its effusiveness – far more gushing epistles were being written in 1823 – but the very clear inference that Robert was chafing to some degree in his relationship with his father, and that he had confided in James, who had five sons, one of whom, George Walter, was just a year older than Robert. James appears to have given predictable if sage and finely put advice; the phrase about 'music at midnight' seems to be his own. Troubled and ill as he was, it is a tribute to James' character that he should have advised the younger Stephenson in such terms.

The link with London business and finance, established through Thomas Richardson, produced more enquiries to Robert Stephenson & Co. than actual orders for engines, fixed or moving. Through the winter of 1823–24, workmen and apprentices had to be taken on and workshops had to be built. The company's first engine was its own, to power the machinery. The possibility of new business was not confined to the British Isles. As the Central and South American states freed themselves from Spanish rule, their legendary mineral wealth, especially in silver, attracted the interest of London financiers and entrepreneurs, who moved swiftly to obtain mining concessions from the new governments. But City magnates were not the fuglemen of this rediscovery; as long ago as 1816 a disillusioned Richard Trevithick had left England for Peru, hoping to make in the silver mines there the fortune which had eluded him with his locomotives. Though many of the workings were disused or derelict, modern technology, it was felt, could extract much that the primitive methods of earlier times had left behind.

Richardson was among the prime movers, and bombarded George Stephenson with enquiries relating to the Mexican Mining Company. To Richardson and his colleagues, one mine was presumed to be very much like another, and George Stephenson was a man who knew about mines.

Very little of Robert's attitude at this time is discernible in letters which have been preserved, and his own actions represent the prime guide to what he was thinking. At first, there was a suggestion that he should make a comparatively short visit to Mexico to ascertain the type of machinery needed by the mines there; and it was made via George. In London, while the Bill for the new Stockton & Darlington branch was in Committee, in case he should be required to give evidence (he was not), Robert received a letter from Longridge on 14 February 1824 to say: 'You would learn from your Fathers letter that we are making arrangements for your going out to Mexico ...'

Longridge asked him in the meantime to consider:

1. if you would like to go
2. how long you would probably be from home
3. what you should have for salary, and
4. how are we to do at home for want of you.[21]

At the same time, the other Robert Stephenson, George's brother, was also being canvassed for the Mexican job, as his nephew soon found out. In early March both Roberts went on a tour, at Richardson's behest, to examine engine sites and mining installations in South Wales and Cornwall. Uncle Robert would later decline the opportunity, but the younger Robert, returning from Cornwall to London, where his railway Bill was still under discussion, met J.D. Powles, chairman of the Colombian Mining Association. The projected American visit had taken on a new aspect. The promoters of the Association knew that mechanical and mining knowledge on the spot was essential; it was obviously impossible to get George Stephenson, who had committed himself to the Stockton & Darlington project, so why not his able young son? Even before his visit to the south-west the prospect of going to Colombia may have been raised with Robert, since he wrote to his father from Okehampton in Devon on 3 March 1824, stressing the value of gaining experience elsewhere:

When one is travelling about, something new generally presents itself and though it is perhaps not seperior to some schemes of our own for the same purpose, it seldom fails to open a new channel of ideas ... this I think is one of the chief benefits of leaving for awhile the fireside where the young imagination receives its first impression.[22]

South America is not mentioned in the letter, which ends: 'Whenever the

Factory comes to my recollection I wish sadly to be at home. I know you will find a want of me …' By 9 March, a letter from Robert to Longridge treats as *fait accompli* that the Stephenson Company is to build a railway or road from La Guaira, on the Colombian coast, to the capital, Caracas: 'We are to have all the machinery to make, and we are to construct the road in the most advisable way we may think, after making surveys and levellings.'[23] His seniors had not anticipated that the young managing partner would shed the leading strings to the extent of making such deals himself, including his own services in the package. While George's reaction has not been directly preserved, there is no doubt that he was very much against the idea of Robert going. Many father–son confrontations centre on a family business or an estate, and in this case, although there was no patrimony or tradition to be conserved, there was a vision of the future to be secured. George, not a harsh father, though he may on occasion have been an overbearing one, was not immune from the paternal desire to see in his son a more successful version of himself, nor from the paternal anxiety that the son might not achieve this without help. He was always a man who wanted results quickly, as the High Pit engine, the safety lamp experiments and even the sundial at Paradise Row, all show. At this opening stage in his son's professional career, George had laid out, with good and firm intention, the track for Robert to follow. With all the strength of his own powerful personality, he busied himself to secure a position for Robert that really required a far more experienced man, and in which guidance would obviously be necessary. To Robert it was plain where his guidance would come from. His own life was at a most delicate and difficult stage – he had tasted independence, made new friendships outside the family, encountered new ideas and new places, and found, perhaps, that his wonderful father did not have all the answers to everything. Had he been less ambitious and less clever he might have gone obediently to Newcastle. But he was, in his own lower-key way, as determined a character as his father. A deeply disingenuous letter went to George in April, marshalling his case:

> But let me beg of you not to say anything against my going out to America, for I have already ordered so many instruments that it would make me look extremely foolish to call off. Even if I had not … it seems as if we were all working one against another. You must recollect I will only be away for a time; and in the meantime you could manage with the assistance of Mr Longridge … And only consider what an opening it is for me as an entry into business; and I am informed by all who have been there that it is a very healthy country. I must close this letter, expressing the hope that you will not go against me for this time.[24]

Robert knew his father well enough to touch the most responsive nerves. George wanted to maintain good relations with the London capitalists, and he certainly wanted Robert to be successful. Robert was clearly intending to come back to rejoin the family business. Mention of his health was shrewd: Jeaffreson asserts that at this time 'the threatening symptoms of pulmonary disease … seemed decidedly on the increase'.[25] Robert does not seem to have taken much trouble about his health: the person who cared more was his father. George took Robert to see Dr Headlam, Newcastle's leading physician, who opined that a temporary stay in a warm country would be beneficial.[26] It may have helped that Robert was to have a travelling companion, Charles Empson, a family friend from Newcastle, eleven years older, who seems to have gone along for his own diversion and edification.[27] In the end, George Stephenson gave his reluctant agreement. His fundamental common sense, as well as his love for Robert, can be given credit for this. A more obstinate parent might have provoked an irreparable breach.

It looked as though Newcastle would be George Stephenson's base of operations, and his new status in that town was indicated by his chairing a meeting for the establishment of a Mechanics' Institute in 1824. Though the Loshes and their friends stayed away, the aim was achieved and the Institute 'struggled into existence'.[28] Robert returned to Newcastle for only a couple of weeks in April before leaving again for London, where he took lodgings: first at the London Coffee House on Ludgate Hill and later at 6 Finsbury Place South. Through May and June he had much to do, including taking a crash course in Spanish and another in mineralogical chemistry, as well as finding out just what was expected of him, what equipment he might need to order or specify, and, not the least difficult, obtaining precise terms of employment from the concessionaires. In the end, he found he was to be the employee, not of the Colombian Mining Association, but of their agents, Messrs Herring, Graham and Powles. His salary was a handsome £500 a year (of which he arranged for £300 to be paid to his father) plus expenses. In typical style, Herring, Graham and Powles instructed him to depart from Falmouth; then, even as he was boarding the coach with a vast amount of excess baggage, he was told that embarkation would now be from Liverpool. Transferring his 21 hundredweights (1,066kg) of luggage to the Liverpool coach, in two instalments, he arrived in the northern port on 8 June.

The fortuitous diversion enabled him to see his father before he left the country. While Robert had been busy with his own avocations in London, George had received a letter sent on 20 May by the promoters of the Liverpool & Manchester Railway, informing him that subject to his acceptance, he was appointed as their engineer. His acceptance was prompt, and when he came to Liverpool on 12 June, it was to take up the reins of his new post and incidentally to wish his son Godspeed. Though he now had even more reason to want Robert to remain in England, he did not make a show of it. Much socialising went on in Liverpool and he wrote to Longridge on 15 June to describe their entertainment by the city's leading men: 'I was much satisfied to find that Robert could acquit himself so well amongst them. He was much improved in expressing himself since I had seen him before; the poor fellow is in good spirits about going abroad, and I must make the best of it.'[29] However, George himself was acquiring touches of polish. The Liverpool magnates lived in grander style than the Peases or Longridges, and he may have benefited from some advice offered by Edward Pease, who wrote to Richardson: 'I have pressed on GS – he should always be a gentleman in his dress, his clothes neat & new, & of the best quality, all his personal linen clean every day. His hat & upper coat conspicuously good, without dandyism.'[30] Robert wrote to his step-mother from Liverpool that: 'Glad would I have been to have joined my father in his undertaking at Liverpool but I do not even now despair of taking the chief part of his engagements on myself in a year or two ...'[31]

For Robert, going to Colombia was meaningful in many ways, of which 'making a fortune' was probably the least important. It was an adventure, with all the lure of far-off travel, exotic places, other climes, new people and unforeseeable events. There was a job to do, for which he knew he possessed the skills. Above all, perhaps, it was his own venture, the first in his life that he had created for himself. In Colombia, to be the son of Mr Stephenson of Newcastle would count for little or nothing – Robert Stephenson would be on his own. Whether he felt stifled by his father's proximity and authority, or whether his share of the family determination made him want to assert his independence, or both, the Colombian Mining Association gave him the chance and he seized it. On 18 June he sailed for La Guaira on board the sailing vessel *Sir William Congreve*. What he had told no one was that he had contracted to go not for one year, but for three. Knowledge of that would have caused a far greater row, and almost certainly would have resulted in the closure or drastic restructuring of Robert Stephenson & Co. By the time his partners found out, he was in Colombia.

# 6

# 'DID ANY IGNORANCE EVER ARRIVE AT SUCH A PITCH?'

George Stephenson was offered the crucial post of engineer to the Liverpool & Manchester Railway without formally applying or even being interviewed. Nothing could say more about his status as the country's prime railway engineer. In May 1824 four members of the Provisional Committee, Joseph Sandars, Lister Ellis, Henry Booth and John Kennedy, crossed the country to look at railway development in the north-east, visiting the Stockton & Darlington line – still under construction – and Stephenson's other railways at Killingworth, Bedlington and Hetton. Sandars had met Stephenson on an earlier visit to Killingworth, probably arranged by William James, but on this occasion George was away, though at all these places his name loomed large. On returning to Liverpool, the delegates reported favourably on the value and utility of railways, and of steam locomotives. At this meeting, on 20 May 1824, it was formally agreed to proceed with the formation of the Liverpool & Manchester Railway Company as a public company with capital of £300,000, and to appoint George Stephenson as engineer, to make a definitive route survey and prepare plans to be submitted to Parliament in the session of 1825. His salary would be £1,000 a year.

The proposal to link Liverpool and Manchester by railway already had a considerable history. These two old market towns were now fast-growing industrial cities whose populations exceeded 100,000 and were rising steeply, with Liverpool primarily focused on its port and Manchester as the centre of the cotton-spinning trade. Raw cotton was being imported from North America in very large amounts: 573,512 'bags' of cotton were imported via Liverpool in 1823; 87 per cent of Britain's total cotton imports, and most were shipped on to Manchester, either by the Mersey & Irwell Navigation (river transport) or the Bridgwater Canal. Both these systems carried coal and much else in the way of goods; both were increasingly overloaded, charged high rates and were very slow. The cost and delays of inland water transport were felt by the merchants and manufacturers to be a major hindrance. To William James, this was one of the most obvious cases in the country for a railway line, and when, in 1821, he was introduced to Joseph Sandars, one of the leading Liverpool protesters against the water-borne carriers, the idea was reinvigorated.

James, still ebulliently active, offered to make a preliminary survey for a line at his own expense. This was a very sketchy one. Soon afterwards he made his visit to Killingworth and became an advocate of locomotives as well as railroads, and the more detailed survey of autumn 1822, in which Robert Stephenson participated, followed. The self-interest of the canal proprietors, and the unwillingness of local landowners to see their grounds traversed by a railway, generated far more hostility to the Liverpool & Manchester scheme than there had been to the Stockton & Darlington. But the Provisional Committee was determined, well resourced financially and well backed. Headed first by John Moss, a prominent banker and lumber dealer, and then by Charles Lawrence, Liverpool's mayor, it had in Sandars a vocal and effective publicist, and in the 35-year-old Henry Booth an effective treasurer and organiser. Their collective determination and skills were opposed by the formidable Robert Haldane Bradshaw, Superintendent of the Bridgwater Trust, whose policy was described as 'Profit-extraction to the utmost limit, regardless of the feelings and interests of the users of the canal',[1] and who rightly saw the railway project as a vital threat.

On George's appointment the Liverpool Committee severed its connection with William James. Sandars wrote to him on 25 May to intimate that: '... the Committee have engaged your friend Mr G. Stephenson ... I very much regret that by delay and promises you have forfeited the confidence of the subscribers. I cannot help it. I fear now you will only have the fame of being connected with the commencement of the undertaking. If you will send me down your plans and estimates I will do everything for you I can ...'[2] James, who had spent some months in prison on account of his bankruptcy, had failed to send in the survey plans, for which he had been paid £300, in time for the 1823 parliamentary session, and still had not done so by May 1824. In an attempt to get back into the game, he asked his brother-in-law, Paul Padley, who had assisted him in his survey, to go to Liverpool on his behalf. But Padley, following what may be regarded as an astute, or only a prompt, move on Stephenson's behalf, had already been recruited with his two colleagues for the new survey. Sad and angry, James believed for the rest of his life that George Stephenson had deliberately schemed to oust him and put himself at the head of the project. He did not yet know that Stephenson was also being approached by the committee of another of 'his' projects, the Birmingham and Liverpool Railway.[3] In fact, Stephenson was simply the beneficiary of James' misfortune, but when he became aware of the other man's resentment, he returned it in full measure. No more would be heard of the agreement made in September 1821. As with Losh, it had been a short partnership. Bankruptcy did not necessarily mean disgrace or the end of a career; Marc Brunel spent time in a debtors' prison in 1821, but was rescued by friends who mobilised government support. James, despite his many contacts, was left in poverty. Stephenson has been criticised for dropping James, but theirs was a commercial relationship and if James had lost the confidence of the promoters, then he was an encumbrance.

Stephenson now had a great deal of work to supervise and to do. 'I have quite come to a conclusion that there is nothing for me but hard work in this world therefore I may as well be as chearful as not,' he wrote to Longridge on 7 June 1824. Somehow, two days later, he had '... been through the whole of the Birmingham Line ... On my arrival at Birmingham a meeting of the most wealthy of the subscribers took place, who would not hear of any other executing their work but myself ...' They accepted his charges even though '... it made me blush to ask it but the hardening lessons I had got from you made me stand to it'. Sometimes details were overlooked, and he was nettled when reminded of it. Twitted by Longridge for sending off an undated letter, he fired back on 11 July: 'You speak of ... meeting me at Manchester, but the time and place you have not mentioned which is equally as bad as an undated letter and rather worse.'[4] The Stockton & Darlington line, with its many sub-contractors, most of them needing close watching and frequent chivvying, was still being built, and without his dynamic presence, construction slowed down.[5] On 16 September the S & D ordered two locomotives from Robert Stephenson & Co. and four stationary engines were already on order from November 1823. A full survey for the Liverpool–Manchester line had to be completed in time for plans to be prepared for the next parliamentary session; and cost estimates also had to be worked out. But interest in railway projects was increasing fast in 1824–25, to an extent that has been described as the 'first Railway Mania'. In *A History of the English Railway*, John Francis reckoned their number as 59, and the potential capital required as £21,942,500. The agreement George had made with Losh and James was a dead letter now, but anyway that only related to locomotives. Though many of the proposed schemes were shadowy or tentative, some, like the Newcastle & Carlisle, were being actively planned, and there were many signs that a vast market for railway surveying and construction was about to open up. The legal profession had also begun to notice the potential of railways, and: 'The ranks of lawyers, more than any other occupational group, contained men with the blend of technical skill, business savvy, commercial contacts and raw avarice needed to promote bubble railway companies at a profit.'[6] George Stephenson had no truck with bubble companies, but by the end of 1824 he had accepted commissions to survey not only the Liverpool & Birmingham, but also the London & Northern and London & South Wales. William James wrote to a Liverpool & Manchester director on 5 September: 'I do not quarrel with Mr Stephenson's good fortune, although, through his fraud, I have lately lost the Birmingham appointment.'[7] But James had lost credibility with railway projectors and was condemned to watch as new lines were proposed.

This embracing of new and substantial responsibilities alarmed Edward Pease, who had already been thinking that George was over-extending himself. On 23 October he had written to Richardson that the engineer's execution 'is torpid, defective, and languid as to promptings',[8] and in December he wrote to Richardson and Longridge to propose that '... we really ought, some way or other, to engraft ourselves on to GS emoluments so far as to indemnify us from loss ... he should not place our property at risque by the application of his time and talent to other objects'.[9] A frank exchange of views between his partners and George took place, and on 30 December 1824 a new company was formed, George Stephenson & Son, with the specific brief of surveying and building railways. Like Robert Stephenson & Co., it had ten shares, of which four were held by Edward Pease (two of them on behalf of Thomas Richardson), two by Michael Longridge, two by George Stephenson and two by Robert Stephenson, who had not been consulted. Around the same time, it was decided that a foundry should be set up at the Forth Street Works; until then, Burrells had supplied castings. On this, George sold his quarter-holding in Burrells.[10]

In contrast to the view that George Stephenson was pushed into railway surveying by his Quaker partners, it is plain that he had taken the initiative and that they were scampering after him. It was the company structure, not the work of surveying, that was wished upon him. The new company was a vehicle for selling his expertise. He and Robert would each receive a salary, after operating expenses and before profit distribution, of £1,500 annually. Michael Longridge was to handle correspondence and accounts. Pease and Richardson had no executive roles. Clause 6 of the agreement provided that Robert Stephenson could, within three months of his return to England, give three months' notice of his intention to terminate his part in the arrangement, if he should wish to do so. A letter from Longridge to Edward Pease of 18 January 1825, with a schedule of 'Proposed Regulations', adds important details. James Adamson is named as secretary at a salary of £100 and seventeen other individuals are named as surveyors or apprentices at salaries ranging from £300 to nothing, or, in the case of four apprentices, board and lodging only. Daily coach hire and travelling expenses were allotted from a guinea for those earning £100 a year or less, up to 5 guineas for George and Robert. Survey teams were allocated to four railways: London & Northern, Liverpool & Birmingham, London & South Wales, Liverpool & Manchester. Robert Stephenson's name heads 'London & Northern', with Joseph Locke as his number two. These were not actual railway companies, but at best organising committees: apart from the Liverpool & Manchester they represent a schema of trunk railway development, as thought desirable by George Stephenson. But by February 1825 Locke was busy surveying a real line, the Leeds & Hull Railway. Longridge notes: 'Some few are already employed by Geo. and he proposes to engage the others at the salaries mentioned if he should be appointed engineer for those Rail-ways.'[11]

William James and another crusader for a national rail system, Thomas Gray, had already, in 1820, provided outlines of what James had called The Central Junction Railway or Tram Road.[12] George Stephenson, already being approached by promoters of railways in different parts of the country, was keen, for reasons of good order as well as personal ambition, to see a systematic approach prevail. It was amply evident that men with access to money, the Peases and their connections, and also the Liverpool merchants, were willing to invest in railways. Since the future of the railway industry was based on two prime elements: the preparation and laying down of a good track, and the provision of locomotives to run on it, why should that future not be divided between two Stephenson enterprises, one concerned with making railways and the other with building locomotives? After the previous joint ventures, here was a much more satisfactory solution that put the Stephensons in the driving seat, and could be expanded as commitments increased. It was a beguiling vision, but in fact, George Stephenson & Son, even though it 'virtually cornered the market',[13] never really got under way as an organised business. It had no property or plant to be managed, unlike the Forth Street Works, just a headstrong and self-opinionated engineer who did not relish any touch of harness and is unlikely to have felt that Pease, Richardson and Longridge deserved any profits from his survey and engineering work. He continued to act as the independent operator that he was and wanted to be. His vision would be achieved in his own way.

With the rush of new railway projects and the need to deploy and supervise his team of engineers, George Stephenson had scant time to give to the Forth Street Works, even though it was a critical moment for the locomotive business.[14] The Liverpool & Manchester directors had briefed the engineer Charles Sylvester to examine the question of using locomotives, and on 15 December 1824 he reported favourably, on the basis that further improvements were likely. In the following January, locomotive trials and demonstrations were carried out at Killingworth in front of Sylvester and other engineers. At the end of 1824 it had been agreed that George would be relieved of his 'ostensible share of management' of Forth Street for the year 1825 and that Michael Longridge would take over at a salary of £200. Timothy Hackworth claimed that overtures were made to him at this time to take over the management of Forth Street[15] and that Stephenson even offered to split his investment with him, but Hackworth refused. As George had known Timothy since they were boys, and had respect for Hackworth's ability, an offer of the foreman's job, at least, is perfectly likely, and would have ensured supervision without prejudice to Robert's eventual return as head of the enterprise. Born in 1786, Timothy was the son of the Wylam colliery blacksmith, and had long been in charge of Hedley's engines at Wylam. It is likely that he did work for a time at Forth Street. Longridge, with his own business and now his role in George Stephenson & Son, was not over-happy with the task of watching over the locomotive works. His association with the Stephensons had become a close one, especially with Robert, who while still in America became godfather to his daughter. It brought business to the Bedlington Works, but it brought him headaches as well. On 7 March 1825 he wrote to Thomas Richardson that:

> It was against my wish they [George and Robert] commenced engine Builders, but after they had begun, considering it to be beneficial to the Bedlington Iron Co. and that Geo. and Robert would benefit from my habits of Business, in which they were both deficient, I offered to take a part with them.[16]

Some special pleading is going on here. Longridge had been in the project from the start, but his letter was in response to some anonymous criticism of delays and poor workmanship at the Forth Street workshops, passed on by Richardson.

George Stephenson had to find a work force for his survey teams, and he used his old contacts and friendships as much as possible. A friend from Killingworth days, Joseph Cabry, had moved to Ness colliery on the Wirral and Stephenson had been trying to get him and his sons back to Killingworth since around 1819.[17] Though Cabry never came back, his son Joseph may have worked on the Liverpool–Manchester survey and two other sons, Henry and Thomas, became apprentices at Forth Street in late 1823. By employing the sons of friends, Stephenson was both indulging in a form of nepotism typical of the time (and perhaps showing his new status to their fathers), and more usefully taking on people whom he felt he could understand, trust and talk to plainly. But also, these young and inexperienced men were not likely to criticise his methods or his knowledge.

His worst problems were with the Liverpool–Manchester survey, where concerted opposition by three big landowners, Bradshaw, Lord Sefton and Lord Derby (and many lesser ones), was making it physically dangerous, if not impossible in some places, for a proper survey to be made, as well as creating serious delays. At this time surveyors had no legal right to enter private ground, and the hostility of the proprietors meant they had to work at night, or in quick 'raids'. George reported to Pease in November 1825 that Bradshaw had guns fired across his property during the night,[18] and that Sefton had a hundred men on hand to chase off the survey parties. Such opposition may not have been the only reason the job was skimped. Although he had some talented men among his staff, George also allowed some very inexperienced or incompetent people to work out levels on some sections, which remained unchecked by himself or anyone else. He probably did not worry too much about this, relying on practical ability in the actual construction to sort out any difficulties. But some of his own estimates were so sketchy as to be useless. In its nature and scale, the Liverpool & Manchester project was a very different proposition to the Stockton & Darlington, and he failed to come to grips with this. Longer, requiring more substantial engineering works, double-tracked, it was of much greater strategic importance and public interest, and, not least, faced intensive and high-level opposition.

Since the L & M Committee had made no secret of its intention to use locomotives, much of the opposition's publicity campaign was directed against this still largely unfamiliar form of traction. Very few people had ever seen locomotives in action, and they were certainly noisy, smoky things. Unbridled by actual experience, technical knowledge or the nature of probability, all sorts of objections, criticisms and prophecies of doom came in flights at every level, from the bailiff standing drinks in the village pub, to writers in newspapers and journals. Cows in calf would abort, game preserves would be disturbed, smoke would kill birds and destroy crops, high-pressure boilers would explode, causing death and wrecking property. Other contributors to the debate claimed that locomotives would not work anyway, or broke down too often, or would not survive Lancashire's 'mizzling weather'.[19] Others yet, like a contributor to the *Quarterly Review* in 1825 against the proposed Woolwich Railway in London, expressed opposition to the notion of being 'whirled along at the rate of eighteen or twenty miles an hour'.[20] As so often in such propaganda campaigns, apart from the pure fictions, elements of true information were distorted or turned towards false conclusions, and out-of-date facts were brandished as new. Stephenson had been criticised before, but to work in a consistently hostile climate was a new experience.

Promoters and opposers all knew that the propaganda war was merely the preliminary barrage: the vital part of the struggle was the parliamentary hearing, whose yea or nay was the decider. Parliament had an established procedure for dealing with Bills for canal and railway projects. The first stage was the deposition of a petition at the Private Bill Office: a set of detailed documents showing the purpose, benefits, cost, funding and ground plan of the project, with a list of the landowners and other interests affected. Each House had standing orders related to the drawing up and presentation of railway Bills, and a committee would scrutinise the petition to ensure that it complied with these. Following that, a Bill could be prepared and considered by the House of Commons in a first and second reading, to be at least seven days apart. By vote of the House, the Bill could be referred to a committee for further discussion, or sent back to its promoters. In the committee stage, members could question the supporters and opponents of the Bill directly and then finally vote on whether to commend the Bill to Parliament, to amend it or to reject it. Promoters needed detailed knowledge of parliamentary procedures and legal requirements, and a parliamentary agent was invariably employed. This was a man of legal training, accredited by the Private Bill Office, and he would be provided with a team of solicitors and barristers to help in presenting the case.

Sandars and Booth had worked hard to provide members of each House with positive information, and the petition stage was passed without trouble. A Bill, 101 pages long, was prepared and given a first reading in the House of Commons on 18 February 1825. Before the second reading, the opponents pointed out that the plans submitted to the Private Bill Office were not identical to those deposited for public inspection with Lancashire's Clerk of the Peace. Despite this, a second reading took place and the Bill passed to the committee stage without a vote. So far, the principle of the project had been undefeated; now it was to be examined in detail. The committee sat from 21 March to 31 May, and eight counsel retained by the opponents faced the legal team of the proposers. These men, trained in the adversarial tradition of the law court, were early examples of what would become a numerous species: the specialist parliamentary advocate on railway projects. Not engineers or businessmen, it was their task to tear holes in the opponents' case by whatever means they could.

It was not until 25 April that George Stephenson was called as a witness. By this time a great deal of evidence had been given and the general feeling among observers was that the case for the railway had neither been fully made nor undermined. It was expected that the engineer's evidence would be crucial. According to Smiles, George 'well knew what he had to expect'.[21] If so, he made remarkably little preparation. Some coaching had been done: on discussing speed with William Brougham, solicitor to the railway committee, George had proposed to claim that locomotives on the L & M could attain 20mph, but Brougham 'emphatically demanded that a lower speed be cited as attainable', insisting that otherwise Stephenson would 'invariably damn the whole thing' and would himself be regarded as 'a maniac fit for Bedlam'.[22] Already some expert witnesses, including Nicholas Wood, had testified as to the capacities of the steam locomotive, and on his first day, Stephenson dealt authoritatively with questions relating to the power, speed and tractive ability of his engines. He was relaxed enough to joke; when asked by a committee member if it would not be an awkward circumstance if a cow were to stray on to the railway line and get in the way of an engine: 'Yes,' replied the witness, with a twinkle in his eye, 'very awkward – *for the coo.*'[23] On the next day, Edward Alderson, for the opponents, shifted his line of questioning to the actual plans, and very soon George Stephenson was wreathed in confusion and embarrassment. His neglect of the survey work was now exposed and he was reduced to blaming his apprentices for errors he should have corrected at the time.[24] Well briefed, the lawyer caught him out in a string of errors, neglects and inconsistencies. Levels were shown to be wrong by up to 10ft. On the bridge across the River Irwell – part of one navigable waterway to Manchester – he had allowed for a 10ft height above normal water level instead of the necessary 16ft 6in. Alderson pressed on:

'What is the width of the Irwell there?'

'I cannot say exactly at present.'

'How many arches is your bridge to have?'

'It is not determined upon.'

'How could you make an estimate for it, then?'

'I have given a sufficient sum for it.'

Stephenson had to confess that he had not set a proper base level for measurements taken along the line. When he finally left the witness stand, Alderson drove home his advantage:

Did any ignorance ever arrive at such a pitch as this? Was there ever any ignorance exhibited like this? Is Mr Stephenson to be the person upon whose faith
this Committee is to pass this Bill involving property to the extent of £400–
500,000 when he is so ignorant of his profession ... I never knew a person
draw so much on human credulity as Mr Stephenson has proposed to do in the
evidence he has given.

The hearings went on for another six weeks, during which the objections of
landowners were considered. The opposition's canal engineers, Francis Giles and
George Rennie, gave their view that Stephenson's intention to take the railway
across the great morass of Chat Moss was a piece of folly which, even if feasible,
would cost in itself more than £200,000 to accomplish. In his final summing up
for the antis, Alderson returned to a personal attack on Stephenson:

I say he never had a plan – I believe he never had one – I do not believe he is
capable of making one. His is a mind perpetually fluctuating between opposite
difficulties ... When you put a question to him upon a difficult point, he resorts
to two or three hypotheses, and never comes to a decided conclusion.

Writing from London on 8 May to Joseph Cabry, Stephenson merely mentioned
'hard fighting in Parliament' in a letter trying to entice Cabry to Forth Street 'to
look after the whole of the men in our manufactory'.[25] On 31 May the committee voted against Clauses 2 and 3 of the Bill: these were substantive clauses and the
promoters withdrew it. The Bill was lost. Even if George Stephenson had made
a better survey, there was no certainty that it would have succeeded, but with the
defects in his work so bitingly exposed, the promoters' case was gravely damaged.
George, boldly self-styled as a civil engineer, had been publicly rubbished and
treated as hardly more than a charlatan. It was the nadir of his professional life.
The grand vision of George Stephenson & Son suddenly looked like a punctured
bladder. William James, the original proposer of the Chat Moss route, might have
been forgiven for a bitter smile. George may have felt relieved that his son was
far away from the scene of his humiliation, though Robert heard of it eventually
in letters from Joseph Locke and Michael Longridge. Replying to the latter, on
15 December 1825, he wrote: 'I still anticipate with confidence the arrival of a
time when we can see some of the celebrated canals filled up. It is to be regretted that my father placed the conduct of the levelling under the care of young
men without experience.'[26] On 17 June the Liverpool Committee met to consider their position. Disappointed but not despairing, they resolved to renew their
application for an Act in the next session of Parliament, with a revised survey. In
the meantime, all agreed to dispense with the services of Mr Stephenson. Even
those who still believed in George's ability to do the job, including Moss – who
wrote to George to say: 'Your talents are of a much more valuable nature than

that of a witness in the House'[27] – Sandars and Booth, had to accept that it was impossible to envisage him presenting their case again.

Fortunately perhaps for Stephenson at this time, he was required to be extremely busy elsewhere. Forth Street, which had been proceeding somewhat lethargically with the two engines for the Stockton & Darlington, and two others for the Mount Moor colliery near Gateshead, needed him badly. With the S & D line itself nearing completion, work on its first engine was speeded up, and the colliery engines, though ordered before it, were set aside. George also received a boost in an unexpected way. Nicholas Wood had been writing a book, *A Practical Treatise Upon Railroads*, which was published in 1825 and was by far the most substantial and authoritative volume so far to appear on the subject. Naturally, it dealt extensively with such topics as rails and steam locomotives, and Wood gave full credit to his friend and mechanical mentor, George Stephenson, as the leading engineer in the field. The book was in immediate demand, all the more so because the Liverpool & Manchester hearings had provided a national forum for discussion of the value and potential of railways. Longridge, a careful business-man, felt it ought to be censored in case it gave away too many trade secrets.

At the beginning of September, George reported to Pease and his board that the new railway would be ready for traffic on the 26th of the month. On the 11th, the first of the locomotives to be completed, and the first to be built at the Forth Street Works,[28] No 1, named *Locomotion*, was tested and proved satis-factory. Stephenson referred to it as 'the Improved Travelling Engine'. It had a single-flue boiler and four wheels, and is the first locomotive definitely known to have outside coupling rods joining its wheels.[29] Dragged from Newcastle on a heavy road wagon by horses, it was set on the rails at Aycliffe. At the same time, the first railway carriage was delivered, named *Experiment*: a coach body with doors placed in the ends, set on an unsprung four-wheeled frame also built by Robert Stephenson & Co. On the 26th, Jemmy Stephenson drove the engine from Shildon to Darlington, with Edward Pease and three of his sons among those travelling in *Experiment*. That night another Pease son, Isaac, died, and the bereaved family did not attend the opening festivities.

On the opening of the Hetton railway, fifty men had dined as the Coal Company's guests at Mrs Jowsey's Bridge Inn in Monkwearmouth,[30] but 27 September 1825 was an event of a different order – it was the first great public railway occasion; harbinger of many to come as new railways opened across the world. The full repertoire of motive power was on display. At the western end, horses brought ten loaded coal wagons from Witton colliery to the foot of the Etherley inclined plane. The wagons were hauled up by the winding engine and rolled down the other side of the ridge by gravity, being braked to avoid too great a speed. At the foot they were hitched to more horses, which drew them over Stephenson's iron bridge to the low end of the steep Brussleton Bank. Many people clung on perilously as they were hauled up to the top. On the far side,

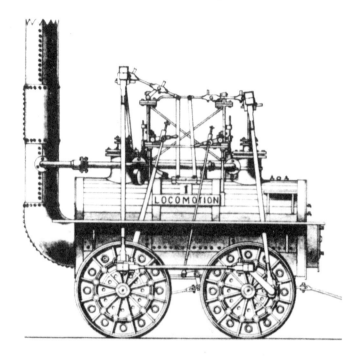

No 1, *Locomotion*

at the foot, *Locomotion* waited with a train of thirty-four vehicles, headed by six trucks laden with coal and flour, then the carriage and twenty-one coal trucks temporarily fitted with seats for invited guests, and six more truckloads of coal at the end.[31] Three hundred shareholders and other favoured persons had tickets, but far more invaded the train, squeezing in and clinging on – as many as 600, it was claimed, adding up to some 30 tons of humanity. George Stephenson was driving, and a horseman preceded the train. The run was largely, but not wholly, trouble-free: a defective wagon had to be detached and *Locomotion* was once brought to a stop when the feed-pump to the boiler became clogged and had to be cleared. At the junction of the short branch into Darlington, a truck with the Yarm town band was joined to the train and six of the coal wagons were detached, their cargo to be donated to the town's poor. Forty thousand was the estimate for the crowd waiting at Stockton Quay as the train steamed in, about 45 minutes later than expected. Amidst wild cheers and the pealing of the church bells, seven cannon fired a twenty-one gun salute, and a waiting band joined the train-borne one. The world's first public steam railway was in business. Stockton Town Hall had a municipal banquet awaiting the guests of honour, in the course of which twenty-three toasts were proposed, the final one being to the line's engineer. Five months after his humiliation, he could savour triumph as the central figure of an event never seen before in the world's history.

# 7

# TESTING TIMES

Robert Stephenson was not without passionate thoughts or self-probings, but did not confide such things to his personal log or diary. Like many other engineers, he treated it more as a technical day-book, recording weather, temperatures and minor scientific experiments; on the very first day of his voyage, he tried to take the temperature of the sea at various depths with a 'Register Thermometer', but found the speed of the ship too great for the instrument to sink. For the rest of the passage he continued his meteorological notes until the Colombian coast came into sight on 23 July 1824: '… at two o'clock cast anchor opposite La Guaira; observed with silence the miserable appearance of the town. The hills behind the town rise to a height that gives a degree of sublimity to the scenery in the eyes of a stranger.'[1]

Following its liberation from Spanish rule in 1819, after the campaigns of Simon Bolivar, the colony of New Granada had formed a union with Panama, Venezuela and Ecuador under the name of Gran Colombia. Still recovering from the turmoil, devastation and divisions of the liberation struggle, the country was destitute and armed gangs made life dangerous. A civilian government was struggling to keep its own cohesion and security, quite apart from its authority in the country at large. Robert had no direct concern with the government or the congress. The Colombian Mining Association and Messrs Herring, Graham and Powles were his authority, and he was as detached from local affairs as a Roman engineer sent to organise the building of Hadrian's Wall. His first project was to consider how the exposed roadstead of La Guaira could be made into a proper harbour, and how it could be linked by railway to the city of Caracas, only 7 miles inland, but 3,150ft (1,000m) above sea level. Sea works were a new thing for him and he went into the question of a possible breakwater with care before recommending against building it. He did propose the construction of a pier, which itself was partly a breakwater, with a seaward sloping edge to break the force of waves. Mountain slopes rose directly behind the shore. For a young railway engineer accustomed to the modest heights and gentle topography of England, a route climbing more than 3,000ft through tropical mountains, in earthquake and landslip-prone territory, was a huge challenge. Working out the likely construction costs, and balancing them against what he could find out about traffic levels from Caracas to the coast, he finally advised against the building of a railway – a sad conclusion for the managing partner following his earlier excitement, but a prudent one.

Having completed his work at the coast, Robert set off for the interior.[2] The next stage was to locate, investigate and report on mines, with the centre of operations being Mariquita, on the Magdalena River, about 65 miles (104km) north-west of Bogotá. A small team of experienced miners had already been sent there and more were to follow. Between late September 1824 and January 1825, Robert travelled on mule-back, with a servant and an interpreter, from Caracas to Bogotá, across rugged terrain of jungle, swamp and rivers. His brief was to look out for mineral deposits which might be viably exploited, and this meant many scrambles with his hammer to collect samples. Daily he was 'accosted by strangers, ready to mislead him with false information. More than once he was induced, by misrepresentations, to ride more than a hundred miles after a mare's nest.'[3] Everything was new and strange, and Robert 'used afterwards to speak in glowing terms of this his first mule-journey in South America'.[4] Over his white cotton suit he wore a hat of plaited grass, with a 9-inch crown and a 6-inch brim, and 'a *ruana* of blue and crimson plaid, with a hole in the centre for the head to pass through', which served as cloak and blanket 'in the net-hammock, which is made from fibres of the aloe, and which every traveller carries before him on his mule, and suspends to the trees or in houses, as occasion may require'. At Bogotá he made contact with R.S. Illingworth, the Colombian Mining Association's resident, and his immediate superior, and also sent reports both to London and to his father. Here, or at Mariquita, a letter from Michael Longridge awaited him, brimming with reproach about the three-year contract:

> When your Father & your other Partners consented that you should go out to Colombia it was with the clear understanding that your engagement was only of a temporary nature: and that as soon as you had informed yourself about the practicability of forming a Rail Way & had made your geological enquiries you should then return to England to make your Report.
> *On no account* would we ever have consented that you should become *the Agent of Messrs Graham & Co. for three years.*[5]

Longridge was trying to persuade Powles, via Richardson, to release Robert from the contract, but without success. Robert trekked on to the dilapidated town of Mariquita, and from April 1825 he and Empson made their base at the Santa Ana silver mine, some 12 miles away. To his dismay, Robert found that numerous items of machinery, which had been shipped from England and brought up the

Magdalena River to the head-point of navigability at Honda, were too heavy and bulky to be transported up the steep tracks onto the plateau where the mines were situated. There was no possibility of using them, so he would have to improvise. From Santa Ana he made another wide-ranging tour on mule-back, looking for mine-workings and mineral deposits, covering about 1,100 miles in the Andean foothills. There was ample time for introspection. Like many other exiles, he wrote more letters than he received and felt sharp pangs of homesickness, writing in August 1825 to Betty that, 'you may easily conceive how often I think of an English fireside and the joys that spring around it'.[6] But he continued to collect, not only geological samples, but also seeds, fruit and flowers, intending to bring them all with him when he eventually returned home. At Santa Ana he had a cottage built in the style of the country, of bamboo, reeds and palm leaves, set on a wooden platform, with its roof overhanging at the front and supported on poles to make a verandah. From here he could watch a population of birds far more numerous than today: golden orioles, humming birds, toucans and warblers, and admire the splendid sunsets. Monkeys, parrots and a big mule named 'Hurry' all formed part of the ménage; 'we eat parrots very generally they are fine well roasted'.[7]

In October 1825, more serious business began with the arrival of over 100 miners from Cornwall. The advance party had renewed the semi-derelict mine-workings and installations, and regular operations were now to start. Far from home, well paid and free from restraint, the miners behaved abominably. Their carousing and fighting on arrival at Honda brought an immediate formal pro-test to Stephenson from the local governor. But he found it very hard to bring them under control. Perhaps his status as an employee of Herring, Graham and Powles rather than the Mining Association reduced his authority with them, so that they acted as if he were merely the wages clerk. But it appears to have been his youthfulness, his mildness of manner and his not being Cornish that provoked their insubordination and intimidating behaviour. Matters came to a head in early December, when Robert had made some changes in their working arrangements, and his cottage was invaded late one night by a mutinous band:

… of the rascals, who, mad with liquor, yelled out their determination not to obey a beardless boy. For more than an hour he lay on his bed, listening to the riot – fearful that the disturbance might lead to bloodshed, and prudently anxious to avoid personal collision with the drunken rabble … But when the insurgents proposed that the 'clerk' should forthwith be taught his proper place, he rightly judged that it would not do for him to remain longer in his private room when his presence might still the storm, and could not aggravate it. Rising, therefore, from the bed, he walked into the midst of the rioters – unarmed, and with no more clothing on him than his trousers and shirt … Taking his place in the middle of the room, he drew himself up, and calmly surveyed them. Silence having its effect, he said quietly, 'It won't do for us to fight tonight. It wouldn't be fair; for you are drunk, and I am sober. We had better wait until tomorrow. So the best thing you can do is to break up this meeting, and go away quietly.[8]

They went, but on a subsequent night they returned, and Robert on that occasion left his cottage 'and, accompanied by two friends, found refuge in the house of a native'. Messages from Illingworth, demanding obedience to Stephenson, were eventually reinforced from London, ending what seems to have been a campaign of deliberate testing of the young man's nerve, though Jeaffreson suggests his life was at risk. Twenty-two years before Joseph Conrad's birth, it was a Conradian episode – the tropical setting, night-time, the degraded, drunken but dangerous assailants, the pensive, solitary young man, self-exiled, finding what he has let himself in for. Did Robert wonder what his father would have done? One can hardly doubt that George would have been speedily into that other room to single out a ringleader, challenge him to wrestle and duly pin him to the ground. But Robert was not George, and had to find his own way of meeting the crisis.

When his rowdy crew found that their young boss was not going to run away, things cooled down. But, 'To the last, he could never get from any man more than half a day's work each day, and he always had nearly a third of his hundred and sixty subordinates disabled by drink'.[9] If it was not a glorious captaincy, it was not a failed one either, and this was of crucial importance to Robert Stephenson. Having got himself into this situation, he had to emerge with credit. From his still not very distant childhood, he had known himself to be someone of whom much was expected, not simply in terms of hard work, but of success in what he undertook. However good he might have been at getting round his father, there was no ignoring the great central precept – like George himself, but even more so, Robert must 'get on' and be successful. Out in front of the real Robert, an ideal-Robert always beckoned him on. If he failed in Colombia, whatever the reasons, and however sympathetic his father might have been, not only would his independent venture have been a ruin, but the matching of his own reality to his father's vision would have been forever impossible, and that was unthinkable. Robert's sense of inadequacy was only one aspect of a character endowed also with grit and determination, but it could be discomfiting; a few years later, in 1830, he confessed to a friend over a bottle of wine: '… I sometimes feel very uneasy about my position. My courage at times almost fails me; and I fear that some fine morning my reputation may break under me like an eggshell!'[10]

During the time they had worked together before Robert's departure, he and Joseph Locke, who was only two years younger, had become good friends. Joseph wrote to him on 24 November 1825 to report that he had been surveying a 50-mile railway from Leeds to Hull, had fallen in love with 'one of the most enchanting creatures under Heaven', and was now surveying a line from Manchester to Bolton. He had also been looking with George at the countryside between Hexham and Newcastle, for the promoters of an alternative Newcastle–

Carlisle railway.[11] George Stephenson & Son were not short of work. But by the end of 1825, the steady growth of trade since 1820 had resulted in a financial crisis; too much capital had been borrowed and invested without showing a good return, the cost of living shot up and there were many bankruptcies among those who had borrowed to finance unproductive investments. Interest in building new railway lines fell away, though not all schemes were dropped. During that winter, George Stephenson wrote a letter to Robert that expressed something other than his usual cheerful optimism. For Robert, it sharpened the pangs of isolation and guilt, and in a letter to his step-mother of June 1826, he wrote:

> … when he spoke of his head getting grey and finding himself descending the hill of life, I could not refrain from giving way to feelings which overpowered me, and prevented me from reading on. Some, had they seen me, would perhaps call me childish: but I would tell them such feelings and reflections as crossed me at that moment are unknown to them. They are unacquainted with the love and affection due to attentive parents, which in me seems to have become more acute, as the distance and period of my absence have increased.[12]

Once the Stockton & Darlington line was open, charge of its engines devolved from George Stephenson to Timothy Hackworth, though Stephenson himself continued to act as consulting engineer to the company. Meanwhile, the Canterbury & Whitstable Railway requested his attentions, having obtained its Act on 10 June. This was another of William James' schemes – its line had been surveyed by him – but the proprietors, having earlier submitted James' costings to Stephenson, who criticised them as too low, now appointed George as engineer, further deepening the rift with James. George delegated the work to Joseph Locke, whose outstanding talents as an engineer had become clear.

Fortified by the opening and successful working of the Stockton & Darlington, George Stephenson did not retreat from business or the public view. But prospects at the end of 1825 looked very different from a year earlier. The surge of speculative interest in railways had been brief and 'the great proportion of new joint stock companies, including most of the thirty new railways originated in 1824–25, were completely wiped out'.[13] Money was tight and confidence low, and in the depressed conditions, which continued between 1826 and 1830, there was limited new business to be got for the Newcastle works, and George Stephenson & Son only had a few relatively small jobs in hand, including the 3ft 6in (1,378mm) gauge horse-worked Nantlle Railway in North Wales, built to transport slate from Cloddfarlon quarries to Caernarfon, and which was laid out by Uncle Robert Stephenson, who had declined Mexico on account of his wife and large family.

The only major scheme to be pushed ahead was the Liverpool & Manchester's new survey and Bill, with which Stephenson had nothing to do. The committee had hired George and John Rennie, who, apart from Telford, were perhaps the most eminent civil engineers of the day, and they appointed an able former military engineer, Charles Blacker Vignoles, as chief surveyor. Despite encountering the same kind of opposition as Stephenson, Vignoles expeditiously worked out an amended route, though it still included the traverse of Chat Moss. Vastly more suave and personable than George Stephenson, Vignoles and George Rennie saw the revised Bill successfully through its parliamentary hearings. Locomotives were scarcely mentioned; horse traction or fixed engines were assumed to provide the power. The railway's Act received the royal assent on 1 May 1826, and on the 30th the directors of the new company met to appoint an engineer. The Rennies were invited to state the terms they would accept for superintending the construction works, and most specifically to indicate how many visits of inspection they would make each year. In the meantime, the board appointed Vignoles to superintend the staking out of his surveyed route. The Rennies' reply made it clear that they could not provide the time and attention to be sole engineers, and on 9 June the board decided to call on either John Urpeth Rastrick or George Stephenson, or both, to consult on the general plan, and then to ascertain whether the Rennie brothers would be willing to act as supervisory engineers to either or both of these two. The Rennies declined absolutely to be involved with either, though they accepted that George might be allowed to run a separate locomotive department.[14]

Haggling went on for a month before the directors agreed to abandon trying to bring in the Rennies, and, following the receipt of testimonials from Joseph Pease of the Stockton & Darlington and Michael Longridge (who both might have been considered interested parties), decided on 3 July to appoint George Stephenson as engineer.[15] Stephenson agreed to spend nine months out of twelve on the L & M, and his salary was set at £800 a year. Josias Jessop, one of the few other civil engineers who had built railway lines, and who had testified helpfully in the second set of parliamentary hearings, was appointed consulting engineer, and so nominally the arbiter of George's actions and judgements, along with Jesse Hartley, a Liverpool engineer.[16]

Stephenson's appointment to the Liverpool & Manchester post marked the end of his residence in Newcastle and the reconfirmation of his position as the No 1 railway engineer. Though a remarkable turn of events, this was much less than a resurrection. Even after the debacle of April 1825, the powerful John Moss had been sympathetic and Henry Booth was firmly on his side. A majority of the directors believed he could do the job and Rastrick never seems to have fully entered the frame: Sandars went to see him and reported that his conditions for accepting the post were unacceptable. With the summer season well advanced, George acted quickly to get work under way, changing to 'his' gauge of 4ft 8.5in (1,435mm) from the Rennies' 5ft (1,530mm). Joseph Locke, William Allcard (aged only 18) and John Dixon were appointed as engineers, with superintendents under them for specific projects or sections; in all, a staff of around twenty-four engineers. Jessop, whose every recommendation had been contested

Olive Mount cutting,
Liverpool

by Stephenson, died in October 1826 and the board did not seek to replace him.
Hartley, a competent engineer rather similar to George himself in temperament and personal style, was content to be supportive.[17] Vignoles, who had been
retained directly by the board, was in charge at the Liverpool end, where his own
plan provided for a deep cutting and a 2,250yd (2km) tunnel under Edge Hill,
taking the line down to the waterfront at Wapping Dock. A dictum of George's
was that, 'The great object in the construction of a Rail-road is, that the materials
should be such as to allow the greatest quantity of work to be done at the least
possible expenditure; and that the materials also be of the most durable nature':[18]
a worthy view, but one which does not touch on management and supervision.
This time, he kept an eye on everything and everyone, and interfered unhesitat-
ingly on matters of small detail. His method irked Vignoles, who was not prepared
to be one of the 'young men', considering himself, at first at least, to be 'co-engineer', while George did not regard him as part of his team at all. On 14 January
1827 a thoroughly disaffected Vignoles wrote to a friend:

I ... consider (and that too in common with almost all other engineers) that the
mode Mr Stephenson proposes to put the works into the course of operation is
not the most eligible; and because it appears to me that he does not look on the

concern with a liberal and expanded view; but considers it with a microscopic eye ... pursuing a petty system of parsimony, very proper in private Collieries or small undertakings but wholly inappropriate to this National work ... I also plead guilty to having neglected to court Mr S.'s favour by sycophantic expressions of praise ... or by crying down all other engineers, particularly those in London ... All these circumstances gave rise to a feeling of ill-will towards me in Mr S.'s mind which he displayed on every occasion.[19]

Through the criticism, an indication of Stephenson's management approach is evident: he demanded personal loyalty and commitment. Following a dispute with Stephenson about the alignment of the Edge Hill tunnel, Vignoles resigned on 2 February 1827.

The major works required at the Liverpool end and at Chat Moss were the only real difficulties on the line. Joseph Locke was drafted to Edge Hill to take charge of the tunnel. Chat Moss lay waiting for George like Grendel's mother in the mere, but he was a confident and determined Beowulf. He knew how he was going to deal with it, and, even if he could not express the reasons, he was sure that the unorthodox strategy would work. Instead of the massive task of digging out, embanking and filling, he proposed simply to cut two drains on each side along the 4-mile bog section, and to place his line on the 48ft (15m) surface between the drains. Cut and dried peat formed the foundation of his road-bed, with further layers of brushwood and heather faggots, which were then packed around with earth, sand and gravel, and finally a thick cinder bed on which the wooden sleepers and the rails were placed. A special board sub-committee was formed to check on progress. Looking back in 1837, Stephenson typically made it seem a victory of one man's confidence over general doubt:

> After working for weeks and weeks, in filling in materials to form the road, there did not appear to be the least sign of our being able to raise the solid embankment a single inch ... Even my assistants began to feel uneasy and to doubt the success of the scheme. The directors, too, spoke of it as a hopeless task ... There was no help, however, but to go on. An immense outlay had been incurred and a great loss would have been occasioned had the scheme then been abandoned, and the line taken by another route. So the directors were compelled to allow me to proceed with my plans, of the ultimate success of which I never for a moment doubted.[20]

Apart from the last dozen words, this was simply not true. In October 1826 and again in November, the board made inspections of the Chat Moss workings and expressed satisfaction with their progress; by that time a temporary track had already been laid on part of the embankment to help with the works. Robert Stannard, one of the larger contractors on the line, was responsible for the proposal to 'float'

IRWELL BRIDGE.

Bridge over the Irwell, Manchester (from a Bury engraving)

the line on a foundation of brushwood hurdles laid herringbone-fashion over the deepest parts of the Moss.[21] An independent inspector, Mr Wilson, from the Kilmarnock & Troon line, made a very favourable report on the work in April 1827, telling the board that there was 'no doubt of the railway being carried over the Moss in a substantial and satisfactory manner'.[22] John Dixon was the engineer in charge. In general, the Liverpool & Manchester board kept a close eye on the construction work and on Stephenson's involvement. In 1827 his salary was raised to £1,000 a year, on condition that he took on no new projects until the Manchester line was finished. As no new railway schemes of any importance seemed likely to arise in the current economic climate, this was not a problematic requirement, and it was obvious that the successful outcome of the L & M project would be decisive not merely for George Stephenson, but for the future of railways.

Construction went on at a steady pace, and George had to take account also of changes of plan by the management, including an extension of the line across the River Irwell, right into Manchester (the Rennie-Vignoles line ended at Salford). He gave out contracts for wrought-iron rails to his own very clear specifications, including a 1,000-ton order for Longridge at Bedlington (who gave the lowest estimate). George had made a significant alteration to the Rennie-Vignoles plan by incorporating two inclined planes on which trains would be rope-hauled. Sylvester's report had stated that locomotives could not pull a load up a gradient steeper than 1 in 360[23] and George's own view was equally conservative[24] – but the basic mode of traction had yet to be formally discussed by the directors. Another line requiring a fixed engine was the Canterbury & Whitstable, whose board got a sample of Stephenson's high-handedness when they wanted to obtain competitive quotes before placing an order, but he went ahead and ordered it from Forth Street, excusing himself in a letter of March 1827: 'I have (although you may think it premature) wrote to Newcastle, to order them to get forward with the cylinder and boiler giving them the proper dimensions. The Company have certainly a right to receive orders from where they think proper, but I feel conscious of giving a just tender.'[25]

However sincere he was in getting the company a good job at a competitive price, George would not have got away with this on the Liverpool & Manchester. While his father thus had much to occupy his mind in a satisfactory way, through 1826 and 1827 Robert Stephenson was feeling increasingly frustrated. The amount of precious metal extracted by his troublesome Cornishmen was small, and it is clear with hindsight that the whole Colombian Mining Association venture was a product of the febrile financial climate of the early 1820s, based more on rumour and optimism than on solid knowledge of the Andean reserves. On 15 December 1825 Robert wrote to Longridge that he hoped to be back in England before his three years were up, in order to manage the production of machinery that could actually be got to the Mining Association's sites: 'Much money has been spent in this speculation, chiefly from bad management.'[26] Among the unusable equipment were two steamboats, which drew too much water for the depth of the river. The investment of £200,000 was certainly not going to produce any dividends,[27] and Robert wrote home that 'all the high-flown descriptions which you have heard in England about the riches of Colombia have been written by superficial observers'.[28] Apart from Empson's company, there were visitors from time to time, some of whom made lengthy stays, including two Frenchmen, M. Boussingault, a geologist and chemist, and M. Roullin, a mathematician, from both of whom Robert gained useful knowledge; while also, 'With characteristic simplicity he begged the few English gentlemen of his acquaintance to correct him whenever he used the diction, idioms, or intonation of North-country dialect'.[29] With the opening of the Stockton & Darlington Railway, and his father's reinstatement on the Liverpool & Manchester, Robert felt very cut off from new developments and inventions. On 23 February 1827 George wrote to Robert summing up the state of his activities, in one of his longest personal handwritten documents to be preserved.

> … Yore mother is getting her tea beside me while I am riting this and in good spirits. She has been in Leverpool a bout a fortnight. We have got a very comfortable home, and a Roume set aside for Robert and Charels [Charles Empson] when they arrive in England.

It records that they are getting on well with the tunnel under Liverpool and with Chat Moss, and notes the very recent departure of Vignoles:

> … a finishing blay to Renny … My assistance is now all of my own chosen.
> The Bolton line which was clandistanly got from me when we were in parliament with the Liverpool bill has been given to me – a welsh line 9 miles long has been put into my hands. a line at Canterbury is put into my hands likewise – we have a most magnificent Bridge to build a cross the sankey valley near newton it will be 70 feet high so as to cross the masts of ships that navigate the

Sankey Viaduct

canal. I have drawen a plan on the gothick principal, there will be 20 arches
of 40 feet span. It will be quite a novel in England as there will be a flat arch
sprung between the centre of the tops of the gothick and so on. it has a fine a
pearance in the plans ...

we have just advertised for 400 waggon wheels and 200 axels and strange to
say Robert Stephenson & Co. offer was lower than any other house and we
have had offers from almost all the best houses in England – my plan of wheels
is now put up like the maile Coach axels but still fast so that one greaseing per
day is a nough – the locomotive engine is working very well at Darlington –
and a great many Coaches on the line I think about 6 and each drawen one
horse which take a bout 30 passengers ... it is expected that a line will be made
from Darlington to York and I have been asked to take the survey – but hope
it will be cept untill you come back to england – this line will suit Mr Charels.

I think the proposed tunnal under the thames was talked of before you left –
it is now got a good way under the river but will cost a great deal more money
than was expected. This is however a very common case with engineers – the
estimate for this concern [Liverpool & Manchester] is 500000£, and I daresay
it will require it all.

the line passes Rainhill very near the same place where Jameses line passed. we shall want one steam engine at that place and a nother at near parr moss also one at the top of the tunnal. I want these engines to be constantly moving with an endless Rope so that the locomotive engines can take hold of the Rope and go on without stoping, the Incline plane will only be ¾ of an inch per yard so that the poor [power] of the locomotive assisted by the perment ones [permanent engines] will get the traffict on in grand stile. we most go at 10 miles per hour. I think I told you about my new plan of locomotive it will be a huge job the Cylinder is intirely within the Bolior and neaither Cranks nor cham [chain?] will be wanted ... you will think I have some mistaken ideas about this but I think not.

you may depend upon it that if you do not get home soon every thing will be at prefection and then there will be nothing for you to do or invent – however we hope that some usfull ideas will be brought from the western world. – the coal trade has been very bad in Newcastle last year ... Mr Wood is expected to be maried very soon of a young lady with a great posithen. she be long to Alnwick. he got a quainted with her at the election.

your mother expects you will not forget the presents. you must bring more than one as Mrs Robert Stephenson will want one by & by – and we expect Mr. Charels will bring plenty of amarica plants seeds for our [garden]. cannot you bring your favorate mule with you. I trust letter will just catch you befor living the country. my kindest love to Charels I am my Dear Dear Robert your affectionate father Geo: Stephenson.[30]

George's reference to the Bolton & Leigh Railway is not clear. He himself has been accused of shady doings with respect to this line, which, originally proposed and surveyed by James, received its Act on 31 March 1824. Its moving spirit, William Hulton, was the unrepentant Tory magistrate who had read the Riot Act to a radicals' meeting at St Peter's Fields in Manchester on 16 July 1819, precipitating the 'Peterloo Massacre' in which eleven people were shot dead by troops and around 400 wounded. Paul Padley had copies of James' plans, which 'found their way into Stephenson's hands. They joined those of the Birmingham and Liverpool in the portfolio of surveys which Stephenson would now adopt as his own and were soon to be joined by those for the Canterbury & Whitstable.'[31] His reference to Rennie is a reminder that George was intensely competitive, and while deliberate, even malevolent purpose can be read into his mopping-up of James' projects, these were lines ready to start and free to choose their engineer. It is most unlikely that copies of the Bolton & Leigh plans, which had been presented to Parliament, were not in possession of the directors. Confident and optimistic, Stephenson's letter sets out a tray of goodies to tease the imagination of a young engineer, and there is no suggestion that anything is amiss at the Newcastle works. But Longridge and Pease were both direct in their warnings that Forth Street was suffering from lack of proper management. Pease had written: 'I can assure thee that thy business at Newcastle, as well as thy father's engineering, have suffered very much from thy absence, and, unless thou soon return, the former will be given up, as Mr Longridge is not able to give it that attention it requires: and what is done is not done with credit to the house.'[32] Around 1825 a general manager, Harris Dickinson, had been appointed, but he was not an engineer.

Herring, Graham and Powles were sufficiently pleased with Robert's services and reports to ask him to extend his contract, but, having served his three years, he declined the offer and refused to wait until a replacement arrived. Illingworth came over from Bogotá to supervise the operations, and Stephenson and Empson set off down-river to Cartagena, loaded with rock samples, seeds, plants and souvenirs. They were close to the Isthmus of Panama, long the subject of debate about the feasibility of a canal, and Robert was very much in favour of the idea. His final letter from Mariquita, written to Longridge on 16 July 1827, enthused: '… how the magnificence of such a work augments in our ideas when we consider the advantages which would arise from it – how it would influence commerce in every quarter of the earth!',[33] but he also noted that the disturbed political situation, likely to lead to the break-up of Gran Colombia, made any action unlikely. He hoped to reach the isthmus from Cartagena and get at least 'a very general idea' of a canal plan, but lack of time, together with transport and perhaps political difficulties, made him abandon the project.

The two young men were not returning home directly. Robert had said in a letter to Longridge in December 1825 that he wanted to visit North America in order to look at steam engines in ships: 'I have heard they have the finest steamboats in the world.'[34] At Cartagena he found two other Britons waiting for a passage: '… in the public room at the inn, he was much struck by the appearance and manner of two tall persons speaking English; the taller of them, wearing a large-brimmed straw or whitish hat, paced restlessly from end to end of the room.' Stephenson spoke to the other man, a Scot named James Gerard, then the restless one joined them, who turned out to be Richard Trevithick. Such was Stephenson's account of their meeting, as passed on by Sir Edward Watkin, a railway magnate of the next generation who had been gathering material for a life of Trevithick, which in the event he did not write.[35] During his years in South America, Trevithick's fortunes had varied dramatically, but now, after a succession of vicissitudes and disasters, he was almost penniless and had very few possessions. He had been rescued from a capsized canoe in the Magdalena River and the imminent attentions of a large alligator, by a British officer in the Colombian military service, Bruce Napier Hall, who happened to be out on a pig-shoot. Almost forty years later, a friend of Hall's, James Fairbairn, read a newspaper report of a talk by Watkin, in which the meeting of Stephenson and Trevithick at Cartagena was described as 'an accident'. Fairbairn wrote to Watkin to say that 'the meeting was not an accident, although an accident led to it', and went on to describe the circumstances of Trevithick's rescue by Hall, who:

> ... took him on to Carthagena, and thus it was he fell in with Mr Stephenson, who, like most Englishmen, was reserved, and took no notice of Mr Trevithick, until the officer said to him, meeting Mr Stephenson at the door, 'I suppose the old proverb of "two in a trade cannot agree" is true, by the way you keep aloof from your brother chip. It is not thus that your father would have treated that worthy man, and it is not creditable to your father's son that he and you should be here day after day like two strange cats in a garret; it would not sound well at home.' 'Who is it?' said Mr Stephenson. 'The inventor of the locomotive, your father's friend and fellow-worker; his name is Trevithick, you may have heard it,' the officer said; and then Mr Stephenson went up to Trevithick. That Mr Trevithick felt the previous neglect was clear. He had sat with Robert on his knee many a night while talking to his father, and it was through him that Robert was made an engineer. My informant states that there was not that cordiality between them he would have wished to see at Carthagena.[36]

It should be noted that the author of this account was not himself at Cartagena; he was merely passing on the reminiscence of his friend Hall. Then on 16 December, Bruce Hall himself wrote to Watkin in reply to a request for corroboration. He confirmed the story of the rescue at the mouth of the Magdalena, and added:

> I will just say that it was quite possible Mr R. Stephenson had forgotten Mr Trevithick, but they must have seen each other many times. This was shown by Mr Trevithick's exclamation, 'Is that Bobby?' and after a pause he added, 'I've nursed him many a time.'
>
> I know not the cause, but they were not so cordial as I could have wished. It might have been their difference of opinion about the construction of the proposed engine, or it might have been from another cause, which I should not like to refer to at present; indeed, there is not time.[37]

These scant notes are all the information we have about this 'small world' meeting. Though other biographers of Robert Stephenson have taken them at face value, it must be said that they are of dubious accuracy. Fairbairn's letter is pure hearsay, only partly backed up by Hall. The most striking part of Hall's statement is that Stephenson found himself referred to as 'Bobby' and that the haggard inventor, still only 56, claimed that he had 'nursed him many a time'. Such dandlings could have taken place while the second Trevithick locomotive was being built at Gateshead in late 1804 and early 1805, when Robert was still an infant. The only known indication that Trevithick visited the north-east at that (or any other) time is a letter he wrote in January 1805 expecting he would '... go to Newcastle-upon-Tyne in about four weeks … I expect there are some of the travelling engines at work in Newcastle'.[38] If he did go, especially when his locomotive was arousing local interest, it is surprising that his visit was not reported in the local press. More strangely, if he did make such a visit, there is no apparent reason why he should have got on to close personal terms with George Stephenson, who was not involved with the building of the locomotive, and was still an obscure brakesman. Possibly, if Stephenson was already friendly with John Steel at that time, Steel might have brought Trevithick to Killingworth in the course of a visit by the inventor, though Robert could hardly be expected to remember it. If Trevithick came after George's return from Scotland, Robert was by then past 'nursing'. George Stephenson never mentioned any personal link with Trevithick, though one might not necessarily expect him to do so, always being keen to protect his own reputation for originality.[39]

At the time of the Cartagena encounter, Robert Stephenson was unknown and George, though with a reputation in mining and railway circles, was not yet a national figure. By the time of the letters to Watkin, they were both famous (and both dead). Fairbairn's second-hand dialogue can be dismissed as a pious fabrication. Hall's first-hand account is more of a puzzle. Robert clearly had no recollection of having ever encountered Trevithick, which seems to surprise Hall, though not many people retain memories of strangers who dandled them when they were very young. The last two sentences quoted from Hall are also enigmatic. The 'proposed engine' referred to may have been something to do with the harbour works at Cartagena, while the other reason for the lack of cordiality remains a mystery. But young men of 26 don't necessarily take kindly to complete strangers exercising claims of consideration from out of the mist of babyhood. It seems most probable that Hall, looking back four decades on what had only later emerged as a striking coincidence, thought that Trevithick and George Stephenson *must* have been friends (Robert's boyhood by-name of 'Bobby' was well known to the public), and allowed some additional colour into his memory of the event. It remains a curious episode, the details of which will probably never fully emerge. The association with Robert's infancy, real or imaginary, has an odd parallel, in that Robert's father might be said to have taken possession of Trevithick's 'baby' in the form of the locomotive. In a much more real manner, Robert Stephenson gave, or loaned, £50 to Trevithick at Cartagena – either way it was a generous gesture. After returning to England, Trevithick maintained an interest in locomotive engineering, and some remarks made much later by Robert Stephenson, relating to the '*Planet*' locomotive type of 1830, confirm that there had been some exchange of ideas between himself and Trevithick.[40]

Trevithick's companion, Gerard, joined Stephenson and Empson on the voyage to New York, which met with drama and ended in disaster. It was the hurricane season and in the passage of the Caribbean Sea the ship encountered first one, then a second drifting, dismasted hulk, each with a starving, desperate crew, in one case reduced to cannibalism:

The one had been nine days without food of any kind, except the carcasses of two of their companions who had died a day or two previous from fatigue and hunger. The other crew had been driven about for six days, and were not so dejected, but were reduced into such a weak state that they were obliged to be drawn into our vessel by ropes.

In a letter of 18 March 1828 to his former colleague Illingworth[41], still at Bogotá, Robert was recalled the events. 'To attempt any description of my feelings on witnessing such a scene would be useless. You will not be surprised to know that I felt somewhat uneasy when I recollected that I was so far from England, and that we might also be wrecked.'[42] Almost at the end of the journey, the vessel ran into a storm and in darkness was driven on to a reef. The masts and rigging were cut away, but the height of the waves made it impossible to launch a boat and passengers and crew spent a night fearing the imminent breaking up of the ship, before the gale abated enough for a boat to be launched. No lives were lost, but Robert lost most of his baggage – except, bizarrely, a case of mineral samples – and his money, arriving in New York with only the clothes he stood up in.[43] The combination of London and Liverpool commercial connections enabled him to get credit, and, once supplied with cash, and disliking New York – '… we soon found out the characteristic impudence of the people. In many cases it was nothing short of disgusting' – he set off almost immediately, with Empson, Gerard and two other companions from the voyage, on a walking expedition inland.

On the Hudson River he could have seen steamboats, but these are not mentioned in his letter to Illingworth. Coming so fresh from Colombia, he was impressed by the populous and well-farmed countryside, which 'affords to an attentive observer a wonderful example of human industry; and it is gratifying to a liberal-minded Englishman to observe how far the sons of his own country have outstripped other European powers which have transatlantic possessions',[44] and he liked the hospitable country-dwellers much more than the New Yorkers. Crossing into Canada by way of the Niagara Falls, which he found less striking than the Tequendama Cataract, a 515ft (157m) fall on the Bogotá River, though 'their magnitude is certainly prodigious', they reached as far as Montreal, presumably by steamer along Lake Ontario and down the St Lawrence River. Although now in a British colony, he was unimpressed, finding it 'far behind the States in everything. The people want industry and enterprise.'[45] According to Jeaffreson, he had up until now still worn his Colombian poncho and wide parramatta hat, with no more than a single change of linen, but at Montreal he, 'equipping himself in the ordinary costume of an English gentleman, went into the best society of the city'.[46] From Montreal they returned to New York, and departed for Liverpool on board the *Pacific* sailing ship, arriving at the end of November 1827.

# 8

# 'WORTHY OF
# A CONFLICT'

Letters from Longridge, as well as from George himself, had warned Robert that he would find his father changed. In the three and a half years of his son's absence, the elder Stephenson's hair had gone white and his face was deeply lined. He was still only 46, but had been driving himself hard for many years, quite apart from the recent pressures of his business life. But he was physically fit, optimistic and greatly cheered by Robert's safe return. George found himself confronting a son who had gained in experience, self-assurance and authority, and was used to making decisions and seeing them through. Robert was now a smooth-spoken man of business, outwardly calm, who commanded respect, and George could only have been pleased and perhaps even a little awed, as much at his son's now refined diction as at his maturity of appearance and manner. Instead of a pupil-protégé, he had a partner. The partner very soon went off, not at first to Newcastle, or to the railway construction, but to London, to give an account of himself to his erstwhile employers, sort out the completion of his contract, establish what he was owed and see what business could now be transacted with them.

Since George had committed himself to the Liverpool–Manchester project, and since his method of project management in any case demanded a very high level of personal attention, he was relieved to be able to transfer some responsibilities to Robert. His salary on the L & M was raised by the board to £1,500 a year in April 1828. On 7 June the final cut was made in the Edge Hill tunnel, 22ft (6.7m) wide and 16ft (5m) high, on its great slope down to sea level. Here was to be no Stygian darkness: negotiations were in hand with the Liverpool Gas Company for interior lighting. Proud of their achievement, the company and the engineer were happy for the public to come and inspect the tunnel.

At this point, Stephenson had men working along the length of the line, but their total number is unknown. Choosing not to follow the example of his own work on the Stockton & Darlington, which he had apportioned mile by mile to sub-contractors, he was effectively acting as his own contractor, hiring and firing

Interior of Edge Hill tunnel (from a Bury engraving)

workmen, directing the works with his own team of engineers to give orders on the spot and moving teams about. Apart from the supervision of board sub-committees, he was also his own inspector. Criticising this procedure, L.T.C. Rolt blames George's 'shrewd Quaker partners' for making him do it and goes on to say that 'He was not the first, and by no means the last, eminent engineer to be beguiled by the sophistries of so-called astute business men'.[1] This was quite mis-taken – to work in this way was George's own plan, and he very much preferred it because it suited him personally, enabling him to embrace control of the whole operation. To combine project management with surveying was fully within the remit of George Stephenson & Son. As a way of working, it was intensely demanding on Stephenson himself, requiring him to be master of almost every detail, to know almost every face and to be, as far as anyone could be, everywhere at once. Far from complaining, he obviously relished it. As a general method of managing a large-scale project, it was undoubtedly a perfect opportunity for cor-ruption and for letting bad workmanship slip through. But Stephenson in this respect was a man of solid integrity, intent on providing a quality job. Another might have used the opportunity to line his pockets, or let his concentration slip on some vital aspect, or even let the whole scheme slide out of control into chaos. Applying massive concentration, tenacity and energy, Stephenson was determined to make it *his* line, in every way. This is not to decry the essential contribution of Locke, Dixon, Gooch and others, but only George had the full picture.

Consequently, when the L & M board sought a loan from the Exchequer at the end of 1828, having already borrowed £100,000 from the same source, Stephenson was deeply displeased when the Exchequer loan commissioners, as a preliminary to granting the loan, asked the doyen of civil engineers, Thomas Telford, to report to them on the works. Telford sent an assistant, James Mills, to carry out the inspections, but he arrived to find that Stephenson and Locke had gone off to the Stockton & Darlington line to look at locomotive operations there.[2] Mills persevered in his mission, talking to such people as would give him information, and piecing together enough details on Stephenson's procedures to send a rather horrified report back to his master. Telford, who in his canal works had effectively created the system of contracts and supervision, was disturbed, and came to Liverpool himself in January 1829. George Stephenson could not ignore *him*. In some ways, these two great men resembled each other, both having risen from humble beginnings, both with a visionary capability and both believers in hard work, but Telford, now aged 72, had taken a more orthodox path and, as president of the Institution of Civil Engineers, represented exactly the school of 'London engineers' which, as Vignoles had observed, George actively despised.

Stephenson conducted Telford and Mills over the line and, though deploring the system and finding it a 'tedious and laborious task',[3] Telford had few serious complaints about the work that had been done, except for over-steep cutting sides and a lack of stout retaining walls.[4] However, he noted that of fifty-eight bridges, thirty-nine were still to be built, and estimated that over £200,000 would be needed to complete the works; 'I am also at a loss ... as to when this communication will be open to the public, or how it is to be worked.'[5]

At the end of 1828, a total of £461,899 19s 6d had been spent, against a share capital of £510,000 and an Exchequer loan of £100,000. In the circumstances it seems rather unfair of Telford to advise against the additional loan, but he could not possibly approve of the way in which the project was being managed. The report was contested in detail by the L & M board,[6] prompting a sarcastic riposte from Mills: 'There was never any great public work before, I apprehend, where one Person was appointed to act as Engineer and to execute the work under the sole contraul of himself, his agents, and own apprentices, hired as resident Engineers at a large salary.'[7] By Mills' reckoning, the work could have been achieved at a third less cost. Ultimately, however, the commissioners allowed the railway company to make a further call of 10 per cent on Exchequer funds.[8]

Ever since the proposal to use locomotives had aroused hostile criticism before and during the first parliamentary hearing, the Liverpool & Manchester board had remained tight-lipped on the subject of traction for their line. There was another reason for this reticence: the board itself was divided on the subject. Henry Booth and some other influential members, including Joseph Sandars, shared George

Stephenson's view and were very much in favour of using locomotives. Other directors, including James Cropper, a member of a leading Quaker family in Liverpool, were sceptical about the value of locomotives and favoured the installation of a series of fixed engines along the 30-mile route. In January 1828 the board began to discuss the matter seriously. At least among mechanically-minded members of the general public there was keen interest, and in his *Account* of the railway's inception, Booth wrote: 'Every scheme which the restless ingenuity or prolific imagination of man could devise was liberally offered to the Company.' He himself was working on a means of smokeless combustion.

Adverse publicity came directly and indirectly from the canal and stage coach operators, the turnpike trusts and the horse dealers, and all the fears and prophecies of disaster of 1825 were resurrected. Naturally, great attention from both sides was focused on the Stockton & Darlington Railway, where locomotives, fixed engines and horses were all in daily use. By the end of 1827 it was operating six locomotives, five of them built by Robert Stephenson & Co. Genial strangers bought drinks for enginemen in local pubs and asked them how frequently their machines broke down and had to be replaced by horses, and how often broken rails had to be replaced. Every accident and incident found its way into public currency and became inflated by rumour. At this time, incidentally, the general public had no opportunity to sample travel behind a locomotive: all passenger services on the S & D were horse-drawn, run by private contractors paying a toll to the railway company. The line was operated like a public toll road, open to anyone with a vehicle of suitable gauge and with flanged wheels to keep it on the rails.

George Stephenson was concerned about the way in which public perception of the locomotive was being manipulated. In Timothy Hackworth on the S & D he had a useful ally, but also a troublingly independent one – the sixth locomotive at the Shildon depot was one that Hackworth had designed and built himself *in situ* in 1827, using parts of an engine that he had persuaded the board to buy in from a 'Mr Wilson, of Newcastle' but which had proved ineffective. Grandly named *Royal George*, it was undeniably the most powerful and efficient on the line. This alone was enough to put both Stephensons on their mettle.

At a directors' meeting on 8 January 1828, George was authorised by the L & M board to build a locomotive, using Booth's smoke-consuming device, for testing to see if it 'will be effective for the purposes of the company, without annoying the public'.[9] In the same month Robert had returned to Newcastle to take charge at Forth Street, though he was also supervising construction work on the Bolton & Leigh Railway; letters were exchanged between father and son on what was needed to make the new locomotive the most improved yet. Well before this, George had realised that his single-flue boiler was inadequate and that a new approach to boiler design was essential. In 1827 he had found time to design the locomotive mentioned in his letter of 23 February. Tubular bars under the fire, through which water passed, and a 'water drum' inside the single flue were intended to expand the heating surface and improve steam-raising capacity. Known as the 'Experiment', this was also the first locomotive to have two horizontally-placed cylinders, which were mounted within the boiler at the back above the fire. Delivered to the Stockton & Darlington in January 1828, its axle weight was found to be too heavy for the rails and it was taken off the road until it could be mounted on six wheels rather than four.

An American visitor in 1825, William Strickland, emissary of the Pennsylvania Society for the Promotion of Internal Improvements, had returned home to publish a detailed *Report* in 1826, which did much to increase interest in English locomotives and railways. Some dynamic Americans made visits to Killingworth and Forth Street in 1828, and were ready to place orders. Horatio Allen, delegate of the Delaware & Hudson Canal Co., was first, ordering one engine from Forth Street and three cheaper ones from Rastrick's foundry at Stourbridge. Next was Lieutenant George Washington Whistler of the Baltimore & Ohio Railroad, father of the famous painter, with two colleagues, Knight and McNeil, who also ordered a single engine. Whistler met both Stephensons, and according to his biographer:

> He liked the elder man. A clever man he was, a wonder, to rise from common digging in the Newcastle pits to engineering – observe how highly! But his son Robert was another matter. Whistler did not prize this young man who had been educated at the University of Edinburgh and talked in highfalutin' terms because he felt the necessity of impressing people.[10]

The reference to George as a one-time coal-digger does not inspire confidence, but contemporary comments on Robert in 1828 are rare enough for this one to be noted.

New business, and the prospect of more, helped raise the pressure on the Stephensons – whether self-induced or not – to a pitch greater than that inside the boilers of their locomotives. Suddenly, there was a great amount to be done. George had already written to Robert on 31 January 1828 to say: 'I trust the locomotive engine will be pushed. Its answering is the most important thing to you, and recollect what a number we shall want – I should think thirty':[11] this was his estimate of the L & M's locomotive requirement. Apart from this major opportunity to be worked for at home, the visitors to Newcastle from the United States, Germany and France were harbingers of demand from beyond the British Isles, sketching the promise of a world market for the British invention. One French visitor in 1828 was Emile Martin, later a pioneer of steel-making; sent to England to learn about steamships, he acquired a keen interest in railways from his visit to the Stephensons. A repeat visitor was Marc Seguin, who had looked over the Stockton & Darlington Railway in the winter of 1825/6, and was in England again from December 1826 to January 1827.[12] An accomplished inventor himself, he had already experimented with boilers and had just applied for a

French patent for a design in which narrow 'fire tubes' drew heat from the fire through the water, greatly increasing the steam-raising capacity of the boiler. He ordered two locomotives from Forth Street for delivery to the St-Etienne-Lyons railway, then under construction. These first export models, delivered in 1828, were of the old standard single-flue, vertical-cylinder design – the new developments going on at the time were too untried for commercial use. On 1 January 1828 Robert had written to Longridge:

> I have been talking a good deal to my father about endeavouring to reduce the size and ugliness of our travelling-engines, by applying the engine either on the side of the boiler or beneath it entirely, somewhat similarly to Gurney's steam coach. He has agreed to an alteration which I think will considerably reduce the quantity of machinery as well as the liability to mismanagement. [13]

George had been consulted and his agreement obtained, but Robert was now taking the lead in locomotive design. His reference to 'ugliness' is interesting. Both he and George were sensitive to the appearance of machines as something distinct from their efficacy. Efforts to reduce mechanical complexity were of obvious utility, as the letter notes, but clearly simplicity was also equated with beauty. The less 'mechanical' a machine appeared, the less ugly it was perceived to be. In these thoughts lie the developing British tradition of locomotive design, which sought to enhance the 'effortless' effect of movement, as if the engine were impelled by some invisible force. Some later designers would be obsessive in their avoidance of any external detail beyond the bare essentials – in stark contrast to the French approach.

Goldsworthy Gurney was a Cornish inventor who had given up his profession of surgery in 1823 to concentrate on building steam-powered road coaches, and who of necessity had to develop relatively lightweight engines. Mention of his coaches shows how aware Robert was of other work on steam traction at this time. The work put into the 'Experiment' was valuable in building the 'Liverpool Travelling Engine', which incorporated a whole package of innovative elements and details, including, as Robert had told Longridge, two cylinders fitted on the outside of the boiler. Completed in June 1828, it was sent for testing on the newly-opened Bolton & Leigh Railway in July, but though George wrote to Hackworth on 25 July to say it worked beautifully, it did not perform so well as to show any real degree of improvement. Meanwhile, 'Time was running short; the period for opening the new line was fast approaching, and yet George Stephenson and his son had not hit on the way to build such an engine as should sweep the ground from under the advocates of stationary machines'. [14] The 'Liverpool Travelling Engine' was given the name of *Lancashire Witch* by Mrs Hulton at the opening of the Bolton & Leigh Railway on 1 August 1828, [15] and Robert used its arrangement of external cylinders and sprung axles on other locomotives built in 1828–29. Each one had different internal boiler arrangements, reflecting the struggle to get this aspect right.

Good diplomacy by the Liverpool & Manchester board enabled it to get a second Exchequer loan without Stephenson having to yield too much. But the directors picked on another aspect of his management – since they were paying a handsome salary of £400 for the services of Joseph Locke, they objected to Stephenson's detaching of him to do a survey of the Manchester & Stockport Railway. (Here was one of the benefits of the George Stephenson & Son business – the Manchester & Stockport were able to imply to potential investors that they had the great man himself, without George having to break his personal commitment to the L & M board not to take on new business.) Stephenson's response was intransigent, insisting that his terms of employment allowed him full control over his assistants, without interference.

In petulant fashion, George withdrew Locke altogether from the Liverpool & Manchester project[16] and replaced him with a younger apprentice engineer, Thomas Longridge Gooch, scion of a Bedlington family and a relation of Michael Longridge, who had been working as secretary-cum-draughtsman in his house in Upper Parliament Steet, Liverpool, since October 1826. Along with his brothers, Tom Gooch would become prominent in railway affairs.

Once again, the Stephenson business was spreading itself too thinly over too many projects. Unfortunately for George's grand ambition, the idea of a regulated, planned national railway system proved to be only a vision. Only the intervention of government could have provided for the kind of system that William James had outlined, and no British government in the 1820s or the decade following had the political will, executive authority or administrative resources to do this. Countries which achieved it, like Belgium, were governed in a far more centralised way than Great Britain. Instead, railway development in Britain was left to private initiatives and grew in piecemeal fashion, subject only to the controls of submitting a Bill to Parliament; whose acceptance or rejection had nothing to do with government policy, and often as little with the merits of the proposal. There was to be no official great Cham of railways. George Stephenson & Son were in the fortunate position that the proposers of new railways very often came first to them, but the company did not have the structure or managerial resources to provide its surveying, plan-drawing and engineering package for a large number of unrelated and geographically separate railway companies or would-be companies; nor could it very well work for rival projects.

Robert Stephenson's oscillations between Newcastle and Liverpool or London might have soon become wearisome if he had not also a personal reason for visiting the capital. In March 1828 he wrote to a friend that: 'If I may judge from appearances, I am to get the Canterbury Railway, which you know is no inconvenient distance from London. How strange! Nay, why say strange, when all my arrangements instinctively regard Broad Street as the pole.'[17] Broad Street, in the City of London, was where John Sanderson, a City businessman, lived, and Robert was courting his daughter Frances. He had met her before he had gone to Colombia, and appears to have lost no time in renewing the acquaintance-ship. By October of that year it had become a romance, and the young lady was introduced to George, though apparently in a fortuitous way: 'When in London I met my father by pure chance,' Robert wrote to a cousin of Fanny's, 'and as he remained a day I had him introduced to Fanny. He likes her appearance and thinks she looks intelligent. I took him to the house without her having the most distant idea of his coming. She did not appear confused, and the visit went off extremely well.'[18]

Occasionally, signs of impatience ruffled his normally cool and equable approach to life. A letter of 27 August 1828 shows irritation and a touch of prig-gishness with the directors of the Bolton & Leigh Railway, who demanded his presence at short notice to discuss some alteration:

> This is one way of doing things, but, proud as I am, I must submit. I have tried in my cool and solitary moments to look with patience on such proceedings, but, by heavens, it requires a greater store than I have. I would patiently bear this alteration if they did it from principle; but knowing, and indeed hearing, them say from what the alteration does really spring, I cannot but consider it unwor-thy of Liverpool merchants. I plainly perceive a man can only be a man. As soon as ever he aspires to be something else he becomes ridiculous ...[19]

This enigmatic last phrase could be a reflection on himself or on the cheap-skate directors. Robert spent enough time at Broad Street to draw some indirect criticism from Thomas Richardson, the only person involved both with his Colombian connections and the two Stephenson enterprises. Richardson had made some comment to John Dixon about Robert's neglect of his duties, which Dixon duly passed on, prompting a direct reply:

> Dear Sir, You do me an injustice in supposing that the ladies in Broad Street engross the whole of my time; I am at present so ardently engaged in the Carlisle opposition [the case against the proposed route of the Newcastle & Carlisle railway] that I have neither time to visit Broad Street nor the Hill [Richardson's house at Stamford Hill] ... You are really too severe when you imagine, or rather conclude, that I neglect business for considerations of minor importance. I am well aware that it is only by close attention to my business that I can get on in the world ...[20]

By this time, 31 March 1829, he was engaged to Fanny Sanderson, but his work ethic, at least as spelled out for the austere Richardson, suggested he was already married to his job, or rather, jobs. He was too conscientious a person for anyone to suppose he neglected the Canterbury & Whitstable line for Fanny, but he did neglect it. He was recruiting miners and designing portable machinery for the Colombian and Mexican mining companies, as well as helping on several railway projects and he had also been appointed engineer of a 16-mile railway in Leicestershire, linking the collieries at Swannington in the west of the county with the city of Leicester. And, of course, he was always returning to Forth Street and the vital development work going on there. Left to the superintendence of a not particularly talented deputy, in the person of Richardson's son Joshua, the Canterbury line was not to be a good advertisement for George Stephenson & Son. At Newcastle, Robert was fortunate to have two men of real ability: William Hutchinson, the works foreman, and G.H. Phipps, a draughtsman, both of whom entered fully into the spirit of the quest. Robert was able to discuss some design alteration on one visit and see the finished piece on his next. Michael Bailey explains his procedure in terms that make it seem perhaps too close to those of a modern industrial workshop: 'He pursued four development programmes: boiler design, to increase steam generation; thermal efficiency, to reduce heat loss between boiler and cylinder; transmission design, to simplify conversion of reciprocating piston action into vehicular movement; and suspension design, to reduce dynamic forces between locomotive and track'[21] – but which emphasise the systematic nature of the work. *Lancashire Witch* became a boiler test bed and ended up with a straight double-flue boiler, the fire-tubes being of 18-inch diameter.

Although the L & M had authorised Stephenson to build a locomotive, strong opposition remained among some members to the idea of travelling engines. Among them was Thomas Shaw Brandreth, who still believed in the future of horse-haulage, but anchorman of the antis was James Cropper. His objections were framed in economic terms, adversely comparing the cost of locomotive haulage with that of fixed engines, though personal opposition to the line's engineer became increasingly evident. Cropper and others were well aware that George Stephenson stood to gain if locomotives were chosen, and therefore considered him a biased adviser; the Newcastle works might also get L & M orders for fixed engines, but would at least have to compete in price against other builders. George Stephenson was well aware of the views of the Cropper camp. For those directors who shared neither Booth's and Sandars' enthusiasm, nor Cropper's hostility, a choice still was necessary, and after several exploratory missions made to see different engines in action, two independent engineers, John U. Rastrick of Stourbridge and James Walker of Limehouse, London, were asked to make a thorough investigation of the alternatives and report their findings. This was duly done in the course of January and February 1829, and the two reported simultaneously with the same conclusion.

Rastrick himself had set up as a locomotive builder and constructed engines to Horatio Allen's order, which went to the USA in 1828; but he and Walker, despite noting the performance and expressing their belief in the potential of locomotives, came down narrowly in favour of the stationary engine system. The verdict was not unreasonable: for all their disadvantages, fixed engines were reliable and proven, whereas locomotives were still very clearly at a developmental stage. Robert Stephenson's reaction again showed that powerful emotions underlay his reasonability and methodical approach. With information supplied by George as well as their own researches, he and Joseph Locke set about compiling a swift counter-report, and he wrote to a friend that he believed locomotives 'still will ultimately get the day, but from the present appearances nothing decisive can be said [by the L & M board]: rely upon it, locomotives shall not be cowardly given up. *I will fight for them until the last. They are worthy of a conflict.*'[22]

These militant words have helped to promote the idea of a 'battle for the loco-motive', but what ensued was not so much a battle as a hectic struggle to produce a locomotive that would silence the opposition. Like Sisyphus of ancient legend, the team at Forth Street had pushed a succession of heavyweight objects up the hill; each had got a little farther, but none had reached the top. The counter-report, full of detail in favour of locomotives, was submitted to the Liverpool & Manchester directors on 20 April 1829. By now, those not committed to either side could only feel thoroughly confused, and James Walker's suggestion for a locomotive trial, made in his report and formally proposed to the board on 29 April 1829 by Richard Harrison, was gratefully taken up. It was agreed that the railway company would offer a prize for 'a Locomotive Engine which shall be a decided improvement on those now in use', and over the following few days the rules and conditions were rapidly worked out.

The trials were to be held in October, and it was evident, if unsaid, that if no locomotive met the conditions, the company would proceed to install fixed engines and work the entire line in that way. To the hard-pressed Stephensons, and Hutchinson and Phipps, the imposition of a six-month deadline cannot have been welcome. Still grappling with boiler design, they were on the verge of completing a second locomotive for the L & M to help in construction work: six-wheeled with two vertically-placed boilers, it was called 'Twin Sisters'. Delivered to Liverpool on 10 June, it worked well enough, but was clearly not the answer to the board's requirements, now set out in the stipulations and conditions for the trial. These numbered eight, with the key requirement in the second: 'The engine, if it weighs six tons, must be capable of drawing after it, day by day, on a well-constructed railway, on a level plane, a train of carriages of the gross weight of twenty tons, including the tender and water tank, at a rate of ten miles per hour, with a pressure of steam on the boiler not exceeding fifty pounds per square inch.'

A week after 'Twin Sisters' began moving wagon-loads of clay for embankment building, Robert Stephenson was married to Fanny Sanderson at Bishopsgate parish church in London. The junior Mrs Stephenson 'was not beautiful, but she had an elegant figure, a delicate and animated countenance, and a pair of singu-larly expressive dark eyes'. Perhaps by the same source, 'a near relation', Jeaffreson was also informed that Fanny was 'an unusually clever woman, and possessed of great tact in influencing others, without letting anyone see her power. To the last her will was law with her husband; but, though she always had her way, she never seemed to care about having it.'[23] Robert was a devoted and indulgent husband, but Fanny nevertheless had to accept that much of a young engineer's life was spent away from the comforts of home and the arms of his wife. Their honeymoon was no more than a few days in Wales, en route for Liverpool, where Robert had business on 24 June. Michael Longridge had pressed them to live close to him at Bedlington,[24] but in the end they took a house in Greenfield Place in Newcastle; as the name suggests, it was then almost in the country, with an outlook over the Town Moor, but much more convenient for the managing partner of Robert Stephenson & Co. to get to his factory than Bedlington would have been. They set themselves up comfortably, with a piano and a 'sofa à la mode' in the drawing room, but only one servant;[25] a Newfoundland dog completed the ménage. The Sandersons were not wealthy, and Fanny 'had no fortune', but Robert was by no means badly off, with his railway earnings as well as his salary from the Forth Street Works.

By then, a new locomotive was taking shape inside those works. No railway or colliery company had ordered it: jointly commissioned by George Stephenson and Henry Booth, its whole purpose was to meet all the conditions of the Liverpool & Manchester trial. Due note had been taken of what new ideas had been effective, or not, on every engine from George's 'Experiment' up to 'Twin Sisters'. Every piece of knowledge, every observation of locomotives at work, was brought up for possible application. Booth, whose interest in mechanical matters was very much a hands-on one, had made a suggestion which George immediately took up. If two wide fire-tubes through the boiler provided more heating surface and so more steam, more quickly, what if there were a larger number of smaller tubes, each with heat flowing through, providing an even greater heating surface? This thought had already occurred to the mind of Marc Seguin, but there is nothing to suggest that Booth could have known of Seguin's idea, or his French patent.[26]

A patent for a boiler with vertical fire-tubes had been granted to James Neville of London in 1826, but application to a locomotive was not envisaged. In any case, it was one thing to envisage a boiler filled with a multiplicity of fire-tubes, and something else to build one that worked, especially on a vibrating, moving frame. In those days, even at the relatively modest pressure of 50 pounds to the square inch ($3.5\text{kg/cm}^2$), the aim was to keep the boiler as integral as possible, and the twenty-five copper tubes, each of 3-inch (7.6cm) diameter, piercing each end of the 'Premium Engine's' boiler, presented a whole new technical problem. Booth's brainwave was matched by George's conception of how the tubes should be fitted, and the technical ability of Robert and his team to fit them. Not only did this locomotive have the first multi-tube boiler, but it was also the first to have a separate firebox. Previous locomotives had integral fire-grates and boilers, which was impossible with the tubular design; the firebox 'had to be designed from first principles and this Robert Stephenson undertook personally, evolving the water-jacketed box with internal stays'.[27]

By 3 August Robert was able to report to Booth that 'the Body of the boiler is finished and is a good piece of workmanship'.[28] Despite his comings and goings, he managed to keep Booth fully briefed on progress. In these letters he also comments on Timothy Hackworth's design. The enterprising Hackworth was also building a 'Premium Engine', but, with only the most basic facilities at Shildon, he had to get the boiler constructed at Bedlington and to obtain cylinder castings from Robert Stephenson & Co.,[29] enabling Robert to study his plan: 'I will write you in a few days detailing Hackworth's plan of boiler, it is ingenius but will not destroy the smoke with coal, which I understand is to form a portion of his fuel; coke will be the remainder – he does not appear to understand that a coke fire will only burn briskly where the escape of the carbonic acid gas is immediate.'[30] A detailed description and rough sketch of Hackworth's boiler design followed on 12 August. Hackworth had to resign himself to the fact that the Stephensons would know the main dimensions of his engine; this knowledge, obtained with Longridge's connivance, reassured them that, in this crucial respect at least, he was still working on traditional lines. Though there were no qualms about perusing Hackworth's plans, snooping by others was different.

Another declared entrant for the trial was Timothy Burstall of Leith, known for his work on steam road coaches. On 31 August Robert wrote to Booth to say that Burstall's son '… walked into the manufactory this morning and examined the Engine with all the coolness imaginable before we discovered who he was' – Burstall junior was speedily shown the door, and Stephenson consoled himself with the thought that there was scarcely time for the Burstalls to take advantage of anything that might have been seen. The engine was by then ready for the road and was tested at Killingworth in the first days of September, with Robert reporting to Booth on the 5th that it was capable of doing everything that was set out in the stipulations, if not more. To any engineer's eye it looked quite different to all its predecessors, with a square firebox riveted to the boiler. It ran on four wheels, but the cylinders, mounted at an angle on the side of the boiler, drove the front wheels only, which were considerably larger than the rear wheels, whose axle was set behind the firebox. A tall smokestack was the only vertical element, secured by two long iron stays to the boiler. Feeling that he had put a great deal of his own expertise into the design, Robert had asked to be included as an equal partner in its ownership. Booth was reluctant to agree: 'I had no wish to dilute my interest in the little speculation I had entered upon',[31] but succumbed to the persuasion of father and son.

Hauled across country to Carlisle in parts, taken by canal from Carlisle to Bowness and from there by ship, by 18 September the engine was delivered to Liverpool. Brightly painted in yellow and black, with the chimney in white, 'the Wheels of the Engine are painted in the same manner as coach wheels, and look extremely well'.[32] It now had a name: *Rocket*. Even its four-wheel tender, or 'convoy cart', had novel features, being the first vehicle definitely known to have axle-boxes with bearings fitted outside the wheels.[33] *Rocket* was the first contender on the scene, almost three weeks before the appointed date of 6 October. By the end of September others had arrived. Burstall had brought his lightweight steam engine, *Perseverance,* down from Edinburgh; Hackworth was there with his just-finished *Sans Pareil*; and a locomotive of which the Stephensons knew nothing, *Novelty*, was entered by the Swedish engineer John Ericsson and his English colleague John Braithwaite. The conditions allowed for any form of power, and apart from the four steam-driven entries, there was the horse-propelled *Cycloped* entered by Thomas Shaw Brandreth. Two horses were placed side by side on a treadmill, whose motion drove the wheels. An American visitor, Ross Winans, with links to the Baltimore & Ohio Railroad, informally entered a humanly-worked 'frictionless' railcar which he called the 'Manumotive', but as it plainly was even less capable than the *Cycloped* of fulfilling the conditions, it was not considered by the judges. In France, Marc Seguin had a tubular-boilered engine almost finished, but even if he had wanted to enter it, it would not have met the criteria.[34] People have often wondered why there were not more entrants: some who might have built a challenger, like Goldsworthy Gurney, could not get the capital together, but in any case, in 1829 the number of people who could design a workable locomotive and get it built was very small.

# 9

# RAINHILL
# AND AFTERWARDS

Huge public interest was aroused by the contest and a large crowd converged on Rainhill for the opening day of 6 October. Not all were local folk or mere spectators – London newspapers sent correspondents and several American engineers were present, including Horatio Allen and E.L. Miller, projector of the South Carolina Railroad, as well as representatives from many British railway companies or provisional committees. A section of straight, level track 1.5 miles (2.2km) long, marked by white poles, had been prepared, with an overlap at each end to allow the trains to come to a stop. Provision was made for supplying fuel and water. A blacksmith's shop had been set up at one end of the test track and a weighing machine had also been installed. Three judges had been appointed: Nicholas Wood, John Urpeth Rastrick and John Kennedy, a Manchester cotton manufacturer who was an enthusiast for machinery and an inventor in his own right. Several hundred constables had been recruited to patrol the unfenced track and keep the crowd, estimated at between 10,000 and 15,000 people, well back.

On that first day, *Rocket*, *Sans Pareil* and *Novelty* all made several demonstration runs, and *Novelty* won the most admiration. Lightest and quietest, and the least smoky, it seemed to skim along the rails, and on one run reached almost 30mph (48km/h). Braithwaite had offered to stake £1,000 on his engine's ability to run from Liverpool to Manchester within an hour.[1] *Rocket* made a run with a 13-ton load of stones and human passengers, and was timed at 15mph (24km/h). *Novelty* took no load on that day and George Stephenson, watching keenly, remarked to Joseph Locke that 'her's got nae goots'.[2] The presence of Charles Vignoles in the London team would have sharpened his sense of rivalry.

On the following day, *Novelty*, pulling three times its own weight, burst the bellows which provided draught for its fire. Braithwaite and Ericsson had built their engine in the seven weeks preceding the trials and had the disadvantage of being unable to test it until they brought it to Rainhill – there was no suitable railway in London. Only on this day were the final terms of the trial announced:

each contender was required to run a total of 65 miles, forwards and backwards over the test length, with a load equivalent to three times its own weight, and at an average speed of not less than 10mph. A stop for refuelling would be allowed after half the distance had been covered.

The trials lasted for a week, an unexpectedly long time due to the breakdowns of *Novelty* and *Sans Pareil* and the appeals of their designers for a chance to make repairs and try again. With three sets of contestants, and a large number of more or less knowledgeable observers and commentators, the atmosphere was tense, with more rumours than facts flying around. Public attendance fell off, though crowds returned on the days when *Novelty* was to be tested. However, by the evening of the 8th, one fact was already incontrovertible: *Rocket* had set the mark that the other contestants had to surpass if they hoped to win the prize. On that day, having raised steam from cold, duly weighed and loaded, it started on the test course at 10.30 a.m. and ran the first half, 35 miles, in 3 hours 10 minutes and the second half in 2 hours 52 minutes.[3] Who drove it is not recorded, though it seems inconceivable that anyone other than the two Stephensons would have been in charge. The stipulations of the 'Ordeal' had been not only met, but exceeded. A very significant demonstration had been made on the morning of the 7th, when *Rocket* took a carriage loaded with passengers 'with great ease' up the gradient of 1 in 96 beyond the Rainhill level; a stretch of line which George Stephenson had considered impossible for locomotives, and always intended for fixed engine working, was well within the power of his new machine.

On the fifth day, while *Novelty* was stopped to have a broken pipe repaired, *Rocket*, without load or tender, swept by in a demonstration run at 30mph (48km/h). Hackworth had been struggling with a leaky boiler and it was not until 13 October that *Sans Pareil* made its trial. Though his locomotive was found to be overweight by 5 hundredweights (254kg), and also to have unsprung axles, thus not conforming to the rules, Hackworth was allowed to proceed, but the venture came to an end when the feed pump failed and the boiler ran out of water. Next day, *Novelty*'s final bid ended when its boiler joints gave way, and Braithwaite and Ericsson withdrew it from the trials. *Cycloped* was never a real contender, and Burstall, who had had to make repairs to *Perseverance* following an accident on the way to Rainhill, made only a token run over the test course before withdrawing his entry. The judges confirmed to the Liverpool & Manchester directors that *Rocket* had fully met their stipulations and merited the £500 prize.

The Stephenson-Booth camp seem to have made no secret of their tubular boiler, since John Rastrick noted its details and sketched them at the time,[4] but even while congratulating George Stephenson on 'the perfection to which he has brought the old-fashioned locomotive engine', the correspondent of *Mechanics' Magazine* believed that 'it is the principle and arrangement of this London engine which will be followed in the construction of all future locomotives'.[5] He could not have got it more wrong. *Novelty*, with its tiny boiler and its bellows-worked fire, was the old-fashioned engine, and *Rocket's* tubular boiler and self-induced blast defined the future development of steam locomotives for 150 years to come. Timothy Hackworth, who had invested his own modest funds in his bid, was very angry in his disappointment, and virtually claimed sabotage. John Dixon, a loyal Stephensonian, who had been with the Rainhill party, wrote to his brother James in Darlington to sum up what he had seen:

> We have finished the grand experiment on the Engines and G.S. or R.S. has come off triumphant and of course will take hold of the £500 so liberally offered by the Company; none of the others being able to come near them. The Rocket is by far the best engine I have ever seen for Blood and Bone united … [Hackworth] openly accused all G.S.'s people of conspiring to hinder him of which I do believe them innocent, however he got many trials but never got half of his 70 miles done without stopping … Vox Populi was in favour of London from appearances but we showed them the way to do it for Messrs Rastrig & Walker in their report as to Fixed and permt. Engines stated that the whole power of the Loco. Engines would be absorbed in taking their own bodies up the Rainhill Incline 1 in 96 consequently they could take no load. Now the first thing old George did was to bring a coach with about 20 people up at a gallop and every day since has run up and down to let them see what they could do up such an ascent and has taken 40 folks up at 20 miles an hour.[6]

These feats were not just showing off – the Stephensons were demonstrating that their locomotive could repeat its performance reliably day after day. This was important as they did not just want the £500 prize; they wanted an order for thirty new locomotives. Surprisingly, no attempt was made to patent the tubular boiler. It can only have been a deliberate decision, though the reasoning behind it is unknown. Perhaps they learned of Seguin's version; or it may be that they wanted to keep it right out of the public domain, since the existence of a patent was often seen by others as a challenge to produce something just sufficiently different to evade the claim of infringement. And at that time it was not intended that any other firm should make Stephenson-type locomotives. Already, too, they were planning an improved version with more, and narrower, tubes. The loser in this would have been Booth, though for his investment of around £180 in the building of *Rocket*, he got £330 back from his share of the prize and of the sale of the locomotive to the Liverpool & Manchester Railway. After the intense involvement of 1829, though he kept up his connection with railways, he did not collaborate again with the Stephensons.[7] On the Liverpool & Manchester board, the anti-locomotive camp were silenced. A further four engines of the *Rocket* type were promptly ordered from Robert Stephenson & Co., and the directors also bought *Sans Pareil* and *Novelty*, which cannot have pleased their engineer. Three further engines were ordered from Forth Street in time for the opening of the line. As the news of *Rocket's* success spread across the world in newspapers, reports and personal letters, the name of Stephenson went with it, synonymous with a new stage in locomotive development.

Life could now resume a more normal mode for George and Robert, though George still had a deadline to meet in the completion of the Liverpool & Manchester Railway, a task he was eager to finish so that he could involve himself in a wider range of schemes. His spirit and confidence were high, and when Captain James Chapman, RN, was appointed by the board to examine and report on the state of the construction works, he made no attempt to hide his wrath, especially when it appeared that Chapman considered himself entitled to make decisions which properly were in Stephenson's province, including the dismissal of a Northumbrian engineman who had been impudent:

> This kind of interference with my duties as well as the doubts and suspicions which had been expressed regarding the opinions I have from time to time given on different aspects connected with this work had occasioned me much uneasiness. I have been accused of jealousy and a want of candour in the case of Mr Brandreth's and Mr Winan's waggons as well as in that of Mr Erickson's engine, and of even worse than this in the case of Stationary v. Locomotive engines. In all these instances, instead of jealousy I confidently state that I have been only influenced by a disinterested zeal for the complete success of your work and by a laudable desire to support and establish my own credit.
>
> May I now ask if I have supported your interests or not? Has Mr Brandreth's carriage answered? Has Mr Winan's saved 9/10 of the friction? Was not Walker and Rastrick's report wrong? Has the *Novelty* engine answered your expectations? Have the *Lancashire Witch* and the *Rocket* not performed more than I stated? These facts make me bold, but they also stimulate me to still further improvement. But I cannot believe that you will permit me to be thwarted in my proceedings by individuals who neither understand the work nor feel the interest which attaches me to this railway. Allow me therefore to ask if you intend Mr C. to continue on the works.[8]

Chapman's role had nothing to do with locomotives, but Stephenson made adroit use of the immense kudos his locomotive triumph had given him. Chapman was withdrawn in January 1830. In this, as in many other ways, George Stephenson staked out the ground for those who followed. The relationship between railway managements and their construction engineers was always to be tautened by mutual misunderstanding and varying degrees of anxiety, and even mistrust. But while hot verbal exchanges were frequent, few later engineers would dare to administer such an epistolary dressing-down to their employers. It might have been his railway, not theirs — as in a sense it indeed was — to make or mar. With Locke and Robert busy elsewhere, George did not spare himself in the huge effort needed, not only to supervise 30 miles of intensive activity, but to plan in detail a great range of new items and equipment, to purchase materials and, on top of all that – especially as completion drew near – to spend time escorting local notables on promotional trips arranged by the directors:

> Almost every detail in the plans was directed and arranged by himself. Every bridge, from the simplest to the most complicated, including the then novel structure of the 'skew bridge', iron girders, siphons, fixed engines, and the machinery for working the tunnel at the Liverpool end [where locomotives were not allowed], had to be thought out by his own head, and reduced to definite plans under his own eyes. Besides all this, he had to design the working plant in anticipation of the opening of the railway. He must be prepared with wagons, trucks, and carriages, himself superintending their manufacture.[9]

Again, Stephenson's subordinates, ignored here, should not be forgotten. Most of the draughting work was done by Thomas Gooch, in addition to writing George's letters and reports. In 1861 he told Smiles how he made his drawings by day at the L & M company office in Clayton Square, from instructions given by Stephenson on the previous evening either verbally or as little rough sketches, as well as writing letters at his boss' dictation. Stephenson himself would have been up by sunrise or well before it in winter, visiting the tunnel and Olive Mount cutting works before coming home for breakfast, then setting off on his horse, 'Bobby', to range along the line of construction. Encounters with the engineer were not always amiable: George Stephenson was in a hurry and his temper was short when he met with delays or difficulties, putting Smiles into an agony of circumlocution: 'Mr Gooch says of him that though naturally most cheerful and kind-hearted in his disposition, the anxiety and pressure which weighed upon his mind during the construction of the railway, had the effect of making him occasionally impatient and irritable, like a spirited horse touched by the spur.'[10] Defaulters and problem-makers were perhaps lucky to get away with merely a verbal pounding. Home again, he checked through pay sheets and reports from the section engineers, had a late dinner and perhaps allowed himself a short nap before a session with Tom Gooch or his successor in the post, the 19-year-old Frederick Swanwick, another pupil-engineer pressed into secretarial duty. Finishing late, he went to bed in preparation for another dawn start. It was a punishing schedule, but wholly self-imposed, because it was the only way he knew to tackle the job. Stephenson had the ability to take cat-naps and awake refreshed: 'His whole powers seemed to be under the control of his will, for he could wake at any hour, and go to work at once.'[11]

Another view of George Stephenson in his Liverpool days, and later, was given by Joshua Walmsley, a wealthy merchant (knighted in 1840) and staunchly Reformist politician. At first sceptical of the railway scheme, he was introduced

to George by Sandars and quickly became his friend and one of his most stead-
fast supporters:

> I soon learnt to have entire faith in Stephenson's genius, and better still, I learnt
> to love the man, to revere his truthfulness and honesty, and value his brave tender
> heart … I never met a truer friend, a more consistent man, or a more agreeable
> companion … His speech was sharp and quick, his manner often abrupt. What
> he said he asserted positively, laying down the law. It was the self-reliance of a
> man whom experience had taught to have faith in himself. Sometimes this self-
> reliance might degenerate into obstinacy, but it was the obstinacy of conviction,
> not of conceit.[12]

On each side of the main Liverpool–Manchester axis, a number of local railways
were being surveyed or built under the aegis of George Stephenson & Son, with
George necessarily very much in the background and Robert as the engineer
in overall charge. One of these was the Warrington & Newton Railway, a short
line which linked the town of Warrington to the L & M at Newton-le-Willows.
When the Warrington & Newton directors asked Robert to survey a line south
from their town to Sandbach, as part of a plan to continue it to Birmingham,
the L & M board raised a protest. They intended any line joining Liverpool and
Birmingham to be part of their system. Already the Stephensons had had to
realise the impossibility of imposing any sort of control or monopoly over the
construction of an unregulated railway system, but could George Stephenson &
Son act as surveyors and advisers to two railway companies competing for the
same territory? In this case, they did. In a letter of 17 December 1829 to Thomas
Richardson, which says, 'The trials at Rainhill of the locomotives seem to have
set people railway mad …', Robert states his own position: 'being … engineer for
the Warrington directors, I could not refuse with any appearance of consistency
to attend to an extension of this line – an extension which, if made, will be of
immense benefit to that which I am now executing.'[13] He completed the plans
for the Sandbach extension and put his name to them. Agreement between the
Liverpool and Warrington directors prevented the issue from coming to a head.

Meanwhile, down on the Canterbury & Whitstable Railway, two able young
men from the Forth Street corps, Thomas Cabry and John Cass Birkinshaw (son
of Birkinshaw of the malleable iron rails), had been building engine houses and
setting up winding engines. Robert appeared on 3 May 1830 to make an inaudi-
ble speech at the opening ceremony. In later years, his public speaking improved
considerably, and, if never an accomplished orator, his speeches were clear, factual
and to the point. Although only a mile of this line was worked by locomotives,
it was the first railway to operate a regular steam-hauled passenger service, using
a single Stephenson locomotive, *Invicta*, sent down by sea from the Tyne. The
Liverpool & Manchester was not far behind: on 14 June the directors held a

meeting in Manchester, to which the Liverpool members came in their own train, pulled by *Arrow*, one of the newly-delivered *Rocket*-type locomotives, driven by George Stephenson. The board passed a unanimous tribute to the 'great skill and unwearied energy displayed by their engineer, Mr George Stephenson, which have so far brought this national work to a successful termination'.

A gala opening was being planned for 15 September, and all seemed sweetness and light between the board and the engineer. Each of the new locomotives was slightly different, in a series of improvements on *Rocket*, which was now revealed as a prototype rather than a perfected machine. The last to be delivered before the opening, *Northumbrian*, no longer had the firebox as a separate piece bolted onto the boiler; boiler and firebox were made as a single unit, with a backplate incorporated to separate them, and with 132 tubes whose diameter was reduced to 1.6in (4cm) providing more than three times the heating surface of *Rocket*. This was the first locomotive to have a smokebox, an enclosed space in front of the boiler in which the exhaust steam from the cylinders and the hot gases from the boiler tubes were drawn up into the tall chimney. It also had the first proper tender, with a built-in water tank rather than a cart with a water cask. *Northumbrian*'s precarious footplate was ascended by the mettlesome young actress Fanny Kemble, fresh from her debut success as Juliet at Covent Garden, and she was taken by George Stephenson for a spin on 25 August, during one of the numerous guest displays that preceded the opening. She loved George almost as much as she loved the wonderful feeling of speed, and her reaction was worth any number of directorial encomiums:

> When I closed my eyes the sensation of flying was quite delightful, and strange beyond description; yet strange as it was, I had a perfect sense of security and not the slightest fear … Now for a word or two about the master of all these marvels, with whom I am most horribly in love. He is a man from fifty to fifty-five years of age; his face is fine, though careworn, and bears an expression of deep thoughtfulness; his mode of expressing his ideas is peculiar and very original, striking and forcible; and though his accent indicates strongly his north country birth, his language has not the slightest touch of vulgarity or coarseness. He has certainly turned my head.[14]

Some people were almost as impressed by the railway itself, particularly the Edge Hill tunnel. In *An Accurate Description* of the line (dedicated to George Stephenson), J.S. Walker notes of the 'Great Tunnel' that the visitor 'cannot fail to be impressed by feelings of awe and admiration. It is the most remarkable portion of the whole work. The road being formed of sand is as smooth as a bed left by a summer sea',[15] with the rails rising an inch above the level. Painted on the whitewashed wall, for promenaders, were the names of the streets below which they were walking, and gas jets every 50 yards lit up the scene: 'The effect

The 'Moorish Arch' (from a Bury engraving)

was grand and beautiful ... no great stretch of the imagination was required by
the enthusiast, to entertain the pleasing delusion that he traversed the splendid
passages of a magnificent eastern palace, in which the fair inhabitants of a hundred harams were permitted to ramble in unwonted freedom among the lords
of creation.' At the upper level, the tunnel emerged into a spacious quadrangle,
with the two engine houses joined by a 'Moorish arch': 'The building is in the
Turkish style and is highly creditable to the taste of Mr Foster,[16] who designed
it.' Above, on each side, rose 'two beautiful Grecian columns, made of chequered
brick, with pedestals and capitals ornamented by stone' – but these aesthetic
objects were also useful, being the engine house chimneys. Walker also noticed
the intentional composition of the three tunnel mouths; the large main one, a
side tunnel to the goods station and a third made purely for the look of the thing,
in which carriages were stabled. It was the architecture, design and spirit of the
romantic eighteenth-century landscape garden brought to railway construction
and enjoyed as such by responsive observers, some of whom, as they let their eyes
play over the tunnel entrances or were borne through the deep cutting nearby
at Olive Mount, cannot fail to have murmured lines from *Kubla Khan* (published
in 1816) to one another. But if at the Liverpool end there was a touch of architectural fantasy, at Manchester there was evidence of a new utility. The railway
terminated in an elevated section on twenty-two brick arches, which 'will be

The Manchester terminus (from a Bury engraving)

appropriated to various mercantile purposes; those nearest the river being already employed as a dye-work'[17] – the first of innumerable repair shops, laundries and the like, established under the arches of urban railway tracks in the world's cities.

In 1830, a year of signs and portents, social and political, as well as memorable events, a disreputable monarch, George IV, was succeeded by the equally unprepossessing William IV. The mood for political and economic reform had deepened, and for most people the question was not whether reform was coming, but how much was desirable, or attainable. In that uncertain year, part-optimistic, part-apprehensive, the opening day of the Liverpool & Manchester Railway remained fixed in many people's minds as an indication of the way in which the country had become different, and could no longer be understood and managed in the old ways. Conscious of the 'national' quality of their enterprise, the directors were determined to make 15 September an unforgettable occasion and the engineers made their preparations accordingly.

Almost 800 guests were invited to ride on the inaugural trains – 772 in all. The principal invitee was the Duke of Wellington, still the Prime Minister, though embattled with members of his own Tory Party as well as an increasingly confident, if not united, opposition. National and local dignitaries included Sir Robert Peel and the two Liverpool MPs, General Gascoyne and William Huskisson, who was a former president of the Board of Trade and leader of the House of Commons. Also, there was Lord Wharncliffe, one of the Grand Allies, and a constellation of engineers including Nicholas Wood, Rastrick, George Rennie and Charles Vignoles. Eight trains were lined up on sidings at the top end of Edge

'North Star'

Hill Tunnel, each with a Stephenson locomotive at its head. First to leave was the proudly named *Northumbrian*, driven by George Stephenson, with the duke and his entourage, and a band. Robert drove *Phoenix*, his uncle Robert had *North Star*, Joseph Locke was on *Rocket*, Thomas Gooch on *Dart*, William Allcard on *Comet*, Frederick Swanwick on *Arrow* and Antony Harding on *Meteor*. The regular drivers were probably acting as firemen on this occasion.

The duke's train was on the right-hand or south-side track, while the other seven trains ran on the left-hand track. At Parkside the trains stopped for the engines to be supplied with water and coke. Although passengers had been asked to remain on board their trains, many ignored the request, and among them was Huskisson. Once a member of Wellington's government, he was not on good terms with his leader, and it seems to have been to make a public show of amity that he went round to the duke's carriage, where a crowd had gathered, spilling on to the other line. Seeing a train approaching, some people cried out in warning and there was a general scramble off the track into the duke's train or to the other side, but Huskisson missed his footing and fell across the line. *Rocket*, with the horrified Joseph Locke unable to do anything to prevent it, ran over his left thigh, as did several carriages before the train came to a stop. The injured man was carried to *Northumbrian*, which was detached from its train and, with two doctors from among the guests and the Earl of Wilton, George Stephenson drove the engine faster than any had gone before (it was later calculated to average 36mph (58km/h)) to the nearest place on the line, Eccles. Huskisson was conveyed to the vicarage there.

*Phoenix* and *North Star* drew their trains and the ducal one forward to Eccles, where *Northumbrian* was again attached to its train and the convoy proceeded, in very much muted spirits and without music, to Manchester, running the gauntlet of a hostile rail-side crowd whose cries and banners proclaimed 'Remember Peterloo'. Manchester at that time had no parliamentary representation. James Walker described the scene as '… awfully sublime. Such a dense crowd of human forms lined, nay actually covered the road that it appeared to be impossible to make a passage through them.'[18] The grand day had turned sombre and threatening: it ended with chaotic crowd scenes at Manchester, the news that Huskisson had died, and delay and confusion on the return because the duke's train had left Manchester only to encounter four of the locomotives approaching the terminus on the same track; these had to retreat before it to Huyton, almost back in Liverpool, before going on again to pick up their trains. Heavy rain was falling as they left Manchester. It was eleven at night before the last of the passengers got back to Wapping terminus in Liverpool, but there was still a crowd waiting to greet their return. A final word on that day of triumph, drama and tragedy may be had from Sir Joshua Walmsley:

Tragic as was the occasion, Stephenson could not resist a quiet thrill of satisfaction as he remarked to me, on returning to Liverpool, that the 'Northumbrian' 'had driven Mr Huskisson to Eccles at the rate of forty miles an hour. Five years ago,' he added, 'my own counsel thought me fit for Bedlam for asserting that steam could propel locomotives at the rate of ten miles an hour.'[19]

# 10

# THE UTILITARIAN
# SPIRIT

Four years of intensive work had done nothing to reduce the elder Stephenson's appetite for new tasks. He did not even take a break. Already, before the end of his labours with the Liverpool & Manchester, George Stephenson & Son had contracted in May 1830 to be engineers to the proposed Sheffield & Manchester Railway. Robert was equally busy. In 1830 he was elected a corresponding member of the Institution of Civil Engineers, and so, unlike his father, became part of the 'establishment'. He owed his election to the joint report with Locke (already a member) on the advantages of locomotives over fixed engines, which was presented at the ICE and published for a wider readership after the Rainhill triumph. At this time, all engineers other than those in military service were civil engineers.

Even before the last of the improved *Rocket*-types had been completed at the end of September 1830, a new approach to design had been adopted at Forth Street with *Planet*, intended to be the first of a series of virtually identical loco-motives which could be built more quickly than before, using a standard set of parts. Up to now, each engine had been a one-off, and the decision to establish a 'class' shows that the Newcastle team felt they had reached a significant point in their development work. No one thought *Planet* and its sisters were the last word, or anything like it, but from then on, locomotive building was an industry rather than inspired craftsmanship. A railway company ordering a *Planet* knew what it was going to get, in terms of power, performance and fuel consumption.

By this stage it might be expected that patents would have been taken out, but this did not happen. One suggestion is that the Stephensons were just too busy to attend to this,[1] but it seems unlikely, especially as Longridge and Pease, both with investments to protect, were on hand to help with the formalities. But a patent might have been difficult to claim. Already, in 1830, a new builder had appeared in the person of Edward Bury, a Liverpool foundryman whose works produced a locomotive named *Liverpool*, which was supplied for testing in July

*Planet*-type locomotive. In later versions the springs were mounted above the frame

to the Liverpool & Manchester Railway. *Liverpool* had the distinction of being the first locomotive to have its cylinders placed under the smokebox and inside the frame, with the pistons driving a crank axle between the wheels. George Stephenson put off testing it as long as he could, complaining that the 6ft driving wheels were too big; eventually it was tested on the Bolton & Kenyon Railway, where it justified his concern by derailing, causing two deaths.

Inside cylinders were also a feature of the *Globe*, a locomotive designed by Timothy Hackworth as the Stockton & Darlington's first passenger engine, and ordered from the Stephenson works in February 1830. Though by now Forth Street were capable of building a locomotive in three months or less, they did not complete *Globe* until October, so that it appeared after *Planet*, which was the first Stephenson locomotive to have inside cylinders. Since the placing of the cylinders inside the frame was seen as an important improvement in design, and was to be for many years a British standard, arguments as to whose was the prior design broke out and, for want of contemporary evidence, are likely to be never satisfactorily resolved. Despite the heat it has generated at various times over 200 years, it is a somewhat arid point. The successive post-*Rocket* locomotives built by Stephenson show a steady reduction of the angle of the external cylinders, to the level-set pair fitted to *Northumbrian*. Once horizontal cylinders were seen to work effectively, internal placing, to conserve heat and improve balance, was an obvious step to try out. George Stephenson and his team had always been rather patronising to Hackworth, whom George had certainly not envisaged as a rival when he got Timothy the Stockton & Darlington job; at a busy time it is all too likely that Forth Street would set a one-off Hackworth order aside in favour of a stream of orders from the Liverpool & Manchester. Hackworth understandably preferred to assume a more sinister motivation. In terms of general design, Robert Stephenson & Co. was ahead of any other builder, but Bury in particular was serving notice that they would not have an uncontested monopoly in locomotive construction.

With the opening of the L & M, so many projects were being offered to George Stephenson that even he had to start saying no. To Charles Tennant of the Edinburgh & Glasgow Railway, he wrote on 28 October 1830: 'I find myself with much reluctance under the need of stating to you that I cannot possibly attend to them ... I have already about 300 miles in hand, so you may be sure I have my hands full.'[2] But four days before the L & M opening, a major survey project was offered to George Stephenson & Son, which the apostle of a national trunk railway system was bound to jump at – a line to run from London to Birmingham. First proposed and surveyed in 1825, but dropped because of the financial crisis, at a length of 112 miles (179km) this was the biggest railway scheme yet. In the summer of 1830, George's cup had been sweetened by his being asked to adjudicate on two possible routes for it: one proposed by Francis Giles, one by John Rennie (remarks by Giles and George Rennie in 1825 had certainly not been forgotten). He chose Giles' route, with modifications, then mobilised all his resources and contacts – as well as his own record and his own powerful personality – to procure the surveying contract instead of Giles or Rennie. Some of the directors and large shareholders in the Liverpool & Manchester Railway had strong links with the Birmingham Committee of the London & Birmingham. Once again, the notion of George Stephenson as a naive engineer manipulated by mercenary capitalists is seen to be wide of the mark. He was adept at bringing his business contacts on side, with his friends of the 'Liverpool Party' added to his existing Pease-Gurney allies.

A contract for surveying and laying out the line was agreed on 18 September, and the Bill was to be ready for presentation to the Private Bill Office in time for the spring 1832 parliamentary session. George was to be paid 7 guineas a day for each day actually spent on the project and Robert was to receive 5.[3] In fact, George scarcely involved himself and Robert took charge of the work. The over-stretched team of engineers had to be augmented and he found two good subordinates in Thomas Elliot Harrison, trained by Chapman, and John Brunton. Both were under 30 – George himself was only 49: in 1830 an elderly railway engineer was a self-contradiction; the nearest thing was Trevithick at 59. As the number of proposed railways increased, it became more difficult to find experienced men. With demand exceeding supply, some individuals inevitably found themselves in positions beyond their abilities. Writing in his own hand (a procedure for close friends only) from Liverpool in December 1830 to Longridge, George commented: 'It is relly shamefull the way the countrey is going to be cut up by Railways we have no less than 8 Bels for Parliament this sissions';[4] most irksome was that by no means all those Bills were the work of George Stephenson & Son. Working with great speed, Robert was able to present a draft route plan in December, which was accepted by the committee; more detailed plans of the tunnel sections followed in February, and then the task of preparing and lithographing multiple copies of the maps and plans was begun.

At least for Robert, the northern end of the route was not too far from the Leicester & Swannington Railway, for which they had had the engineering contract since June 1830. News from this site was not good. His manager was Joshua Richardson, transferred from Canterbury, and the construction was apportioned in orthodox fashion among six contractors. But Richardson was not on top of the job, and at least one of the contractors took advantage of his laxity; the directors complained, and Robert sent him back to the Canterbury line, found a replacement and restored proper order among the contractors. Small contractors were a constant source of problems: they ran out of money and went bankrupt, or tried to use inferior materials, they frequently fell behind on their schedules and their labourers were often farm boys, unskilled in the work and without the stamina of the professional navvy. On the Leicester & Swannington, two were deprived of their contracts and another had a fatal fall down a tunnel shaft. In digging the tunnel itself, at Glenfield under the Leicester Forest ridge, over a mile long at 1,796yd (1,618m), a huge bank of soft sand was encountered which had not been identified in the trial borings. Instead of a clean bore through hard rock, a brick tunnel had to be made, a more difficult and expensive affair.

Near the Swannington end of the line, the country estate of Snibston, close to Coalville, was advertised for sale in early 1831. Coal had already been dug there from seams close to the surface, and Robert, scanning the topography, formed the view that further deposits lay beneath at a deep level and that the estate would be a good purchase. His father duly came down to have a look, agreed and bought it on 26 July 1832,[5] in partnership with Joseph Sandars and Joshua Walmsley. George liked the district so much that he bought a house, Alton Grange, between Coalville and Ashby-de-la-Zouch, and removed there from Liverpool later that year, driving down in easy stages in a gig with Betty.

Work began on sinking a pit at Snibston, but called for frequent utterance of George's favourite slogan, 'Persevere!' Water-logged Keuper marl would make the brick walls of any pit bulge and give way, but Stephenson brought Northumbrian mining techniques and had the pit lined with cast-iron plates. Then a 22ft layer of extremely hard greenstone was encountered, which had to be cut through at a daily rate measured in inches. Sandars lost heart, but when Walmsley, trusting in George's confidence, offered to buy him out, he hung in with them.[6] Under the greenstone, at the considerable depth of 600ft (200m), was a rich seam of coal, conveniently close to the railway line, but it had taken two years of hard work before the colliery could begin operations and see a return on the investment.

George's country house provided him with a comfortable and centrally located base, but he was very much on the move. Apart from the not-yet abandoned Sheffield & Manchester Railway, there was also a request from the promoters of a London to Brighton railway to survey a route, though this proved to be another abortive speculation. Liverpool was still the main focus of his attention. After the infighting of 1829 on who should provide a railway between Liverpool and Birmingham, the issue was still to be decided. For one body of promoters, Charles Vignoles made a draft survey of a direct line south via Runcorn with a bridge over the Mersey estuary; another group proposed a tunnel under the Mersey and a line to Chester. A potent little band, Charles Lawrence, John Moss, Joseph Sandars and their colleague Robert Gladstone, formed the nexus of a committee well endowed with finance, drive and railway expertise to promote a line which would link up with the Liverpool & Manchester at one end, and the London & Birmingham at the other. Such a line would be the backbone of the new industrial England, and they named it the Grand Junction Railway. They retained the services of George Stephenson & Son, and George sent his most gifted subordinate, Joseph Locke, to survey the route, starting from the end of the Warrington & Newton Railway, which the new, larger concern proposed to purchase.

Political events in the stormy years of 1830–32 delayed their plans. Parliament had other things on its mind than private bills for railways. A general election in 1830 was followed by another in 1831, as the governing class nerved itself to permit enough political reform to stave off the threat of serious unrest, riots, perhaps even revolution, amongst an increasingly impatient population. Newcastle buzzed with radical and pro-reform agitators, many of them not qualified to vote but calling for the franchise to be extended to all adult men. Debate among the members of the 'Lit and Phil', and of the Mechanics' Institute, would have been intense. Robert Stephenson's views on popular opinion and on democratic government had been influenced by his Colombian stay, where he had been more struck by the general ignorance of the people and the corruption and demagoguery of the politicians, than by the idealism of a young republic.

George, in his new rural home, was not far from the mining settlements, and the farm workers, vastly more numerous then than now, were also among those shouting for reform: ten were killed in riots in 1830 and 450 were transported to the Australian penal colony. The National Union of the Working Classes was set up in 1830 – George's response would no doubt have been that they should get on with their work: that was the way to 'get on' in life, as his own career showed. Most of his business associates backed reform, often for business reasons. Taxation through the 1820s had borne much more heavily on commerce and industry than it had on those whose wealth came from landowning and agriculture. Beyond that, they were strongly influenced by Utilitarian ideas, which had 'a very powerful following among merchants and manufacturers and in the professional classes'.[7] In the same groups, evangelical Christianity, whether Anglican or Nonconformist, was also strong. Both of these elements were involved in the long campaign to win public opinion over to the abolition of slavery in the British colonies, now reaching its peak, along with the parallel demands for domestic reform of the electoral system and the Poor Law. The range of new ideas energised on a national scale by William Wilberforce, Lord Ashley and Edwin

Chadwick was being ardently debated and demonstrated for from one end of the country to the other.

These great and urgent issues did not capture the Stephensons. They were not alone, of course. Many people were hostile to agitators who seemed to be shaking the social fabric, and many more felt that none of it mattered to them. Practical issues were what engaged the Stephensons and the application of ethical or metaphysical questions to human society did not interest them; in as far as it meant change, they disliked it. If Wilberforce felt that there was such a thing as a nation's conscience, and that it ought to be stirred, the Stephensons were among those who thought he would be better off minding his own business. George, though conservative by instinct, was essentially apolitical; his friend Joshua Walmsley later summed up his views: 'On the subject of politics he was generally reticent. He had a certain disdain for it as a hopeless confusion, void of any law that he could grasp.'[8]

In 1830 Robert's political views had scarcely moved beyond a few prejudices. Extrapolation of their general point of view from their comments and behaviour might produce something like this: 'Work is the key to achievement and prosperity. Anyone can succeed who makes the effort. If they do not make an effort, it is their own fault. When genuine misfortune happens, family and friends should care for one another.' An example of this comes from January 1831, when the collapse of a shearlegs hoist at Forth Street killed one workman and injured another. The dead man was John Stephenson, George's youngest brother, an unambitious member of the family. He left a widow and five children, and George took personal charge of ensuring they were provided for.[9] That was a special case, but it was axiomatic to George that an employer should provide for the basic welfare of his workers. Old-fashioned individualism and paternalism – as a social and political outlook it was narrow, with its eyes on the ground, and its inability to deal with wider issues was hidden only by the fact that its natural tendency was to avoid confronting them. Inconsistencies of attitude were inevitable. Both Stephensons were all for government intervention when it came to setting out a national railway system, but Robert would be fiercely against the government setting standards for iron bridge construction.

Yet, more clearly than anything else that was happening at that time, the work of the two Stephensons, and their personal approach to it, revealed the Utilitarian idea in action. The steam railway was already showing how it benefited commerce and industry by speed and cheapness of bulk transport, but, in an unforeseen aspect, it was also providing pleasure and convenience for large numbers of people who used it to travel. The passenger receipts of the Liverpool & Manchester Railway were more than ten times what had been estimated. In these respects, the greatest good of the greatest number was being furnished on a daily basis, and on the felicific calculus the railways would have rated highly. This was not an outcome that the Stephensons had anticipated, but they accepted it and built it into their future plans, and seem to have ignored the irony of their siding politically with those who, like Colonel Charles Sibthorp, MP for Lincoln, were the most outspoken and intransigent opponents of railways; while their personal friends were mostly of Liberal and sometimes Radical persuasion.

Though presented for the 1831 session, the Liverpool & Birmingham Bill was abandoned with the dissolution of Parliament on 22 April. The proposers took the opportunity to have a more intensive survey done and to recapitalise the venture. Robert Stephenson was also looking to the future. From late 1830 he had decided that the capacity of the Forth Street Works was too small to cope with the potential worldwide demand for locomotives. Expansion was a possibility, but Stephenson does not appear to have discussed this with his partners. Instead, he and his father planned to start a new partnership with Charles Tayleur, an early backer of the Liverpool & Manchester Railway and on its board in 1829–30, and whose son, also Charles, a contemporary of Robert's, would manage the new works. These would be in or near Liverpool. When Michael Longridge was informed, he was immediately anxious about the future of Forth Street, and in a series of letters and meetings through the first half of 1831, he and Robert discussed the implications of the proposal. Quite apart from whether the new venture would draw business away from Newcastle, there was the question of joint ownership of technical drawings and patents. At this time, Forth Street was just beginning to show a return on the investment made by Longridge, Pease and Richardson. Why should the Tayleurs benefit from designs which they had not paid for? Stephenson's partners allowed themselves to be persuaded, but they bound their managing partner to some clear conditions:

> Should I become connected with another Manufactory for building Engines in Lancashire or elsewhere I have no objections to bind myself to devote an equal share of my time and attention to the existing establishment at Newcastle. I will also pledge myself *not to hold a larger interest in any other factory*, than I have in Forth Street and *to divide the Locomotive Engine Orders equally.*[10]

It was also formally agreed that the name of Stephenson should not be attached to any other works, and that neither Robert nor George should withdraw any of their money put into Robert Stephenson & Co., whether as capital or as loans, without the consent of their partners.[11] The Stephensons' action in this case sweeps out the last vestige of suggestion that they were pawns of the Quaker money-men. George was playing the capitalist's game with a confident hand. Longridge, Pease and Richardson had to accept that he and Robert could now raise funds and find new backers on the strength of their name and fame. Agreement having been reached, the plan for Charles Tayleur Jr & Co. went into temporary abeyance with many other projects at a time of great political uncertainty. Some were still going ahead, and in the early summer of 1831 Robert was engaged as consulting engineer, at a fee of £300 a year, to a new coal and mineral line, the Stanhope & Tyne Railroad, intended to cross the high Durham moors from a limestone quarry at Stanhope and the Pontop collieries to the Tyne at South Shields. Its road had been acquired, not through Act of Parliament and land-purchase, but by the old Northumbrian way-leave system, whereby a rent for the track was agreed with the landowner for a period of years at a time. In the old days these rents had been very small, but landowners had become more conscious of railways as a cash-cow to be milked hard, and the agreed rates were very high. The line was worked by a mixture of fixed engines, horses and locomotives, with a wagon lift at each side of a steep ravine at Hownes Gill. Unusually for him, Robert accepted £500 worth of shares in the company rather than cash payment for his work. The resident engineer on this difficult project was Thomas Elliot Harrison, whose father was one of the promoters. Harrison senior was something of an opportunist, who sold out his interest when things began to look bad, but Thomas was both able and honourable, and had already embarked on a long and friendly professional association with Robert Stephenson.

Though he had not done the survey for the line, Robert accompanied the Sheffield & Manchester Bill to Parliament in the spring of 1831, to face hostile questioning from the canal interests, but he had schooled himself for these sessions and was well able to stand up to the parliamentary lawyers. As if positioning himself on his father's failure as a parliamentary witness, and setting himself to be the opposite, he was always fully master of the plan; crisp, firm and positive in his replies and with a metropolitan polish as shiny as theirs. The Bill was passed, though the project was later abandoned. A further test came when the London & Birmingham Bill began its parliamentary course in February 1832. Powerful opposition was unable to stop the Bill either on standing orders or on any discrepancy or error in the measurements and plans – a tribute to Robert's professionalism considering the length of the proposed line – and it proceeded to the Lords. Here, however, it was rejected, on the grounds that the proposers had not sufficiently proved the necessity of 'forcing the proposed railway through the lands and property of so great a proportion of the dissentient landowners and proprietors'.[12] Robert was depressed by the failure, but Lord Wharncliffe, who had chaired the Lords' Committee, told him privately not to take it to heart: '... you have made such a display of power that your fortune is made for life.'[13] Wharncliffe presided at a meeting called a week after the rejection of the Bill, which decided to renew the application. In the course of the autumn, while Robert and his lieutenants made a third survey, avoiding the territory of the objecting landowners, the principal opponents of the railway were visited and 'the bribe succeeded where reason could find no entrance'.[14]

While Robert was facing the parliamentary counsel, George, with Joseph Locke, paid his second visit to Ireland, in connection with the Dublin & Kingstown Railway, in February 1832. The first section of the Leicester & Swannington

Railway was opened on 5 May 1832, with a locomotive sent down from Forth Street by sea and canal. On the rest of the line, including the long Battle Flats cutting, work continued, and Robert Stephenson still had management problems. Richardson's replacement, John Gillespie, also proved unsatisfactory, reporting directly to the directors instead of to his employer, which provoked an outburst from Robert to George:

> What a grievous thing it is to witness such a wanton breach of faith on the part of Gillespie – Why did he not let me know before it was discovered by the Directors that the cutting was likely to be behind hand – Means might have been taken to avoid the unpleasant triumph which our enemies will have … The receipt of the letter from Gillespie actually unhinged me, by the anxiety which it occasioned …[15]

Gillespie was relieved of his position and young Birkinshaw was made resident engineer. Such vehemence shows the extent to which Robert lived on his nerves, the strains heightened by the actions of 'our enemies'. The *Edinburgh Review*'s October issue had contained a long, detailed, anonymously written article on railway development, which included an examination of George Stephenson's conduct as engineer of the Liverpool & Manchester Railway. According to its author, the directors of the company had been so 'fascinated' by Stephenson as to allow him a monopolistic control. He was said to have imported workmen from the north-east and to have condemned any engines or rolling stock which did not come from the Stephenson works in Newcastle: '… their engineer is an extensive manufacturer in articles, of which, when manufactured by others, he is the arbiter', and that consequently, 'The mechanical ingenuity of the country is excluded'.[16]

The implication was clear – Stephenson was using his position for his own benefit in a way which was unethical if not illegal. Evidently, someone on the L & M board had provided information to the writer, Dr Dionysius Lardner, an Irish academic, first Professor of Natural Philosophy at London University. Lardner had established himself as a populariser of mechanical and scientific ideas, though his habit of rushing into controversy with unconsidered and often blatantly unscientific assumptions has eclipsed whatever merits he had; leaving him remembered only for such claims as that people travelling in trains at 30mph would be asphyxiated. But the *Edinburgh Review* was one of the prime journals read by people of influence, and to be publicly attacked in this way was galling and potentially damaging. A detailed rebuttal of the claim was made by another L & M director, Hardman Earle, a Stephenson supporter, published in Liverpool by Charles Lawrence, pointing out that only 10 per cent of the construction workers were from the north-east; that two *Novelty*-type engines bought by the board had been failures, that *Sans Pareil* had had to be sold off for £110; and that the

*Planet* class was undoubtedly 'nearer to what we consider perfection (relatively, of course) than any other'. But, coming on top of the London & Birmingham Bill's rejection and the difficulties with the Leicester & Swannington directors, the attack had a strong effect on Robert, and through the summer and autumn of 1832 he was often ill.[17] In the on-going process of self-understanding, he now knew it could be easier on the mind to confront mutinous miners than to live from day to day with the worries of what dilatory or disaffected subordinates might be doing in his absence.

By December 1832 the long national political crisis was at an end, with the Reform Bill passed and a new Parliament elected under new rules. The resultant change in political complexion, though it may have gone against the Stephensons' personal grain, was to their commercial benefit: 'The middle classes, or rather the Whig aristocracy governing on their behalf, had freedom to rule Great Britain in full accordance with the requirements of advancing Capitalism.'[18] With the new session of Parliament, the Bills for railways between Liverpool and Birmingham, and Birmingham and London, were again presented. Both were passed in May 1833. Disappearance of most of the objections to the London & Birmingham line was explained by the tripling in land purchase cost from the original estimate of £250,000 to £750,000.

The Grand Junction Railway already had its chief engineer ready to start work, or so George Stephenson considered. He had friends on the board, he – or rather his employee – had surveyed the line, and he was free and willing. But difficulties arose. The directors had been greatly impressed by Joseph Locke, and Locke's contract with Stephenson had just expired. As a free agent, he too sought the post of chief engineer. From Locke, there was the promise of detailed planning, efficient management and proper reporting; as far as George Stephenson was concerned, no one doubted that the job would be completed, but with all the possessiveness, lack of reporting and consultation, obscure payment systems and general hand-to-mouth style of operating that even his friends had found hard to put up with. But this was an 80-mile line, more than twice the length of the L & M. To George's intense annoyance, it became clear that the board wanted Locke to play a major role and that Locke, despite his respect and real admiration for his former boss, was not willing to work to his orders.

After much wrangling, Stephenson was appointed engineer to the entire line, at a salary of £2,000 a year, on 25 September 1833. But just a month later, Locke was appointed engineer, 'under Mr Stephenson with the special care of the Liverpool end of the line', but also superintendence of the whole works in the absence of George. His salary was £800.[19] Stephenson's overall responsibility was titular only, with executive control only over the southern half, from Newton to Whitmore, while Locke, as engineer in charge from Whitmore to Warrington, was answerable directly to the board. George took this with ill grace. With his son busy negotiating for sole responsibility as chief engineer on the even longer London &

Birmingham, he felt it an indignity to be given half a loaf by the Grand Junction. During the surveys and the parliamentary hearings, both L & B Committees had been very pleased with Robert, and Richard Creed, the London-based company secretary, wrote to him following a meeting shortly after the passing of the Bill: 'Nothing is said as to appointment of engineer or solicitor, but I think *you* may be easy on that head. You have friends here and in Birmingham who appreciate your merits and services.'[20]

But the first trunk railway into the metropolis was a mighty prize, and just because Stephenson had done a survey did not mean it would fall into his lap. With a powerful committee at each end of the line, election of a board took time. While waiting for a decision, Robert went on a sales and 'customer relations' trip to Germany with Longridge's 18-year-old son, James.[21] Eventually, on 19 September, his appointment as engineer-in-chief was confirmed, at a salary of £1,500 a year plus £200 allowed for expenses. Prior to that, the board had paid John Rastrick and another engineer, H.R. Palmer, to inspect Robert's plans and report on them, which they did most favourably. Neither his contract nor his father's with the Grand Junction was made on behalf of George Stephenson & Son: each was appointed in his own right. While between them they had all of one, and half of the other, biggest railway projects in the country, the partnership established to dominate, if not monopolise, railway construction was effectively terminated. They would often consult each other, and did combine on certain projects, but generally from now on both Stephensons would take on work as individuals. For George, it was a setback – for Robert, a step forward.

# 11

# LONDON
# AND BIRMINGHAM

In 1833 George Stephenson was 52, and could look back on a unique range of triumphs and tribulations in the course of a strenuous half-century. On the whole, it was a very satisfactory perspective, right back to the young man who, overwhelmed by the awful gap between ambition and its realisation, had wept on the roadway at Killingworth. He was famous, very comfortably off, with a large house; he was happily married, his son was prospering in the path he had laid down; he had several different enterprises going ahead, with his mining interests as well as the locomotive works and his railway projects. His advice was constantly and deferentially sought on topics to do with railways and loco-motives, his health was good, and he could feel there were more years of work left in him – a necessary consideration for someone who had come to identify life with work. But in 1833–34, at least in connection with his work on the Grand Junction Railway, his grip slackened, and the sense of a great task to be mastered, the close attention to details, the creation of a team spirit among his young assistants, the boundless energy and personal involvement that character-ised his work on the Liverpool & Manchester, were far less evident. In effect, he spent two years in a massive sulk about not getting the whole job. He had the chance to prove that he, the old master, could do his part better than the man he had trained up from a gangling teenager; better even than Robert on his great undertaking. But he did not rise to that challenge and his work on the Grand Junction showed only the worst aspects of his earlier approach. Among his old team the best men were no longer willing boys, but experienced engineers, and they preferred to sign up with Robert or Locke, leaving him to recruit new men with scant experience. Locke and Robert found it hard enough to com-plete their teams. While construction of the Grand Junction's northern section was speedily put in hand, George Stephenson's activity on the southern section was sporadic and unco-ordinated. But from his camp came a constant stream of sniping against Locke and his methods.

Robert tried to poach his old Stockton & Darlington associate, John Dixon, from the Liverpool & Manchester Railway, where he was now on the staff, but to no avail.[1] However, he was able to assemble a strong team, which included Thomas Gooch and John Birkinshaw, and some others who would also rise to eminence, including Charles Fox, an ambitious 23-year-old engineer who already had a group of his own articled pupils to bring in. Among these was Francis Conder, a sharp-eyed youth who would later write an engaging memoir of a life in engineering which began under the chieftainship of Robert Stephenson. George Phipps from the Forth Street Works was one of the draughtsmen. Drawing ruled all their lives in the first months. Having had time to reflect on how he was going to tackle the job, Robert had resolved that every structure on the line should be designed, drawn up and specified, so that the contractor for each section should know exactly what was required of him. It was a method exactly opposite to his father's, but Robert had no intention of following George's style of project management.

With his main base at the London end and a subsidiary office at Coventry, his role as engineer-in-chief was to keep in constant close contact with his four main deputies on four of the five sections into which the project was divided, to maintain contact with the directors, to deal with major purchases and to resolve whatever problems arose. In addition, he took personal charge of District No 1, from the Camden Town terminus north to Harrow. He and Fanny moved from Newcastle to London, first to a rented cottage in St John's Wood, then to a house he bought on Haverstock Hill. Although Robert settled happily in London, he always recalled their Newcastle period with fondness, and it was during that time he made the friends whom he returned to visit in later years. By now, four years into their marriage, it was clear that it was likely to be a childless union, but it remained a happy one. Robert Stephenson was a sociable sort, though he preferred the company of close friends to large gatherings, and he probably had little in the way of casual small talk. Pupil-engineers were also asked to his house for evenings of 'music, talk and cigars'.[2] Fanny was no doubt glad to return to her native London. Robert was on good terms with his in-laws, especially Fanny's brother John.

Four years was the time-frame set by the board for construction of the London & Birmingham Railway, not a long time for a line which required several long tunnels, some very deep and long cuttings, a great many bridges and viaducts, and several aqueducts. As if conditioned by the ultimate speed of the system, railway construction was generally accomplished much more rapidly than comparable lengths of canal. Railways and their rapidity speeded up people's expectations of what could be done within a short period of time. Such a vast project would require the intensive attention of the chief engineer, but Robert also had loco-motive-building on his mind. Late in 1832, with the *Edinburgh Review* controversy buzzing about their ears, the Stephensons and Tayleurs had agreed on a site for the second locomotive works, at Newton-le-Willows, close to the junction of the

Warrington & Newton with the Liverpool & Manchester Railway. The new factory, known as the 'Vulcan Foundry', was being built and fitted out in the course of 1833 and would commence production at the start of 1834. Meanwhile, Forth Street was very busy with orders and enquiries. Harris Dickinson, a dynamic figure who the Stephensons felt took too much on himself, particularly by interfering in mechanical matters, died suddenly in 1832 and was not replaced. Once again, Michael Longridge was expected to take on the managerial tasks which Robert, in sporadic visits, could not fulfil. Robert, however, was very much involved in 'product development' and on 7 October 1833 he took out a patent on an improved type of locomotive: this was a six-wheeled machine, with driving wheels on the middle axle, larger than the fore and aft wheels, and without flanges. The omission of flanges was, according to the patent specification, because experience on the Liverpool & Manchester Railway had shown that: 'The cranked axles of the great wheels have been found liable to break, and then the engines run off the rails, and sometimes overturn, or are otherwise injured. They have also in some cases run off the rails when the crank axle has been only strained, without being actually broken.'[3]

Crank axles were the Achilles heel of early locomotives. American engineers loathed them and soon opted for outside cylinders and cranks, with very few exceptions. British locomotive men persevered with the internal drive and tried to improve its details and metallurgy. But the great advantage of the 'Patentee' was that a larger boiler could be mounted on it. This larger and heavier locomotive type supplanted the *Planets*, and was itself only usable because rails had been undergoing a corresponding process of improvement so that the greater weights could be borne. Railway companies responded favourably, and the demand for 'Patentees' from new lines on the European continent, as well as from British companies, kept the Forth Street order book full and got the Vulcan Foundry off to a flying start.

While George was certainly consulted on these developments and would indeed have expected to be fully informed, his own activity in the course of 1833 to 1835 is hard to trace. He was certainly not devoting himself to the Grand Junction Railway in the way its directors were entitled to expect. At Snibston, he was building a village for the miners with a church, a chapel for Dissenters and a school. The houses were of above average standard for a time when artisan housing was often instant slum-creation: 'At Clay Cross ... the workmen's houses contain four rooms, two on the ground floor and two above, with plots of gardens attached, and other conveniences.' The same writer noted admiringly that the school was put up 'when as yet the concern only paid 3½ per cent'.[4] When the new pit at last came on stream (there were three shafts ultimately) he would certainly have been on hand to supervise. Stephenson sent the first train of 'main coal' (hewn from the main seam) to Leicester by railway. Cheap bulk transport resulted in a fall of the coal price, which Smiles estimated saved the citizens about £40,000 a year, 'or equivalent to the whole amount then collected in Government taxes and local rates, besides giving an impetus to the manufacturing prosperity of the place'.[5] Here again something like the Benthamist 'felicific calculus' was at work, with its mechanistic assumption that the happiness of a community or nation could be measured by material criteria.

Leicester's early advantage would soon be matched in other cities, and cheap coal would fuel England's ever-expanding range of industrial products, from Staffordshire crockery to Birmingham metalware and Yorkshire woollens. George travelled the country by gig or carriage, invited to comment on prospective railway routes or plans, and in this way still contrived to hold in sight his old vision of a national system of trunk and feeder lines. Whatever his abilities as a surveyor of levels and angles in detail, his eye for the general characteristics of a landscape was unmatched. He could indicate the line of a route through complex topography, which the men with theodolites would in due course confirm as the best one; he, of course, prided himself greatly on this facility, once telling a friend that, 'I have planned many a railway travelling along in a postchaise, and following the natural line of the country'.[6]

On the Grand Junction, Locke had completed his detailed plans and had let the contracts for every stage on the northern section by 25 September 1834; on the southern section, George had yet to finalise negotiation of a single contract and his plans were sketchy and costed only in the most approximate way. There was no clear set of specifications for contractors to bid on. Faced with order on the one hand and confusion on the other, the directors asked Locke to assist on the southern section, and he began a thorough revision of the specifications and costings. One of the main structures, the Penkridge viaduct, estimated by Stephenson at £26,000, was contracted for at £6,000[7] (this was the first big railway job of Thomas Brassey, who would, over the next two decades, become the railway contractor par excellence). George Stephenson can hardly have been surprised when, on 26 November, the board resolved to appoint Locke as joint engineer-in-chief, and with 'general superintendence' of the whole line.[8] He continued to ignore his responsibilities.

On the Birmingham line Robert was following his own assiduous methods. From May 1834 he had taken over a large establishment, the Eyre Arms Hotel at St John's Wood, as his headquarters. Military analogies arise readily in narrating the progress of the London & Birmingham's construction. Robert Stephenson, as supreme commander, had two forces to contend with. One was the English terrain and English weather, the other was the board of directors, who, like politicians prosecuting a war, were ultimately responsible for the entire operation and could scarcely bear to leave the professionals to get on with the job. The name of Stephenson was a mixed asset: it bespoke expertise and success in railway matters, but among some people it also aroused suspicion of self-serving decisions and refusal to allow fair and free competition. George Stephenson was not wholly

innocent in this respect. He had seen off Edward Bury from the Liverpool &
Manchester, not just because his own engines were better, but because Bury was
a competitor (whom he resented all the more because Bury's foreman, James
Kennedy, had learned all that he knew at Forth Street).

The 'Liverpool Party', well represented on the London & Birmingham board
and shareholders' committees, were fully aware of this background, and while
some of them were Robert's supporters, others, led by the Croppers, were thoroughly anti-Stephenson and had opposed his appointment. Though George
always got on well with the Peases, he had never been friendly with the Croppers,
a more politically active Quaker family, strong believers in free markets and with
a sharp eye for any kind of commercial monopoly. In the Stephensonian context,
it is easy to see the Croppers as opponents of progress, but they were not mali-
cious in their attitude. Considering it to be wrong that the Stephensons should
build railways and then specify and sell their own locomotives to those railways,
they cared nothing for the personal and external events that had brought the
situation about. Not among the more mechanically-minded directors, they possibly thought that one locomotive was very much like another; certainly they felt
that their friend Bury should be given a full chance to compete for the L & B
locomotive contract. James Cropper, his associate Theodore Rathbone and some
others saw the Stephensons, not without some justification, as trying to exert a
stifling monopoly in the railway world: a force to be resisted.

This division, together with the natural anxieties and concerns of any board
embarking on a vast enterprise – the estimated construction cost was £2.5 million – ensured that the engineering works would be under intensive and critical
scrutiny. As time went on and the engineering team continued to labour over
their drawings and specifications, there were murmurings for more visible action.
In May 1834 the first three contracts were let, covering the line out of London
as far as Kings Langley, and the remainder were let in batches through 1834 and
1835; by the time work began on the northernmost sections, the work in London
was well advanced. In all, there were thirty sections, though some larger contractors took on two or three, and W. & L. Cubitt, then at the start of a long corporate
existence, had four.

A new addition to the headquarters team was made on 17 September 1834.
George Parker Bidder, now aged 27, who had been a fellow student of Robert's
at Edinburgh, had become a civil engineer and had assisted Rastrick and Palmer
in their report on the L & B surveys, and had met Stephenson in the course
of that work. Several writers have assumed that the two men had been friends
since Edinburgh, but Bidder was taken on at the recommendation of Isaac Solly,
the first L & B chairman; a letter from Stephenson to him, in mid-October
1852, refers to them both as 'two who were at Edinburgh together and who
were afterwards thrown together by accident in a course of Engineering which
can scarcely be said to have a parallel'.[9] At the time, Robert was still digest-

ing a depressing piece of news. On the preceding day, all connection between George Stephenson and the Grand Junction Railway had finally been broken off. On 12 August George had been reduced to the status of consulting engineer at £300 a year, and that precipitated his resignation, which was accepted at the board meeting of 16 September.

Joseph Locke, already effectively sole engineer-in-chief on the whole project, had his status confirmed from 12 August. But by now he was so 'harassed by Stephenson's venomous taunts' that he was for a time in a state of nervous prostration.[10] Due to the diplomacy of John Moss, who was chairman of the Grand Junction, the parting with Stephenson was not, in the end, an acrimonious one; George would have known that to make enemies of those members of the 'Liverpool Party' who had been his friends would be disastrous for the locomotive business and for his own participation in future railway projects. But he never had a good word to say for the Grand Junction.

Making an analogy with the building of the Great Pyramid, Peter Lecount's history of the line's construction struck a pattern in his awestruck references to the scale of the London & Birmingham Railway. But though there were new constructions and new solutions to problems, Robert Stephenson was not venturing into the dark. Railway tunnels, cuttings and embankments had all been built previously and he had the experience of others to draw on, as well as the latest machinery (though the steam-shovel and steam-hammer had yet to appear), and very large capital resources. The London & Birmingham was not to be the longest railway in the world – the South Carolina Railroad of 1833 beat it in this respect by some 30 miles – but in terms of cost, scale of construction works and geological problems, the English line was second to none at the time.

Difficulties began not far from Robert's front door with the 1,100yd (1,006m) tunnel under Primrose Hill. Making a bore that could withstand the pressures of the London clay required a complete redesign, with brickwork 27in (69cm) thick using fast-setting Roman cement, at a cost almost double the original estimate. The contractor had to be relieved of his agreement and the railway company took over construction directly in November 1834 – a procedure that was to be repeated seven times on other sections, with two other contracts being given to new contractors. Robert Stephenson, and anyone else who knew about railway construction, expected unforeseen problems to arise, but they came with disconcerting frequency and sometimes in a disastrous and fatal manner.

The long tunnel at Watford, with its deep approach cutting through soft chalk, also gave difficulties. Little was known about soil mechanics and the behaviour of different clays, marls and mudstones, but the engineers of the L & B learned very quickly on the job, and Stephenson revised many of his plans for cuttings to reduce the angle of slope, which resulted in a greater workload for his army of navvies, a large increase in the amount of excavated material and, inevitably, a large upwards re-estimate of costs. Throughout the project, the emphasis was on getting the work done and the section complete: digging and building went on by the light of flares in some sections; the budget was going through the roof, but the project was running more or less on time. Robert Stephenson was far more willing than his father to keep the directors informed, to have them on site and to listen to their suggestions, but his sangfroid was tested at times. In the early stages of building, a group of directors became seized by the notion of the 'Undulating Railway', a fashionable notion which required a line to be laid out in a succession of down-grades and up-grades, with the idea that the momentum attained running downwards would get the train to the top of the next incline and so on for the length of its journey, at a great saving in fuel costs. Dionysius Lardner was one of its advocates.

Robert successfully resisted pressure to incorporate a test section of this see-saw in his carefully drawn-up plans. He was less successful when, at the end of 1834, an issue arose for which the Cropper group had been waiting. The railway company's London Committee ordered two *Planet* locomotives, one from Forth Street and one from the Vulcan Foundry, to work on the construction project. It seems odd that Stephenson did not foresee the reaction, which struck him with all the shock of an ambush. On 26 January he wrote to Longridge:

> Our enemies, viz, Rathbone and Cropper, are raising a hue and cry about our having an Engine to build at Newcastle – they say another article will be brought out by Lardner on the subject. They half intimate that I shall withdraw either from the Railway or the Engine building. The revenge of these people is quite insatiable. This distresses me very much. Can I withdraw temporarily from the Engine building? … the Directors support me, but it makes sad uphill work.[11]

Longridge's reply was supportive, suggesting Robert's shares might be transferred temporarily to his father, and adding: 'I feel very solicitous that you should devote the whole of your faculties *undividedly* to this magnificent undertaking; this being once *well accomplished*, your name and future are built upon a Rock, and you may afterwards smile at the malice of your enemies … when you arrive at the sober age of fifty you will bear these rubs better.' In February, however, the board passed a resolution moved by James Cropper, that '… for the supply of rails, chairs, sleepers or locomotive engines it be publicly known that no Director, Secretary, Engineer, Sub-Engineer or servant to the Company can be a Contractor …'[12] Robert Stephenson, wisely, did not attempt the too-transparent manoeuvre of share transfer, and the issue remained untested until the question of a general locomotive purchase should arise.

From these stressful days, we get the first detailed and objective portrait of Robert Stephenson from the pen of Francis Conder, though he wrote it as an older man, adding hindsight to what he had seen through 18-year-old eyes.

The personal appearance of that fortunate engineer is not unfamiliar to many of those whose eyes never rested on his energetic countenance, frank bearing, and falcon-like glance. It is rarely that a civilian has so free and almost martial an address; it is still more rare for such features to be seen in any man who has not inherited them from a line of gently-nurtured ancestors. In the earlier days of Robert Stephenson, he charmed all who came in contact with him. Kind and considerate to his subordinates, he was not without occasional outbursts of fierce northern passion, nor always superior to prejudice. He knew how to attach people to him: he knew also how to be a firm and persistent hater. During the whole construction of the London and Birmingham line, his anxiety was so great as to lead him to very frequent recourse to the fatal aid of calomel. At the same time his sacrifice of his own rest, and indeed of necessary care to his health, was such as would have soon destroyed a less originally fine constitution. He has been known to start on the outside of the mail, from London to Birmingham, without a great coat, and that on a cold night; and there can be little doubt that his lamented and early death was hastened by this ill-considered devotion to the service of his employers, and the establishment of his own fame.[13]

There are several useful pointers here. In passing, it may be noted that Brunel, too, before he began to use his own carriage, had a reputation for sitting on the outside of stage coaches without enough warm clothing. Conder goes on:

The elder Stephenson had no share in the actual direction of the works ... and as his son found himself sit firmer in the saddle, he showed at times something of his father's determined and autocratic temper. He met his people with a frank and winning smile; his questioning was rapid, pointed and abrupt, and his eyes seemed to look through you, as you replied. Very jealous of anything like opposition or self-assertion, very unjust at times in suspecting such a disposition, he was disarmed by submission, and quieted by very plain speaking. On one occasion he had detected what he thought, and probably truly thought, to be a notorious instance of 'scamping' on the part of one of his contractors. He sent orders to the man to meet him on the works, and coming up rapidly as he caught sight of him, burst out: 'So-and-so, you are a most infernal scoundrel!' 'Well, sir,' meekly replied the delinquent, 'I know I am.' Robert Stephenson's lecture was at an end, and the man was dismissed with something like a friendly warning.[14]

How far Robert was a 'natural' leader of a very large team, and how far he deliberately modelled his style on men who were – his father first and foremost – and played up to it, is hard to determine. Those prominent eyes in his portraits and photographs can suggest the apprehensive prey as much as the questing falcon. George was hot, visceral and engaged, and Robert strove to be cool, cerebral and detached. But then and always, by ambitious engineers and rugged navvies, all equally ready to spot and pick at weakness in the man at the top, Robert Stephenson was, beyond question, respected as 'The Chief'. His tight-sprung determination and intense nervous energy set the whole approach to the great undertaking, and if self-doubts were present, they were not apparent. Demanding total dedication from his deputies, he allowed them far greater responsibility than his father would have done, but every major foundation was inspected by him, and at the problem sites he was not only present, but had solutions, or gave his men the confidence to persevere.

Where the slippery clay slopes of the 1.5 mile- (2,500m-) long Blisworth cutting threatened to slide down, he inserted 'counterforts' of his own devising, deep-rammed dykes of gravel or rubble, to stabilise them. At the Wolverton embankment, 45ft high and over a mile long, the clay constantly slopped and spread out at the foot instead of forming a regular slope. Here was one of the sections where the contractor failed and the company took over, in June 1837. There was no magic solution, and for several months earth was dumped in tens of thousands of barrowloads until at last a stable structure was achieved. Worst, and most celebrated of the line's difficulties, was the 2,398yd (2.2km) tunnel under Kilsby Grange, a few miles south-east of Rugby. The ridge here had already been pierced by the Grand Union Canal at Crick, and sand deposits were known to be present, but trial soundings for the railway had narrowly missed a huge deposit of water-bearing sand close to the southern end of the tunnel's line, which was encountered when digging began in autumn 1835.

Blisworth Cutting

The contractor, James Nowell, struggled on until February 1836 when the railway company took over the tunnel works. Nowell, 'overwhelmed by the calamity', according to Smiles, died shortly afterwards. Smiles records that Robert Stephenson consulted his father, and when George came to view the scene, they both decided that pumping out the water, by engines installed at the top of a series of shafts, was the right solution.[15] Jeaffreson gives credit to Robert alone, claiming that when visited by Captain Constantine Moorsom, the Birmingham secretary, to suggest he should call in other engineers, Robert replied: 'No; the time has not come for that yet. I have decided what to do. I mean to pump the water all out, and then drive the tunnel under the dry sand. Tell the directors not to be frightened ...'[16] Moorsom's visit was at the instigation of the Cropper camp,[17] but he returned to Birmingham a staunch defender of the company's engineer. Thirteen pumping engines were set up above the quicksand, but Stephenson was not simply throwing hardware at the problem. In the course of close observation of the pumping operations, which went on through 1836, Robert made the important realisation that it was not necessary to pursue the impossible task of drying out the entire cubic mass, since the resistance of the coarse sand to the passage of water made it possible to 'establish and maintain a channel of comparatively dry sand in the immediate line of the intended tunnel, leaving the water heaped up on each side by the resistance which the sand offered to its descent to that line on which the shafts and pumps were situated'.[18] By the time the water adjusted its level, the tunnel lining was in position, with 36 million bricks forming the double-track passage, ventilated by two great 60ft-wide ventilation shafts. Kilsby was conquered, giving Robert and others valuable experience in water engineering, but completion of the section was delayed by a year and its cost, originally contracted for at £99,000, was more than three times that amount.

The London terminus was originally intended to be close to the present-day Camden Lock, to the north of the city as it existed in the early 1830s. Robert Stephenson, considering that the utility of the line would be much greater if it went right to the north bank of the Thames, was encouraged by Charles Parker, the company solicitor and a good friend, to make this proposal to the board, which he did, rather diffidently, in early 1835, only to have it promptly rebuffed. But soon afterwards the matter was reconsidered, helped by a complaisant attitude from the Earl of Southampton, owner of much of the ground between Camden Town and Euston Square, who had noticed that land values close to the Liverpool & Manchester termini had gone up. An extension as far as Euston Square was agreed, and an addition to the Company's Act was procured without difficulty. With his cuttings, tunnels and embankments, Stephenson had been at pains to minimise gradients on the line so that locomotives could be used throughout, with their whole power employed in pulling heavier trains. From Euston Square up to Camden Town, however, there was an unavoidable gradient of 1 in 77. This would have been a challenge to the locomotives of the time, but in any case

Ventilation tower above Kilsby Tunnel

the residents of Camden Town protested against the intrusive noise of locomotives, and this section of the line was at first operated by rope haulage from fixed engines, with trains running downhill by gravity under the control of brakesmen. Running most of the way in a cutting with high retaining walls, the line gave space for four tracks, as it was anticipated that the Great Western Railway, under construction from 1835, would share the Euston Square station. This never happened, because of the difference in gauge between the Great Western and London & Birmingham tracks.

By this time, Robert Stephenson and Isambard Brunel, engineer of the Great Western, were on friendly terms with each other. Exactly when they first met is not known: Robert had been a member of the Institution of Civil Engineers since 1830, Brunel for longer and the ICE's regular meetings often ended in informal and convivial discussion. George Parker Bidder noted in his diary for 17 March 1835, '… attending the Engineers Institution with Mr Stephenson', and, two weeks later, 'Attended the Engineers Institution in the evening and argued with Brunel and Vignoles'.[19] But a letter from Stephenson of 11 November 1834 headed, 'My Dear Sir', suggests they were not yet friends;[20] whereas by March of 1835, he is 'Dear Brunel'.

The Stephensons wanted the Great Western Railway to use Whitby stone for its construction works.[21] Since July 1832 George had been involved with the

The fixed engine at the top of the Camden Town incline

promotion and survey of a railway between the Yorkshire port of Whitby and the inland market town of Pickering.[22] Robert and Brunel produced a joint report in the summer of 1835 for the Brandling Junction Railway, formed by the Brandling family to link Gateshead with the Wear estuary;[23] the line had been surveyed by George and this was probably the first occasion on which the elder Stephenson and the younger Brunel met, and one of the few when they were not at loggerheads. Robert and Brunel were also meeting to discuss the possible Great Western terminal at Euston Square. Bidder, still in the employment of the London & Birmingham Railway, was about to resign his post in order to work exclusively for Robert Stephenson: 'to take in hand under him at a salary of one & a half guineas per diem the London and Brighton Railway and also the super-intendence of his private business', and 'To receive all the profits if the produce falls short of £300 per annum'.[24]

This new employment, concluded in the early summer of 1835, was of vital importance to both men. Following their second coincidental throwing-together, they had become fast friends as well as colleagues, and the relationship would remain close and cordial until Stephenson's death. Bidder, son of a Devon stone-carver, had been exhibited by his father as The Calculating Boy before more enlightened patrons had arranged for his proper education. Newly engaged to be married, in March 1835 he was a gregarious and relaxed young man of 29, also both industrious and perspicacious, and his calculating facility never left him. He was the eldest of a large family, while Robert was an only child, and in his inaugural address as president of the Institution of Civil Engineers in January 1860, he would touch-ingly recall their long friendship: '… his conduct towards me was that of an elder and affectionate brother; he has encouraged me in all he thought right, and has not failed to criticise all he deemed wrong, and had the necessity arisen, he would have applied his whole fortune, to his last sixpence, for my benefit.'[25]

But much was to happen before that proud and sad day came. Bidder might not have been offered, or have taken on, such a confidential job if they had not become friends and trusted each other; however, his new role also required his full range of skills from negotiation of contracts to draughting and on-site decision-making. Robert Stephenson's original contract with the London & Birmingham precluded him from work on any other railway. With the pressure of work on that line and his involvement with locomotive works – he was still managing partner of Forth Street – it might be thought he had scant time to take an interest in other concerns, but this was by no means the case. Most of the building went on in a steady manner under the supervision of his able team, and the detailed drawings, which had taken thousands of man-hours at the start, now proved their worth.

In the course of 1835, Robert managed to get the London & Birmingham to concede that he might act as a consulting engineer, subject to permission from the two committees and 'provided such business be strictly confined to Chamber consultations, and in no instance demand his attention in the field'.[26] This could have been an irksome restriction, but with Bidder acting as a virtual extension of himself, it was far less of a problem to work on the other side of the fence. The L & B also raised his salary to £2,000 a year, matching what Brunel was getting on the Great Western; the rise was perhaps also motivated by their anxiety not to lose him to another company. Senior engineers were by now making a lot of money. Samuel Smiles was told by Charles Binns, who was for a time George Stephenson's private secretary, that George used to reckon his time at 10 guineas a day, and to apportion this among the companies retaining him at the time. However:

> When Robert heard of this instruction, he went directly to his father and expostulated with him against this unprofessional course ... George at length reluctantly consented to charge as other engineers did, an entire day's fee to each of the Companies with which he was concerned while their business was going forward, but he cut down the number of days charged for, and reduced the daily amount from ten to seven guineas.[27]

In this respect, as in his somewhat old-fashioned clothes, George was very much a man of an earlier generation.

Though it was surely inconceivable that he would quit the Birmingham line, Robert had made no secret of his interest in the proposed London & Brighton Railway. Linking the capital to a fashionable and fast-growing resort, this 60-mile (96km) line had obvious potential. The Stephensons had already done a general survey for one group of promoters, noting a possible route, but nothing had come of it. In December 1834, a Parliamentary Committee was considering two alternative schemes, and Robert Stephenson was allowed a few days off in order to review them both and write a report. One route, drawn up under the aegis of Sir John Rennie, was 'direct', cutting through the ridges of the South Downs; the other was more circuitous, making use of river valleys wherever possible. Despite making severe criticism of the survey methods, Robert favoured the longer route. Both schemes were withdrawn, but the debate went on, and in May 1835 a new group of proposers employed Robert Stephenson to make a fresh survey. This work was done by Bidder.

Also in May, Robert procured a few days' leave of absence in order to travel with his father to Brussels, capital of the new kingdom of Belgium, for a meeting with King Leopold to discuss the planning of a national railway system. Belgium became the Stephensons' first territory for overseas operations. Three newly-delivered 'Patentees' were on parade at the inauguration of the Brussels–Malines Railway on 5 May, when George took a modest place in a roofless third-class carriage,[28] but the modern-minded Belgian monarch bestowed the Order of Leopold on him.

Robert's arrangement with Bidder was not at all meant to exclude George, who realised its advantages. With his own substantial portfolio of railway applications to Parliament, the combined workload was heavy, and Bidder, who had been using the L & B facilities at the Eyre Arms until the end of 1835, leased an office at 16 Duke Street, Westminster, convenient for the Houses of Parliament and only two doors away from the Brunel office.

In the autumn of 1835, with the opening of a section of the line from Camden to Watford becoming imminent, the question of the London & Birmingham

Railway's locomotive stock arose again. As engineer to the line, Robert
Stephenson had to specify the type of locomotive to be used, and he proposed
the best ones he knew, which were his own 'Patentee' design, recommending to
the London Committee that ten should be purchased. Though he did not specify
a manufacturer and the London Committee put the order out to tender, the
Cropper-Rathbone faction objected since Stephenson had an interest in the roy-
alties that would be payable, whoever built them. A comprehensive bid by Charles
Tayleur for the Vulcan Foundry to build and manage all the company's locomotives was rejected for the same reason, and Edward Bury put in a similar proposal
in December 1835. Bury's engines were lightweight four-wheelers, certainly not
'Patentees', nor did they meet Stephenson's specifications, and Stephenson protested strongly against them while Bury was equally vocal in their defence.

Now some extraordinary manoeuvres ensued. It was agreed to place the order
with Bury, but it transpired that his foundry had no spare capacity to complete
the locomotives in time. Meanwhile, Stephenson, with the help of William
Cubitt, was parading a split-new 'Patentee' on the completed section of line, and
warning the directors that the railway would open with no engines to provide
a service. The London Committee responded by ordering twelve 'Patentees', of
which two would come from Robert Stephenson & Co., and the rest from other
builders. Two days later, on 15 January 1836, the Birmingham Committee met
'as an integral part of the direction of the company' and recommended in the
strongest terms that '... no time should be lost in concluding the agreement with
Mr Bury and no arrangement met which might interfere with it'.[30] The London
Committee was persuaded to rescind its order, and eventually a contract was
signed with Bury in May. Six other builders – not including Forth Street – had to
be roped in to complete the job, and the angry and disgusted Robert Stephenson
was left with the realisation that instead of his own engines adding their lustre
to the splendid line that he was building, the trains would be operated by Bury's
small, and as he correctly foresaw, inadequate locomotives. The affair also seems to
have brought the Stephenson-Tayleur partnership to an end; whether amicably
or not is unclear. But if the Stephensons' involvement was going to mean a block-
age rather than a flow of orders, it was no longer useful and a split was inevitable.

Controversy was hot on the South Downs, too. By the end of 1835, among
several rival London–Brighton schemes, there were only two serious contenders:
the 'Stephenson' line via the Mole and Adur valleys, and the 'Rennie' direct line,
now again brought forward. The old Rennie-Stephenson hostility resurfaced in
a campaign of pamphlets accusing each other of mistakes and false statements.
Involvement in such public disputation about another railway was embarrassing
for the London & Birmingham's chief engineer, but the arguments went on into
the summer of 1836, when the 'Rennie' Bill was rejected by the Commons and
the 'Stephenson' one got as far as the Lords, only to be dismissed there. Since
George Stephenson had made a survey, he was called to give evidence, and cut as

poor a figure as he had in front of the Liverpool & Manchester Committee more than ten years earlier. His survey, made by following the turnpike road in a post-chaise, had been of such a flying nature that he could not even recall some of the place names along his route.

Many other railway Bills were passed during the session, including the Midland Counties Railway, from Rugby to Derby, the North Midland from Derby to Leeds, the Manchester & Leeds, and Birmingham & Derby Junction Railways. To the first two of these, George was chief engineer (with Robert included as nominal joint chief engineer of the North Midland). Though these were only a few of the schemes now going ahead, it was still a strong hand in the developing railway game, and the recurrence of 'Midland' in the names is significant. Even if it had to be built in stages, George Stephenson was envisaging a trunk route that would link the North and West Ridings of Yorkshire with Birmingham and the south on the one hand, and with Liverpool and the Atlantic trade on the other, and he had drawn in some of the 'Liverpool Party' to invest in these lines, along with local capital. Free from what he had considered an ignominious role on the Grand Junction Railway, and perhaps conscious of the fact that he had lost a few laurels in his behaviour towards it, the elder Stephenson was moving into a fresh phase of intense and productive activity.

On a visit to Whitby in the summer of 1834, he had made a new friend, George Hudson, at that time treasurer of a committee set up to promote a railway from York to Leeds. It was a chance meeting, though they were bound to have encountered each other at some point. Hudson, aged 34, had risen from modest beginnings on a small Yorkshire farm, and though a fortunate marriage and a substantial inheritance had provided motive power, there was no denying the extraordinary determination and force of character he displayed in his self-appointed role as a projector of railway schemes, as in everything else he did. Already he had a broad vision of railways traversing the country in an ordered fashion. Vociferously Tory in politics, he enjoyed the good things of life; he had not shaken off the broad rustic accent of his youth and his manner was uncouth, without any sort of polish, metropolitan or even that of a proud provincial city. All these characteristics would have appealed to George Stephenson. Hudson was like a younger, more focused and purposeful William James, and Stephenson saw in him a potential strong ally for the future. In late 1835 Hudson had persuaded the York Committee to drop the idea of its own line to Leeds and instead to link up with the North Midland at Normanton, close to Wakefield, and the proposal was passed by Parliament in 1836 as the York & North Midland Railway, with George Stephenson as its chief engineer.

Up in Newcastle, an older ally was showing signs of wobbling. In the autumn of 1835, Michael Longridge told his partners that he was no longer prepared to act as manager of the Forth Street Works and that, furthermore, he intended to set up his own locomotive construction business at Bedlington. The long-suffering

Longridge had been stretched between his own business and Robert Stephenson & Co.; he had also seen the Stephensons set up another locomotive builder. Now George was living in the Midlands and Robert in London, and both, as ever, were happy to take his willingness for granted. Forth Street had a full order book, and other builders were starting up as the demand for steam locomotives grew ever larger. Why should he not get a piece of the action? Longridge's decision seems entirely reasonable. He was not demanding that Robert should return, but merely saying that he wanted to relinquish his management role.

Robert did not go to Forth Street at all in the year of 1835, and a visit in April 1836 did not resolve the difficulty with Longridge. In a tetchy letter written from Newcastle on 12 April to Joseph Pease, he toyed with the thought of selling the Forth Street Works to Longridge,[31] partly because of a worry about whether a new manager might – like Dickinson – interfere on the design side. As managing partner, it was up to him to take a lead, but he was slow to act, though Longridge's urgings made clear that the problem would not go away. But he had a great deal on his mind. Five separate sets of promoters, including the group which had retained him, were still quarrelling over the Brighton line, and there were all the other railways, at different stages from planning to actual engineering. Kilsby was still a serious worry and Brunel visited the workings in August 1836 to give support and advice.[32]

In the end, Stephenson found the ideal man for Forth Street in Edward Cook, a family connection of Fanny's recommended by her father. Cook took up the post of chief clerk and ran the commercial side with the same efficiency that Hutchinson as works manager brought to the manufacturing side. With 'Uncle Edward' in place from November 1836, Robert evidently felt much better about his absence from Newcastle, and there was a frequent correspondence between them, with occasional brief references by Stephenson to family matters. Michael Longridge remained a partner in the now rapidly expanding business, and George Stephenson continued to correspond with him, but the old intimacy between Longridge and Robert lapsed; Robert now had a closer confidant in George Bidder.

In February 1837, the Stephenson practice moved from Duke Street to a bigger office at 35½ Great George Street, nearer Parliament. Robert was dividing his time between it and his office at the Eyre Arms, where all the business relating to the London & Birmingham Railway was still being done. At one point he had optimistically said that the line would be open by the end of 1836, but it was evident that there was at least a year's work still to be done, and the costs had risen dramatically. As Michael Bailey wryly remarks,[33] this has remained a feature of large-scale civil engineering contracts, and the L & B board, like all subsequent boards, became very anxious. Further loans and calls on shareholders' funds were needed, and in April 1837 directors were asked to exercise supervisory roles on each section of the line. Robert Stephenson was not greatly alarmed by any of this. He knew the line and its problems better than anyone – he reckoned he had walked the whole course of the railway fifteen times in his years with the London & Birmingham. Much of the cost over-run was beyond his control, caused by rising prices of materials and changes of policy by the board itself, including the extension to Euston Square, which required numerous bridges to allow for streets as yet unbuilt, as well as the broad cuttings. With no prospect of seeing his 'Patentees' running on the tracks he had planned and built, he was keen to end his exclusive tie to the London & Birmingham and to involve himself more closely with the many other schemes being dealt with by the Great George Street office. The work at Kilsby and Blisworth went on slowly despite 24-hour activity; but finally, after almost a year of constant pumping, tunnelling at Kilsby resumed in December 1836.

Through 1837 the rash of new railway proposals continued unabated. The undecided Brighton line was still being argued over and, by a House of Commons resolution, an independent military engineer was appointed to review the three main proposals. On 27 June Captain Alderson, RE, reported that though Stephenson's line, 'considered in an engineering point of view, is preferable to either of the others', nevertheless other factors made Rennie's line his overall preference. The Bill for this line was speedily passed and John Urpeth Rastrick was appointed as engineer. Robert Stephenson did not accept the verdict and wrote two papers, one short, the other somewhat longer, in vindication of his own route. More than sour grapes against his 'enemies' lay in this: he was truly convinced that his route was much superior.

In this same year he recommended the London & Blackwall Railway to take a route surveyed by Rennie, rather than one they had asked him to survey (done by George Bidder); admittedly the Rennie route was manifestly preferable,[34] and Robert was angling for the construction contract – and obtained it. The wrangle over the Brighton line shows two schools of thought in dispute. By no means every opinion or prejudice of George's was shared by Robert, but both Stephensons believed strongly in building railway lines that followed, wherever possible, the easiest route offered, even if it meant quite lengthy detours. Where physical obstacles were unavoidable, they resorted to tunnels, deep cuttings and high embankments, to obtain as level a road as possible.

Perhaps the extreme Stephensonian point of view was shown in George's proposed line of August 1837 from Lancaster to Carlisle, where the fells of Grayrigg and Shap lay implacably across the most direct route. His railway would have crossed the shallows of Morecambe Bay on a great causeway, burrowed through the Furness peninsula and followed the Cumbrian coastline round to Carlisle. A number of important seaports, as well as the Cumbrian mining area, would have been served, while the line as was ultimately built – by Joseph Locke – went up and over through unpopulated hill country between Carnforth and Penrith – but was a far shorter, 'direct' route. Even on his Manchester & Leeds line, begun in 1837, with its unavoidable long tunnel beneath the Pennines, George Stephenson took a relatively circuitous path.

Back on the London & Birmingham, a train service was started between Euston Square and Boxmoor (present-day Hemel Hempstead) on 20 July, the same month that Locke's Grand Junction Railway opened in its entirety; and the railway company could start to make some earnings against its vast capital outlay.

Another fundamental railway dispute was slowly warming up at this time. Proposals for new lines in England, Wales and Scotland, the European countries and many in North America, had generally accepted the rail gauge established by George Stephenson on the Stockton & Darlington, and especially the Liverpool & Manchester, of 4ft 8.5in. The half-inch has always been something of a mystery; his original intention was simply for 54in (1.534mm), and a quarter of an inch on each side seems hardly worth anyone's trouble. Brunel, however, was intent on building his railway from his own principles and chose to give it a gauge of 7ft 0.25in (2.15m), his quarter-inch being no less arbitrary than Stephenson's half-inch. Perhaps both indicated a kind of 'ownership'. His broad gauge, Brunel claimed, would provide for greater speed and stability, as well as greater carrying capacity in wider trains. Before accepting his proposal, the Great Western Railway's board made a number of inquiries among engineers, including Robert Stephenson, who refused to make any public criticism of Brunel's gauge, though he had told Brunel privately that he could not support it. On 5 August 1838 he wrote to Brunel:

> As my opinions of the system remain unchanged, you will I am sure readily see how unpleasant my position would be, if I expressed myself in an unequivocal manner in my report; and to do otherwise would be making myself rediculous since my opinions are generally known – To report my opinions fully therefore would do harm instead of good to the cause in which you are interested and this I am sincerely desirous of avoiding.[35]

Had the GWR asked George for his view, they would have been rocked by a blast of rage. The 'standard' gauge was his and for a young engineer, new to railways, to summarily dismiss both it and his method of laying the rails was an affront. The difference in gauges was the principal reason why the Great Western and the London & Birmingham did not share a terminus, and Brunel designed his own station at Paddington, while Robert built the iron train sheds at Euston and the architect Philip Hardwick designed a grand Doric entrance to the station, as well as a handsome classical-style building for the Birmingham terminus at Curzon Street. As long as the Great Western remained a separate system without physical links to any other railway, the gauge difference was not a problem, though the prospect of future difficulties was obvious enough.

Brunel had other reasons to be grateful to Robert Stephenson, who lent him the London & Birmingham drawings to help with the planning of the Great

The train shed at Euston

Western. In fact, the younger Stephenson was remarkably open in all his operations on the London & Birmingham, allowing and encouraging the artist J.C. Bourne to make his celebrated series of drawings and lithographs of works on the line, and T.T. Bury to make a set of coloured prints. Peter Lecount, an assistant engineer on the Kilsby–Birmingham section, wrote a history of the building of the line. As a result, the L & B is the best documented of early railways and became a benchmark in planning and building for thousands of miles of railways soon to spread across many countries. It may be that Stephenson was keen to show Cropper and his associates that he had nothing to hide, but for a man who felt as beset by foes and hostile critics as he did, it was a courageous attitude to take. Though Francis Conder described him as 'a firm and persistent hater', it is notable that he never let professional disputes with men whom he respected interfere with personal friendships, and his sustained hostility was reserved for those who showed personal animosity towards himself and, more especially, towards his father.

On 23 December 1837 George Stephenson attended, as an honoured guest, a dinner given by the assistant and sub-engineers of the London & Birmingham Railway to their 'Chief', at the Dun Cow Inn at Dunchurch in Warwickshire. The line was not yet finished, but it was now only a matter of brickwork, barrowloads and time. All the great obstacles and challenges had been successfully overcome. Other friends were also there, including Tom Gooch who came down from his new posting on the Manchester & Leeds Railway, and the evening was one of the happiest occasions in the Stephensons' lives. A correspondent reported on it for the *Railway Times* – a publication which owed its existence to their work – as a warm and emotional event:

One thing was of universal remark: this was the great alteration for the better in the appearance of the latter gentleman [George], he looked at least half a dozen years younger ... There is the making of a hundred railways in him yet ... It would have done any man's heart good to have heard the deafening applause which followed when the healths of the father and son were drunk; everyone felt they came warm from the heart and spoke of feelings that could not be uttered ...[36]

Robert retired from the table at 2 o'clock in the morning with Frank Forster, engineer of the Kilsby–Birmingham section, who had been chairman, but George was then voted into the chair and kept things going until 4 a.m., when Gooch replaced him until 6. Some of the diners were apparently still there at 8 o'clock in the morning. As a testimonial, his assistants presented Robert with an ornate silver soup tureen, inscribed: 'To Robert Stephenson, Esquire, Engineer-in-Chief of the London and Birmingham Railway, a tribute of respect and esteem from the members of the Engineering Department who were employed under him in the execution of that great work. Presented on the eve of their gradual separation.'

Six weeks of hard frost through January and February 1838 delayed further bricklaying, and a hoped-for partial opening of the railway, excluding the Kilsby section between Denbigh Hall and Rugby, had to be put off from the end of January to 17 September. Against the initial estimate of £2,400,456, the final cost of the London & Birmingham was £5.5 million, or about £50,000 for every mile of the line. In a report to the two committees, dated 17 February 1838, Stephenson set out a defence both for the delays and for the doubled cost:

> ... in cases where the Company have been compelled to take up works the task to be performed has required means hitherto without precedent in engineering and consequently involving expenses which no experience could indicate ... I had no guide – I was thrown entirely upon my own resources and those of my Assistants to devise means for accomplishing that which was not only deemed precarious but nearly impracticable in point of time.
>
> Under such circumstances I must without hesitation admit that my calculations have not been borne out but on the contrary far exceeded. Throughout the course of these works I have in vain indulged a hope that the expenditure might be lessened by establishing a system and adopting suitable plans for procedure pointed out by experience – but, unfortunately the advantages accruing from the employment of proper methods have been counteracted by the pressing necessity for an early completion of all the works ...[37]

It was a shrewd apologia, but in his engineering works, as in his locomotive building, Robert Stephenson was always more concerned to produce a quality job than to meet the estimate or charge the lowest price. On 21 June the

Interior of Kilsby Tunnel, base of ventilation shaft (from a Bourne engraving)

last brick was placed in the wall of Kilsby Tunnel by Charles Lean, the sub-engineer who had supervised the pumps and who, on one occasion, heroically rescued a party of workmen from an almost completely blocked shaft rapidly filling with water. Three days later the first train ran from Euston right through to Birmingham. Four days after that, the young Queen Victoria was crowned in Westminster Abbey, and though the frustrated board had hoped to have the entire line working in time for the event, visitors to London had to be conveyed from Birmingham to Rugby by train, by coach from Rugby to Denbigh Hall, and again by train from there to Euston. The official opening and a full service began on 17 September 1838, when a special train took the London directors and company officials, including Robert Stephenson, over the embankments and viaducts, through the cuttings and tunnels, and into Curzon Street station, Birmingham, for a slap-up dinner at Dee's Royal Hotel.

12

# LIVES OF
# ENGINEERS

To Augustus Welby Pugin, whose *Contrasts between the Architecture of the Fifteenth and the Nineteenth Centuries* was published in 1836, the great Doric portico at Euston Square was a monstrous irrelevance. He might have liked it a little better if it had been in the Gothic manner of which he was the champion, but nineteenth-century British writers on architectural aesthetics generally kept well clear of any sort of railway building. This is a pity, because the more ambitious architecture of the early railways was generally good, and often of necessity original, responding to needs that had never before existed. At least one Victorian observer wondered, 'Why are our architects so inferior to our engineers?'[1] and the definitive buildings of the era were the work of engineers. Robert Stephenson's locomotive roundhouse[2] at Chalk Farm in London, his iron train shed at Euston, Brunel's more ambitious iron naves and transepts at Paddington and the iron railway bridges whose girders matter-of-factly spanned wide gaps in a way which would previously have been thought impossible, were all novel constructions, and exciting to many. Pugin had written: 'It will be readily admitted that the great test of Architectural beauty is the fitness of the design to the purpose for which it is intended, and that the style of a building should so correspond with its use that the spectator may at once perceive the purpose for which it is erected.'[3] He might have had in mind such novelties as the great round towers crowning the vents of Kilsby Tunnel. Sixty feet (20m) high and wide, doorless, windowless, roofless, floorless, they might have seemed the legacy of a departed race of giants, though to the pastoralists of the Warwickshire uplands, memories of the army of navvies and masons who built them were still traumatically recent. In fact, Pugin was thinking, as ever, of churches.

In the 1830s there was no specific architectural style that reflected and responded to the character of the times, and the engineers, guided by their own acquired tastes and their eagerness to reassure an uncertain public, followed the established forms of design: 'Many of the great engineering achievements were

The Doric Entrance, Euston Square Station

partly hidden by a veneer of classic or Gothic detail, applied in the belief that an architectural need was thus fulfilled, without anyone suspecting that a new, authentic architecture was being suppressed.'[4] But the engineers saw the problem, not as the creation of a style, but as the choice of one that would satisfy a committee, like Brunel's design for the Clifton suspension bridge at Bristol:

> I have to say that of all the wonderful feats I have performed since I have been in this part of the world, I think yesterday I performed the most wonderful. I produced unanimity amongst fifteen men who were all quarrelling about the most ticklish subject – taste.
>
> The Egyptian thing I brought down was extravagantly admired by all and unanimously adopted ...[5]

Guidance for the less gifted was readily available. Sandwiching the text of the third, much-expanded edition of Nicholas Wood's *Practical Treatise on Rail-Roads*, published in 1838, are booksellers' advertisements for Caveler's *Select Specimens of Gothic Architecture* and Chambers' *Treatise on Civil Architecture*, with details and illustrations of classical rules and models. An engineer-in-chief could select his mode, and in the architecture of the London & Birmingham Railway, Robert Stephenson opted for neo-classical. Careful attention was applied to Vitruvian detail in the stonework, with mouldings and pilasters even in such hitherto unknown structures as inward-curving retaining walls – the visible portions of inverted vaulting, built *beneath* the track in deep cuttings through the unpredictable and elastic clay.

If the tunnels at Edge Hill and the 'Moorish Arch' had had a touch of eighteenth-century fantasy and orientalism, reminiscent of shelly grottoes and romantic chasms, the monumental architecture and massive engineering works of the London & Birmingham announced an era of imperially confident construction. With these to his credit, even if he felt his triumph diminished by the presence of Bury's underpowered little puffers on the line, Robert Stephenson had also come out completely from his father's long shadow. In his own right, he was now one of the country's leading engineers. George himself had the unusual pleasure of recognition by a national body when the British Association for the Advancement of Science (founded in 1831), having chosen Newcastle as the venue for its 1838 gathering, appointed him a vice-president in its Mechanical Section. He took the opportunity to revisit Killingworth with some of the delegates, pointing out, 'with a degree of honest pride, the cottage in which he had lived for so many years, showed what parts of it had been his own handiwork, and told them the story of the sun-dial over the door'.[6]

In June of the same year, a ferocious attack on George Stephenson was made in the *Railway Times* in the guise of a review of an anonymously written book, *General Observations on the Principal Railways in the Midland Counties and the North of England*. Noting that this work describes any line not associated with Stephenson – even the Grand Junction Railway – as 'certain to fail', and 'those which are equally certain of triumphant success' all being Stephenson-engineered, like the York & North Midland, the Birmingham & Derby, etc., the reviewer comments that among railway investors, there are 'a large proportion who still seem to adopt no other rule for their guidance in the selection of the lines to which they shall subscribe ... than the single consideration of whether or not the services of the once celebrated engineer of the Liverpool & Manchester Railway, Mr George Stephenson, have been secured'. Condemning such folly, the author castigates Stephenson's 'mortal dislike to every railway in any part of the kingdom in the hands of any other engineer ...'; and although acknowledging George's 'extraordinary natural powers, and capacity, and genius', begs him to recall 'once more, and before it is too late, the friends whom he has deserted, cease to be guided by tools and sycophants, and malevolent angry feelings for which there is neither foundation nor excuse'.[7] A footnote adds: 'The public must carefully distinguish between Mr *George* Stephenson and Mr *Robert* Stephenson ... the latter, a gentleman in education, in feeling, and in conduct' – the imputation against George is clear.

In reply, the anonymous author, writing from Liverpool, sought to defend Stephenson from this 'vomit of hatred', but there is no doubt that his book accurately represented George's point of view, and that the equally anonymous reviewer had a case to make. Such denigration of projects with which George Stephenson was not involved (or from which he had excluded himself) made him a controversial or unpopular figure to many people – 'the once celebrated engineer' at this time was not a serene Olympian, but angry and jealous, frustrated at the way 'his' creation was moving away from him.

By the end of 1838, both Stephensons were wealthy men. George was a country gentleman and Robert led the life of a prosperous *haute bourgeoisie* in London. Not much is recorded about George's household at Alton Grange, but evidently he and Betty kept a hospitable establishment. In September 1836 George Parker Bidder met Mrs Stephenson senior for the first time when she was in Manchester with her husband, and reported to his wife Georgey that 'Mrs S appears to be a good-natured & perfectly unassuming old lady'.[8] Soon Georgey and their baby son were regular guests at Alton Grange, while the two Georges were off on railway affairs. Even more commodious was the house on which George took out a lease on 25 March 1838, and to which he and Betty moved in August that year. Tapton House,[9] just outside Chesterfield, was a rather plain red-brick mansion, from whose hilltop site extensive grounds sloped downwards, with big though neglected gardens, and here he proposed to indulge himself as a gentleman farmer.[10] Robert and Fanny's house on Haverstock Hill was made lively by visits from Fanny's younger relations. On 27 February 1837, when Robert had written, with manly reticence, in a business letter to Uncle Edward Cook up in Newcastle, that 'we are all tolerably well at Hampstead', a sprightly young Sanderson lady added a postscript:

My dear Uncle, – Cousin Fanny would have filled up this part, but she is in bed with a sick headache. Tell Mr Hardcastle Mr G. Stephenson's brother Robert is dead, the new groom has been thrown from his horse, and both horse and man are at present perfectly useless. That is what Mr Stephenson calls being tolerably well at Hampstead.[11]

Now in his mid-thirties, Robert's tastes and ideas were formed. An early riser, he normally spent two hours before breakfast reading technical and scientific literature. Mathematics, chemistry, geology and physiology were his favourite subjects and he liked to keep up to date in them. It does not appear that he indulged in light reading, but 'few weeks passed over in which he did not find an hour to devote to an English poet':[12] which ones, we do not know – Browning and Tennyson were already published but Robert may well have preferred the verses of writers like Felicia Hemans, who had died in 1835, but whose *Collected Works*, with 'Casabianca' and other tear-inducers, published in 1839, were vastly more popular than the poems of either of those rather highbrow young men. His friend Joseph Locke (a poetry-reader like his father) never went anywhere without a volume of Byron,[13] but Byron would not have been to Robert's politically and socially conservative taste.

Music, played and sung, was an essential part of home life, but it seems Robert did not bring out his flute in parlour gatherings. The impression is of a solid, cheerful early Victorian household, busy with its own affairs and those of the wider family, well insulated in every way from the poverty and degradation that

Charles Dickens was already portraying in his early novels, and somewhat philistine in its tastes. The business of the head of the household might be thoroughly modern, but he conducted that elsewhere. Only in one respect did Robert and Fanny show a touch of social pretension. The legend of the Stephenson family origins had been told to Fanny and, at her persuasion, Robert made application to the College of Heralds for permission to use the coat of arms of the Stephensons or Stevensons of Mont Grenan, in Kilwinning parish, Ayrshire, with whom Old Bob's father had traditionally been associated, and who had been in possession of the estate between 1683 and 1778. The college borrowed some of the devices from the Mont Grenan arms to incorporate into Stephenson's, and Robert parted with a substantial fee to become armigerous on 21 November 1838. Jeaffreson reports him as embarrassed by this lapse into *nouveau riche* ostentation, saying, not long before his death, to a friend: 'Ah, I wish I hadn't adopted that foolish coat of arms! Considering what a little matter it is, you could scarcely believe how often I have been annoyed by that silly picture.'[14]

But in one important way both the Stephensons' style of living was quite different from that of their contemporaries, and in this they were precursors of later peripatetic captains of commerce and industry. Other men of business were partners or employees in a single, fixed enterprise, went to the counting-house, law office or factory in the morning and returned home in the evening. For the Stephensons, Great George Street was only the base from which they were conducting a range of large-scale undertakings that had no parallel at the time. Some were short-term jobs, others lasted for several years. In supervising or inspecting these projects, they were often only home for brief periods, living in inns – hotel keepers loved the influx of well-paid, hungry and thirsty engineers. Other railway builders like Joseph Locke and his new partner John Errington did the same, but the Stephensons were unique in the late 1830s and early 1840s for the number of projects they managed. In 1835 they had visited Belgium, consulting on a national railway system planned under government control. In 1837 George was back, for the opening of the Brussels–Ghent railway, and was made much of by Brussels society in a way that London's social leaders had never done. The railway-construction business had rapidly become multinational in scope, with contracts also in Germany and Italy, and beyond. Only Robert Stephenson could have put in the grand *non sequitur* at the end of a letter to his colleague Frank Forster, looking towards the completion of the London & Birmingham project:

Dear Forster
At the meeeting of our Board [L & B] today, the importance of opening on the 17th of September was so strongly argued on the score of the immense expense and inconvenience of the coaching [between Rugby and Denbigh Hall] that I felt bound to promise them it should be practicable on that day. I must, therefore, lean upon you again, and I do so with confidence; but if you should find it

impossible to complete both lines, you had better at once consider the propri-
ety of putting in some points and crossings, so that our line may be passable over
that part of the Hill Morton embankment (and through Kilsby) which may,
perhaps, not be quite closed. The state of the permanent road is not of so much
consequence as the existence of it throughout, so that the trains may be able
to pass upon it at a very slow speed. I shall be coming down to Coventry on
Saturday night or Sunday morning. Could you drop me a line by return to say
where I shall meet you? Will you go to Cracow to make a line from that place
to Warsaw? I have an application; but I fear your Liberal principles will give rise
to some objection on the part of the Autocrat of all the Russias. The line is 100
miles in length. More of this when we meet: in the meantime try if you can't
convert yourself into a Tory.[15]

Poland was then part of the Russian Empire, and it would seem that Tsar
Nicholas I's personal interest in railway construction extended to scrutiny of
the political views of supervisory engineers. In September 1836 the first three
locomotives to be imported into Russia had come from Robert Stephenson &
Co., Timothy Hackworth (who had become an independent builder in 1833,
keeping the Stockton & Darlington engine contract) and Charles Tayleur.
Though they did not set up branch offices in other countries, the Stephensons
were dispatching men as well as machines all over Europe, as locomotive tutors
and railway engineers. They were not by any means England's first exporters
of machinery, but the scale of operations, with locomotives and the construction business, put them into a class of their own until the building boom of the
mid-1840s. They were fortunate in that their fame drew inquiries and orders:
for railway planners everywhere, their name was a guarantee both of experience
and of up-to-date technology.

Between April and July 1839, Robert was advising on railway projects in
France, Belgium, Switzerland and Italy, in the process making a number of useful
contacts for future business. In Italy he was appointed to supervise the survey for
a new railway from Florence to Pisa and Leghorn, and later became this line's
engineer-in-chief. Most of the time he was in Belgium, where as engineer-in-
chief of the lines from Ostend through Brussels to Liège, and from Antwerp to
Mons, he had to inspect the work of assistants in residence. As soon as he was back
in England, he was plunged into a round of visits to construction sites, committee
discussions and planning meetings. For weeks on end he was at home only on
Sundays, 'his presence in his own house almost a surprise to its inmates'.[16]

A new role had also come his way: arbitrating between contractors and railway companies on disputes relating to claims for extra payments or inadequate
work. Robert's cool and detached judgement – when not feeling personally
attacked – and his ability to sift through sub-clauses of contracts and conflicting
columns of figures gave him a high reputation in this tricky aspect of business.

Railway contractors and material suppliers organised a testimonial dinner for him at the Albion Hotel, Aldersgate Street, London, on his 'natal day', 16 November 1839, and presented him with a silver candelabrum and dinner service valued at £1,250. George Stephenson was present to witness the honour paid to his son by over 200 guests.[17]

For Robert, with the Birmingham line finished, everything pointed to a future even busier than the previous four years had been, particularly as his father was beginning to cut down on his own commitments. Smiles called the three years ending in 1837 perhaps the busiest of George's life.[18] In August 1837 he had said in a letter to Michael Longridge that he intended giving up business altogether in the course of the next three years.[19] While some men dream of a garden, he, having just been in Cumberland, had '30 or 40,000 acres of land on the West Coast of England' in mind to take in hand. In any case, the intimation of retirement proved a gross overstatement. In 1836 he had been appointed engineer-in-chief of the York & North Midland Railway, of which his dynamic friend, George Hudson (Lord Mayor of York from 1837 to 1839), was now chairman. Construction had been slow, but the northern section was opened with municipal and ecclesiastical pomp and ceremony on 29 May 1839. Addressing the guests at the inevitable gargantuan dinner in York's Guildhall, George Stephenson told them he had started work as a ploughboy at the age of 8 and confided that: 'The fact is, I am going to end my days as a farmer.'[20]

Since August 1837 he had also been involved, as engineer-in-chief, in the construction of the Manchester & Leeds Railway, the first trans-Pennine line; though with Tom Gooch drafted in from the London & Birmingham, he had a thoroughly reliable man in charge. Opened in January 1841, this line included the Littleborough Tunnel under the Pennine ridge, which for a few months, until the opening of Box Tunnel on the Great Western, was the longest in the country. George was in close discussion with Hudson on other projects, but also accepted commissions in 1838 to survey routes for two important new railways, one from Newcastle to Edinburgh, the other from Chester to Holyhead. Engineering work on the North Midland was nearing its end, and, with the southern section of the York & North Midland due for completion at the same time, there would be an uninterrupted line of railway from London's Euston Square, by way of Rugby and Derby, to York. Northwards from York, the Pease family and others were proposing to build what they called the Great North of England Railway, through Darlington to Newcastle: a venture of their own which did not involve the Stephensons or Hudson. Robert was also happy to participate in the schemes of the energetic Yorkshireman. On Hudson's boards there were no carping Croppers and it was fully understood that the best locomotives would be used on the completed line, which, whether built at Forth Street Works or not, meant Stephenson's patent model. Indeed, Robert had been appointed locomotive superintendent of the North Midland Railway on 1 January 1839.

1. George and Robert Stephenson, with a 'long-boiler' locomotive, by John Lucas
(Institution of Civil Engineers)

2. High Street House, Wylam

3. West Moor
colliery, with the
whim in the centre

4. Nicholas Wood

5. William James

6. Bedlington Ironworks
(Northumberland Archives)

7. Edward Pease
(Darlington Local
Studies Centre)

8. Stockton
& Darlington
Railway opening

9. 'Experiment', the first railway carriage

10. Forth Street Works, Newcastle

11. Charles
Blacker Vignoles

12. Liverpool & Manchester Railway share certificate

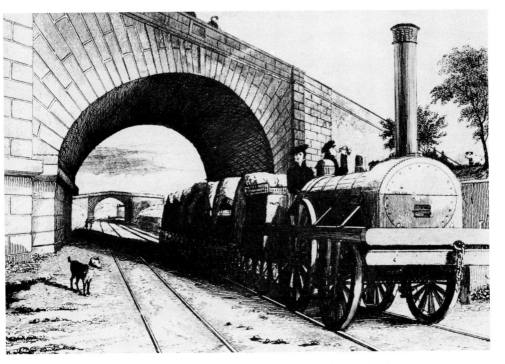

13. The skew bridge at Rainhill (from a Shaw etching)

14. *Rocket*

15. George Stephenson's lime-works at Ambergate

16. Conwy Tubular Bridge, completed

17. The High Level Bridge, Newcastle Upon Tyne. (Institution of Civil Engineers)

18. Britannia Bridge – one of the main tubes on the staging

19. The Great Hall, Euston Station. The statue of George Stephenson was a later addition

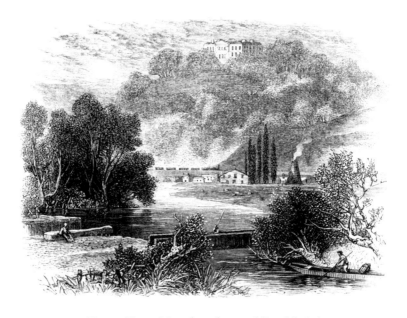

20. Tapton House. Note the railway and lime kiln below

21. George Stephenson.
Engraved by W. Hall, after
Lucas

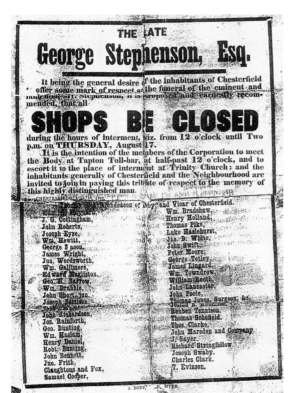

THE LATE

# George Stephenson, Esq.

It being the general desire of the inhabitants of Chesterfield to
offer some mark of respect at the funeral of the eminent and
distinguished Mr. Stephenson, it is proposed and earnestly recommended, that all

# SHOPS BE CLOSED

during the hours of interment, viz. from 12 o'clock until Two
p.m. on THURSDAY, August 17.

It is the intention of the members of the Corporation to meet
the Body at Tapton Toll-bar, at half-past 12 o'clock, and to
escort it to the place of interment at Trinity Church: and the
inhabitants generally of Chesterfield and the Neighbourhood are
invited to join in paying this tribute of respect to the memory of
this highly distinguished man.

Archdeacon of Dby and Vicar of Chesterfield.

| | |
|---|---|
| Edmund Maynard, | Wm. Bradshaw, |
| J. G. Cottingham, | Henry Holland, |
| John Roberts, | Thomas Pike, |
| Joseph Eyre, | Luke Haslehurst, |
| Wm. Hewitt, | Jas. B. White, |
| George Bason, | John Swift, |
| James Wright, | Peter Moore, |
| Jno. Wordsworth, | George Tetley, |
| Wm. Gallimore, | James Lingard, |
| Edward Mugliston, | Wm. Towndrow, |
| Geo. H. Barrow, | William Booth, |
| Wm. Drabble, | John Lancaster, |
| John Short, jun. | John Poole, |
| Joseph Battison, | Thomas Jones, Surgeon &c. |
| | Reuben Tennison, |
| John Richardson, | Thomas Schofield, |
| Jos. Rainforth, | Thos. Clarke, |
| Geo. Bunting, | John Marsden and Company, |
| Wm. Haslam, | J. Sayer, |
| Henry Daniel, | Richard Stringfellow, |
| Robt. Bunting, | Joseph Swaby, |
| John Bennett, | Charles Clark, |
| Jno. Frith, | T. Evinson. |
| Claughtons and Fox, | |
| Samuel Cooper, | |

J. ROBERTS, PRINTER.

22. Notice of George
Stephenson's funeral
(Courtesy of Chesterfield
Local Studies Centre)

23. Robert
Stephenson.
Engraved by W.
Hall, after Lucas

24. Robert
Stephenson & Co.
business card

25. The Victoria Bridge, Montreal, under construction (Institution of Civil Engineers)

26. Commemoration of Robert Stephenson's funeral (*Illustrated London News*)

27. The young George
Hudson, a contemporary
sketch

28. Photograph of Thomas Sopwith
(Royal Meteorological Society)

29. Joseph Locke, by John Lucas
(Institution of Civil Engineers)

30. Sir Joshua Walmsley,
by W. Daniels (Victoria
& Albert Museum)

31. Sir William
Fairbairn, by W. Holman
Hunt (Institution of
Mechanical Engineers)

32. George Parker
Bidder, by John Lucas
(Institution of Civil
Engineers)

Early in 1840 disagreement broke out among the partners in Robert Stephenson & Co. The reason is not wholly clear; some tension had existed since Michael Longridge went into locomotive-building on his own account, but the cause seems to be associated with details of management or financial responsibility. In a letter to Cook sent in February, Robert Stephenson's tone was emphatic:

I shall read Longridge and Pease a lesson by post in a day or two; they shall either rescind the £200 resolution or else I shall go out of the thing altogether. Indeed there is scarcely one thing they have done that I will not undo – the resolutions are conceived and phrased in a style which I will not put up with from those who have done nothing for the concern for many years past and only began to show an interest in our proceedings after I had succeeded by my own resources and those of my father in establishing a character for the firm. The impudence of a Quaker is beyond estimate by Heavens![21]

The Pease mentioned here was probably Edward's son Joseph rather than the old man himself, who felt that Joseph paid too much attention to worldly matters. Whatever the rights and wrongs of the issue, Stephenson was being unjust to Longridge who had carried much work on his behalf for years, until Cook's appointment. The quarrel seems to have died down. One of the cyclic dips in Great Britain's generally upwards-reaching economic trend occurred in the early 1840s, and nerves may have played a part if orders were falling. Forth Street laid off staff and reduced wages in 1840–41. In general, however, Robert Stephenson & Co. was now a substantial and profitable business – twenty-four locomotives had been ordered by the Paris–Orleans Railway alone in early 1841[22] – and it is only to be expected that the partners would take a closer interest and not simply let the absentee managing partner run the show through 'Uncle Edward'.

During the winter of 1839–40, Fanny Stephenson became unwell. At first Robert treated the matter fairly lightly, and in a letter to Edward Cooke, which for once was wholly personal, asking him to order two Newcastle plaids, one for Fanny, one for her cousin, he mentions that: 'Fanny I think is going on well although she is still grazing on macaroni and occasionally a little marine flesh, vulgarly called fish …'[23] Fanny's condition did not improve and she was diagnosed as suffering from cancer. Just as his father had seen his mother gradually waste away, so Robert Stephenson had to witness the steady decline in vitality of the wife he loved. He was able to afford the best medical help the time could provide, but cancer, in its various forms, was little understood. From the few scraps of evidence available, Fanny seems to have borne her suffering with fortitude.

Diversifying from railways, Robert became consulting engineer for the new Bute Docks in Cardiff in February 1840, and in the same year was appointed engineer to the proposed London & Westminster Waterworks. At this time, water supply was a private, not a municipal concern, and the provision of pure water to the ever more populous capital was both a vital necessity for daily life and public health, and an opportunity to make money. Since his digs through the chalk ridges at Watford and Tring, Stephenson had been aware of their huge water reserves, and of the natural filtration provided by the chalk. In the course of 1840 he had an exploratory well sunk in Bushey Meadows to assure the investors of the potential rate of flow. With eight companies already in the business of supplying London with water, there were powerful competitive interests, and though he kept up his connection with the enterprise through the 1840s, nothing came of it.

These were only sidelines, however, and railway development claimed most of his attention. Despite the chillier economic climate, there was a great deal to watch over. In the previous year, interest had grown in a new form of railway traction, which threatened to do away with locomotives altogether. This was the 'atmospheric railway', based on scientific ideas which had been around for over a century, since the Frenchman Denis Papin. In principle it was simple: a tube between the rails had air sucked out of it by a pump driven by a stationary engine. The resultant semi-vacuum drew a train along the rails by means of a piston inserted into the tube. To make this possible, the top section of the tube was made of flexible material and divided, so that the rod attaching the piston to the carriage could push it open as it travelled along. The engineer Samuel Clegg patented a valve for this purpose in 1839 and went into partnership with the Samuda brothers' engineering company to develop the atmospheric railway. A model was set up in London, and George Stephenson was taken by Charles Vignoles (the events of 1827 forgotten or forgiven) to see it. His judgement was prompt but accurate: 'It won't do: it is only the fixed engine and ropes over again, in another form, and to tell you the truth, I don't think this rope of wind will answer so well as the rope of wire did.'[24]

Interest was great enough for a trial section of atmospheric line to be built at Wormwood Scrubs, London, in 1840, to test the idea in practice. Robert had used a 'pneumatic telegraph' tube to send messages from Euston to the Camden rope engine in 1837, but he shared his father's view about atmospheric traction. At this time, however, he did not consider locomotives suitable for all railways. The London & Blackwall Railway, the world's first 'mass-transit' urban line, had just opened in July 1840, and he and Bidder, as consultant engineers, had arranged a rope traction system for its 3.5 mile (5.6km) length. Even four years later, Robert did not believe that locomotive-hauled trains could run at less than half-hourly intervals,[25] despite the fact that, at Bidder's and his recommendation, this was also the first railway with a full electric telegraph system to send messages between stations. Meanwhile, with the completion of the North Midland's link line between Sherburn and Altofts Junctions, a Stephenson locomotive hauled the first train to run from York to Derby on 30 June 1840, with George and Robert Stephenson, George Hudson and their guests; then back from Derby to Leeds. Now the York–London Railway was complete: 217 miles (347km) of iron road, albeit with four different owning companies.

By 1840, the Stephenson works in Newcastle had supplied locomotives to railways in Austria, Belgium, France, the German states, the Italian states, Russia and the USA. In some of these countries, locomotive works were already established or planned, and it could not be expected that customers would automatically look to England in the future. A full-time sales operation was needed and Robert swooped on an experienced salesman, Edward F. Starbuck, a partner with Longridge in a metals-broking business in the City of London, who now also became an agent for Robert Stephenson & Co.[26] Writing to Joseph Pease on 24 October 1840, Robert noted that, 'It is most essential that we should cultivate our continental business and I cannot conceive a better opportunity of doing so than that which has occurred thro' Starbuck'.[27] From now on, if not before, Starbuck was among Robert Stephenson's closer friends. Export business was all the more vital because it helped to even out the impact of the occasional depressions of the British economy.

To add to his business preoccupations and domestic worries, Robert Stephenson found himself threatened from a completely unexpected direction. In 1832, for his work as consulting engineer to the Stanhope & Tyne Railroad, he had accepted £500 of shares in the company instead of cash. Crushed lime was in great demand for iron-smelting, and initial prospects for the line had looked good. Soon it became clear that the difficulty and cost of operating such a steep and exposed railway had been greatly underestimated. Seeing what was coming, the original proprietors sold out and the new management borrowed large sums in order to keep the line running, but also to invest in other railways and to pay fictitious dividends to the shareholders in 1835 and 1836. By 1837, the precarious state of the enterprise was fully apparent, and in 1839 part of the line was closed down, though wayleave payments were still legally due. As it was not a limited liability company, its shareholders were legally responsible, jointly and severally, for its debts. Though Robert knew of the company's difficulties, he was apparently unaware of his potential obligations until, in December 1840, to his horror, he was presented with a large bill which was due for payment and which the company was in no position to meet. Reacting swiftly, he called an extraordinary general meeting of shareholders to determine what should be done.

Meetings on 29 December and 2 January 1841, revealed assets of £307,383, at most, and liabilities of £440,852.[28] The company was insolvent, and with the help of Charles Parker, his solicitor, Stephenson took the lead in a decision to dissolve and reform it as an incorporated company with a new capital of £400,000, taking on the debts and assets of the existing business. His own stake in the restructured concern was £20,000. To raise this sum he had to realise some of his assets, and his father stepped in to buy half of his share in Robert Stephenson & Co. Legal and parliamentary proceedings relating to the Stanhope & Tyne Railroad, now to be renamed the Pontop & South Shields Railway, were put in hand. He wrote revealingly to Edward Cook on 29 December 1840: 'The history of the Stanhope and Tyne is most instructive, and *one miss* of this kind ought to be, as it shall be, a lesson deeply stamped. If the matter get through, I promise you I shall never be similarly placed again.'[29]

Robert's 'Walter Scott' episode passed off without bankruptcy, but it was deeply felt, and on 4 January 1841 he wrote again to Cook:

> The transaction is not intended to be otherwise than *bona fide* between my father and myself. The fact is, I owe him nearly £4000, and I have not now the means of paying him as I expected I should have a few months ago. All my available means must now be applied to the Stanhope & Tyne. On the 15th of this month I have £5000 to pay into their coffers. The swamping of all my labours for years past does not now press heavily on my mind. It did so for a few days, but now I feel master of myself; and though I may become poor in purse, I shall still have a treasure of satisfaction amongst friends who have been friends in my prosperity.[30]

A careful man is more keenly nipped by a slip of attention than one who habitually takes risks, and Robert Stephenson was normally careful to a fault. In one semi-comic episode, having been pushed back from the first lifeboat by the ship's mate in the shipwreck of 1827, he found that priority had been given to a cheaper-rate passenger who, like the mate, was a Freemason. A few days later in New York, Robert joined the St Andrew's Lodge No 7 of the Freemasons, apparently as a form of life insurance, since he is never known to have attended any meetings or been a member of any English lodge.[31] This strain in his character made Stephenson tend towards prudence in his dealings and actions, but some writers have considered that the Stanhope & Tyne experience was so searing as to encourage him into double dealing with his employers – a whiff of brimstone that inevitably heralds the appearance of George Hudson on the scene.

George Stephenson and Hudson kept up a warm relationship, and on 20 January 1841 George (who had invested £20,000 in the company) was elected as a director of the York & North Midland Railway.[32] Through 1841, the elder Stephenson was concentrating more on his own industrial interests and less on railways, but Hudson and Robert Stephenson were in close contact. Robert was head of the country's largest railway engineering practice, bearer of a surname that resonated favourably with railway investors large and small, and managing partner of a locomotive works. By 1841, Hudson's immense energy and resource had made him into a successful railway developer, who saw that railways were already big business and going to be bigger. Shrewd, persuasive in an overbearing sort of way, explosive in temperament, grandiose in his entertainments, crushingly arrogant to those whom he despised (who included most of the directors of his various companies), Hudson was not an unfamiliar character-type, but was able to inflate himself into a hugely dominant figure, who for almost a decade was known as The Railway King, by exploiting two related areas of opportunity. On the one hand, there was the potential for a national railway system, and on the other, the ease with which a determined man could manipulate the finances of railway companies, and Hudson seized both with relentless zeal. Guile, braggadocio and a theatrical ability to wrong-foot his opponents carried him through a succession of crises and challenges, and all the while his shareholders received regular impressive dividends, plus alluring offers for reinvestment in new schemes, without troubling to consider whether the one might be feeding the other.

Hudson himself became immensely wealthy. Apart from political Conservatism, he and Robert Stephenson had practically nothing in common. Robert was a private man, prudent in his investments, reserved in his manner, with no ambition to be a social lion and preferring the company of fellow engineers. But far more than any other single individual, Hudson offered something close to the old Stephensonian dream – a planned national system of railways, powered by Stephenson locomotives. Similar ambitions were harboured among the 'Liverpool Party' of railway developers, but the hostility of the Cropper faction prevented any comprehensive alliance with the Stephensons. Rivalry between Hudson and the Liverpudlians was bound to arise.

At this time, the British government had embarked on one of the first of its occasional forays into railway strategy, and had appointed a three-man commission to advise on the best route for a trunk railway linking London and Scotland. Current assumptions were that only one line could be economically feasible. The Liverpool men were promoting the already-mooted extension of the Lancaster & Preston Junction Railway from Lancaster through the Lake District, while Hudson projected a York–Newcastle–Edinburgh route. Reporting in March 1841, the commissioners supported a west coast route, partly because they thought this line to be closer to completion; they also suggested that if a line was built to Newcastle from the south, it might be continued to Edinburgh. The Stephensons and Hudson were thus presented with a semi-open door, at which they would push in due course.

Far from Hudson's sphere of influence at the time, the Great George Street practice had been active in East Anglia since 1839, when George Bidder supervised the building of the Northern & Eastern Railway to link London and Cambridge. Its progress was slow, hampered by lack of funds. Robert Stephenson was engineer-in-chief to this line, and was also consulted in 1840 by the equally straitened Eastern Counties Railway, anxious about its ability to complete the line via Colchester to Norwich and Yarmouth. Stephenson was retained as engineer-in-chief for its Cambridge–Brandon–Norwich line and several others. He also became consulting engineer to the South Eastern Railway in January 1841, to advise it, not only on new branches, but also on main lines in north-east France, linking the Channel ports with industrial Flanders and Paris. These were important commitments, and also led to locomotive orders, though in the end the South Eastern's French ambitions were not to be realised.

Closer to home, Robert was to find himself in choppier waters. Also at the beginning of 1841, in response to an appeal from the directors of the Great North of England Railway, he took over as its engineer. It had almost completed the line from York to Darlington, but the engineer, Thomas Storey, had resigned, leaving some works unfinished and numerous bridges in an unstable state.[33] Stephenson promptly took the work in hand and, by the end of March 1841, the line was opened to traffic. The company's next task was to extend its line from Darlington north to Gateshead on the south bank of the Tyne, but it had no funds and no starting date for this more difficult section, which traversed hillier country and crossed several east–west railways, including the Stanhope & Tyne.

At the GNER's half-yearly meeting in March 1841, Stephenson's appointment was confirmed and he was unanimously voted engineer-in-chief. His report to the company on this occasion included a comment that went beyond his engineering role:

> I now offer you spontaneously my opinion … Having been for several years alive to the importance of a continuous railway communication, not only between Darlington and Newcastle, but between England and Scotland, to which the public are most anxiously looking forward, I feel that I have yet a duty to perform in reminding you of its influence on the revenue of the Company.[34]

Urging the company to get on with the northern section, he pointed out that, in the approach to the Tyne, costs could be saved by using existing railways to offer 'a continuous and by no means unnecessarily circuitous connexion with the banks of the Tyne',[35] rather than having to build a new line all the way. Among these other railways were the Stanhope & Tyne and the Durham Junction, in which the Stanhope & Tyne company held shares. Though he did not mention his own link with that company, it was no secret, and the GNER directors would have been fully aware of it. Unfortunately, their own company was in disarray,[36] with no prospect of being able to raise the cash to complete the northern section, and with growing dissension between those directors who were also on the Stockton & Darlington board and those who were not. The Stockton & Darlington company had its own ambitions to control access to the Tyne from the south. To an engineer who was raring to complete the east coast route, the Great North of England Railway had scant hope to offer.

Now George Hudson took a hand in the matter. To him, with his eye on a wholly Hudson-controlled railway from the Midlands to Edinburgh, a long Quaker-owned section between York and the Tyne was not a desirable thing, and, for him, economic recession was an opportunity rather than a reason to hold back. He convened a meeting in Newcastle, on 30 April 1841, of representatives from the eight railway companies with an interest in the Rugby–Newcastle route. Hudson was perfectly open about his interest in speedily closing the Darlington–Newcastle gap, and backed the notion of using existing lines as much as possible. It was agreed that an alternative route from Darlington to the Tyne should be proposed, and 'the greatest cordiality and good feeling prevailed'.[37]

L.T.C. Rolt follows Hudson's biographer, Richard Lambert, in accusing Robert Stephenson of bad faith to the GNER by colluding with Hudson to set up this alternative route, but there seems little justification for their severity. At worst, after having been made engineer-in-chief to the Great North of England, he indicated to Hudson that he would switch to a Hudson-managed line, but it is more likely that they had discussed the matter long before. Lambert himself notes that when Stephenson became engineer to the Great North of England, it was '... upon his own terms – which included a strong recommendation to the Company to drop all thought of proceeding further with the second and northern half of the project, the line between Darlington and Newcastle'.[38] For Hudson, always keen to hold down capital costs, Stephenson's zig-zag route was an excellent idea. Mileage charges would be a welcome boost to the finances of the Durham Junction and, beyond that, there was a strong likelihood that the new railway would buy out that struggling line completely, and also the Stanhope & Tyne, relieving Robert Stephenson entirely of his unwanted responsibility. At a further gathering of the interested companies on 6 September, Hudson's method of financing the line was revealed: £500,000 in £25 shares was to be raised from shareholders in the eight companies, who would be guaranteed an annual rate of interest of 6 per cent on their investment, even while the line was being built. Once the line was opened, the companies would jointly lease and run it 'and recoup themselves out of its earnings for their guarantee'.[39] All those present were greatly enthused – Hudson had once again pulled a golden rabbit from a rather shabby-looking hat.

The GNER company knew perfectly well what was going on, since it was one of the eight partners in Hudson's Newcastle and Darlington Junction project. Its half-yearly meeting on 12 March 1842 noted placidly that the Newcastle and Darlington Junction was applying for its Act, and that its own board was applying to Parliament to transfer their approved line to the new company, and to 'abandon altogether' the section replaced by Stephenson's proposed zig-zag.[40] Difficulties were made by the Stockton & Darlington Railway, whose board realised, somewhat late in the day, that it would remain a secondary cross-country line with no share in the North–South trunk route. This sparked argument on the GNER board between those who were also S & D directors and the others. For the latter, to be embraced by Hudson was a highly desirable consummation; for the S & D, it meant being shut out from the party. Rolt's comment on Stephenson's advice to the GNER, '... apparently Joseph Pease and his colleagues did not suspect that

Robert Stephenson had any ulterior motive in giving it', implies that Stephenson was pushing his own interest against that of his employers, but Pease was no fool, and Robert's other interests were public knowledge. Rolt notes that, 'His advice was accepted; land which had been acquired for the railway North of Darlington was sold back to the original owners ...'[41] His suggestion that the Quakers had a right to feel 'tricked and betrayed by Robert Stephenson' has no foundation (coincidentally, Robert at the time was complaining about Quaker 'impudence' in the Forth Street management). Michael Bailey also says, 'he placed himself in an awkward position'.[42] But there is no indication that the Darlington Quakers bore Robert any grudge, and Edward Pease always referred to him in terms of warm friendship.

It was not the Great North of England, but the Stockton & Darlington, which had been outmanoeuvred by Hudson, shattering their plans for an extended system which would control access from the south to Tyneside's fast-growing population and industries. The real conflict of interest lay with the men who sat on both the board of the S & D and of the GNER. The Stockton & Darlington company's own difficulties were indicated by their attempt to get Hudson to use 5 miles of their own railway at the southern end of the new line, and it was only on Hudson's rejection of this – on Stephenson's advice[43] – that they resorted to open opposition. A fierce war of words broke out in the winter of 1841–42, with a Darlington resident, Captain Watts, describing Robert's line as 'an abortion, with a crooked back and a crooked snout, conceived in cupidity and begotten in fraud'.[44] Hudson was more than capable of fighting his own corner, and Robert kept out of this battle.

Lambert goes on to characterise Robert Stephenson as having 'little of the geniality and unpretentious simplicity of his father; he belonged to a harsher, more selfish type which despised the very class from which it had sprung'[45] – this does much less than justice to the complexities of Robert's nature, as perhaps these pages have already shown. John Hart, reminiscing on the 1840s, felt that: 'There was something very attractive about Robert Stephenson – a geniality of address, looks and manner which made him always a favourite';[46] and for Smiles: 'Robert Stephenson had nothing of the snob in him. He was not ashamed of his father having been a working man; on the contrary, he was proud ...'[47] Lambert represents Hudson as abetted and encouraged by others, whose ethical standards were no better than his, but who let him do the dirty work, Robert Stephenson being one of them. But no one pushed Hudson into dubious practices – he leapt into them with all the zest of a bullfrog for the pond. Stephenson had many opportunities to enrich himself as an associate of Hudson, but did not take them, and whether this was because of prudence or a high moral standpoint, it shows that he was too much his own man to let himself become, like so many others, a passive recipient of Hudson's bounty. Hudson and he were very useful to each other, but it was a business rather than a personal friendship.

In a different analysis, Rolt wrote that the Stanhope & Tyne affair 'determined Robert's course of action and drove him, under pressure of necessity, to seek Hudson's powerful aid ... forced by the threat of ruin to play Hudson's game for a while'.[48] But Robert had set about his own solution to the threatened disaster, including the reincorporation of the company and the preparation and submission to Parliament of its Bill as the Pontop & South Shields Railway. Without Hudson it would have been a longer haul, but Stephenson restored the business to stability without Hudson's involvement. And Stephenson, in his own way, had been playing Hudson's game for several years.

Railway promotion and management was a tough, hard-nosed business, and commercial morality was often kicked aside in the hunt for corporate or personal advantage. Company law was still in a primitive form and the ethics of any large business were a reflection of the personal qualities of the owner or dominant figure. In this often unsavoury arena, Robert Stephenson preferred to be a background operator, advising his friends and the companies with which he was connected, watching out for his own interests and always with the strategic intention of making a North–South trunk route. As a role, it may smack of the small boy who explains to his playmates how to make mud pies and then looks on as they begin to throw them at one another. Eventually they are dragged away for a scrubbing, while he, his white sailor suit unspotted, expresses shock at the idea that there could be a link between pie-making theory and pie-throwing practice. Suspicious eyes have been cast on Robert's part in these Hudsonian scrimmages, but have found no evidence of active involvement; on the contrary, he successfully kept a reputation for independence and for honourable dealing, which explains why he was frequently asked to arbitrate in others' disputes.

In August 1841 he made another visit to Belgium, where the works were progressing so well that the king bestowed on him the Order of Leopold, 'as a testimony of His Majesty's satisfaction in the improvements he has made in locomotive engines – which improvements have turned to the advantage of the Belgian iron railways'.[49] Earlier in the year, he had been on a working visit to Italy, as was noted by a new friend he made on his return to Newcastle. Thomas Sopwith, cabinet maker, engineer, surveyor and voluminous diarist, was exactly the same age as Robert. He had grown up in Newcastle, but had not gone to Bruce's school. Sharing interests in geology, meteorology and other sciences, as well as having a strong religious faith, the two were very much on the same wavelength in everything except politics; like most of Robert's friends, Sopwith was of Liberal inclination.[50]

George Stephenson's expressed yen for farming did not mean that his entrepreneurial fires had been banked. The collieries at Snibston were in full production, and in 1841 the Leicester & Swannington Railway carried 55,302 tons of Snibston coal.[51] While supervising tunnelling works on the North Midland line at Clay Cross in Derbyshire, George saw coal seams and decided it was time to dig under new pastures. The *Railway Times* reported: 'Mr George Stephenson, the celebrated engineer, has concluded a contract with the owners of lands on the townships of Tapton and Brimington and Newbold, adjoining the Locksford estate in Derbyshire, for the purchase of the whole of the coal thereunder, and we understand that collieries and limeworks on an extensive scale will soon be in operation.'[52] He set up a new partnership to establish the mines: Robert was included, as were Sandars and Walmsley from Liverpool; a London banker, George Carr Glyn, chairman of the London & Birmingham Railway; and George Hudson. Snibston coal went only a few miles to Leicester, but Stephenson had an altogether larger idea here – he believed that rail-borne coal to London from the Midlands could match or undercut the price of sea-coal from the north-east. This was not generally believed to be possible, but again Stephenson's belief proved itself. He was eloquent on the subject: 'The Lord Chancellor now sits upon a bag of wool; but wool has long ceased to be emblematical of the staple commodity of England. He ought rather to sit upon a bag of coals, though it might not prove so comfortable a seat ...'[53] He is quoted again as saying: 'We want, if I may say so, a stream of steam running directly through the country ...' What he had in mind was a slow-speed line for coal and freight only, and, in conformity with his own firm view, '... if the line be flat, it is not of much consequence whether it be direct or not'.[54]

Economic recession or not, the Stephenson office remained a busy place. Locomotive development was a pressing matter: the 'Patentee' design was six years old and demand for faster and more powerful engines was mounting. If Robert Stephenson & Co. were to remain the market leaders, they had to produce an effective new design. Already the Birmingham & Gloucester Railway had imported American locomotives to work on its steep Lickey Incline south of Birmingham; Norris, the American builder, was a dynamic businessman and his agent was active and persuasive. Writing to Joseph Pease, on 13 October 1841, Edward Cook expressed cautious hope for an upturn of business, noting that, '... the talk here is that Hawthorns have little or nothing to do, but I have no direct means of knowing whether this be so ... we have up to this time in the present year finished and sent away 31 engines, and ... we calculate on the despatch of 6 or 7 more'.[55] Other orders were in suspension 'until the result of our Patent Engine be known'.

In collaboration with Hutchinson, Robert Stephenson had been working on a new design and had taken out new patents on 23 June. With Sopwith, he went to Bristol on 25 September to inspect Brunel's 'stupendous iron steamship' *Great Britain*, which was under construction there.[56] A few weeks later, Robert wrote a letter of introduction to Brunel for an American, 'Mr Stevens, an engineer from New York. He is decidedly possessed of more information regarding American railways and locomotives, engines and steamboats, than any engineer I have met with from the other side of the water'.[57] This was Edwin A. Stevens, joint founder of the Camden & Amboy Railroad, a son of John Stevens, one of America's great steam pioneers.

Meanwhile, Forth Street had been building a locomotive to the new design, and on 12 October a prototype went to the York & North Midland Railway for testing. A six-wheeler, soon known as the 'long-boiler' type, and with a boiler certainly longer than on previous Stephenson engines, its prime characteristic was that all the wheels were placed between smokebox and firebox, so that the latter protruded behind the wheelbase. Two types were patented, one with two sets of coupled wheels, the other with two carrying axles and a single driving axle. Numerous detail improvements were incorporated, the prime aim, according to Stephenson himself,[58] being 'the purpose of obtaining an economical use of fuel, which at that time was considered of paramount importance', and the short wheelbase was the result of most turntables being built for four-wheel engines. Like all contemporary locomotives, they were intended as coke burners; the need to 'consume their own smoke' was still insisted upon. The boiler was supported on a plate frame set inside the wheels (from now on, the inside frame became standard for the great majority of British and other locomotives). This had already been done by Bury on his lightweight engines, though with bar frames rather than built-up plates; these also had overhanging fireboxes and, as Stephenson later readily admitted:

> I originally objected to this projection of the firebox beyond the axle; but after an experiment continued uninterruptedly for a series of years, with an enormous traffic, it would be absurd to reject such practical evidence ... It was this evidence that led me to the construction of the new long-boiler engine, and the abandonment of the objection I originally entertained and acted upon.[59]

The 'long-boiler' answered the need of the time and orders, especially from European railways, flowed in. Soon they were being built in other works, under licence; equally soon, modified versions appeared, some with cylinders placed outside the frames. For several decades to come, the Stephensonian six-wheeler would haul most of Europe's slow goods trains.

# 13

# 'I AM GEORGE STEPHENSON'

The scene is a first-class railway saloon carriage, rocking along somewhere in the English Midlands, around 1845. There are four passengers, 'a gentleman, a young lady – unacquainted these two' – also an active-looking man of 65, wearing 'a broad long-tailed coat of mottled green, black velvet waistcoat, pepper-and-salt trousers, cloth buttoned boots and a white necktie, pinned with a large common brass pin in a way of his own'. He has a young male companion, evidently a clerk or secretary. The man in the green coat

> was seated opposite the lady, a fine, tall, handsome girl, of a very perfect physique, and evidently a *lady*. He eyed this girl admiringly and critically for some time, and then rapped out this: 'You'd make the mother of a fine breed of navvies, my lass!'
>
> Of course she was most indignant, changed her seat to the other end of the carriage, and sat gazing out of the window, her face like a peony.

The other gentleman expostulated on her behalf and said it was the most marked and uncalled-for insult he had ever heard.

> 'Sir,' said the old man with dignity, 'I am George Stephenson.'
> 'Well, George Stephenson, you are no gentleman,' was the reply. After this the solemn silence of insulted English reigned supreme.

The episode is narrated by John Hart,[1] George's one-time personal assistant and one of his warmest admirers. He thought Stephenson was probably thinking aloud, 'as men of sixty-five will'. But it is quite revealing of how George, by now, had become, in his own mind, a sort of national institution. Such recognition was not yet universal, and though the *Railway Times* noted in January 1842 that energetic measures were in hand for a testimonial to George Stephenson, and that one had been proposed some years before, but for some reason 'the matter was allowed to slumber', this new initiative also failed to take off.

By that time there were almost 1,900 miles (3,040km) of railways in Great Britain. Though most were struggling and investment had come to a stop, there were many prospective lines whose promoters were waiting for an economic upturn. In East Anglia, the Eastern Counties Railway was making very slow progress and its future was threatened by a rival proposal to build a line linking Norwich and Yarmouth with the Northern & Eastern Railway at Bishop's Stortford. To counter this move, Robert and George Stephenson did something most unusual. On their own account, they set up as railway promoters, planning and surveying a scheme to build a railway between Norwich and Yarmouth. The magic of their name attracted other investors and the scheme was passed by Parliament in June 1842. Implicit in the strategy was a co-working deal with the Eastern Counties Railway, which had neither the money nor the reputation to defend its own future.

In their zeal to ensure orderly development of railways, the Stephensons were now both rich and influential enough to make things happen their way, at least in one region of Great Britain. Robert was appointed engineer-in-chief of the Yarmouth & Norwich Railway, though George Bidder did the actual supervision. George Stephenson was chairman of the board and, typically, also became involved in plans to develop Great Yarmouth as a seaport. Robert bought no shares in the company; his capital was still tied up in the re-formed Stanhope Railroad, which as the Pontop & South Shields had got its Act as a limited company in May, and he was elected its chairman in August. The Yarmouth & Norwich line was built by a firm of contractors, Grissell and Peto, who had first worked with Stephenson and Bidder on the Blackwall Railway. By now railway construction was mostly in the hands of large contracting companies, and the old system of short sections of line being built by small, often local, contractors was largely abandoned. As the contractors gained in expertise, experience and capital resources, life was made somewhat easier for the supervising engineers, since a uniform approach and standard of work could normally be expected. Among the contractors, the personalities who loomed largest were Thomas Brassey and Samuel Morton Peto. Both built railways for Robert Stephenson, but Peto became a friend as well.

In 1841 George set his mark upon the green slopes of the pleasant Derbyshire valley of the River Amber. Already the Clay Cross collieries were hard at work, and the North Midland Railway added its clashings and clankings to the scene. Now, like a genie of fire and smoke, he descended on the hamlet of Ambergate and set up a great lime-works. In a grimly impressive rank, twenty-one massive kilns were built, each with its chimney belching out a sooty chemical mixture whose toxicity was not even thought about. Limestone quarried at Crich, in the hills, was brought down by tramway and two self-acting inclined planes to the valley, where locomotives pulled the loaded trams to the furnaces. Coal came from his own mines. The output of 200 tons of lime each day made up five trainloads.

George Stephenson by now was an industrialist and capitalist on a substantial scale. Though well away from the fumes of Ambergate, his own Tapton estate was close to the railway, which bordered its edge, and he liked to see the trains go by. He had two dogs, a pedigree spaniel named 'Carlo' and a 'nameless, half-bred, rough terrier' which he used to take on pit visits, until one day 'with a departing yelp' it fell down a shaft.[2] Betty shared his love of animals and one of her two African grey parrots was the only creature that dared to mock George, calling out '*Coom*, Betty!' as soon as it saw him.[3] The household staff was a large one, as befitted a wealthy captain of industry. In charge of the stables was John Wigham, George's friend from Killingworth with whom he had once contemplated emigration, and who could remember the days when George Stephenson was proud to be provided by his employer with a single horse to ride. George himself was happy to remember that time, and was often invited by Mechanics' Institutes and similar bodies in northern towns to address them. Samuel Smiles heard him at the Leeds Mechanics' Institute:

> He was always an immense favourite with his audiences there. His personal appearance was greatly in his favour. A handsome, ruddy, expressive face, lit up by bright dark-blue eyes, prepared one for his earnest words when he stood up to speak and the cheers had subsided which invariably hailed his rising. He was not glib, but he was very impressive.[4]

The unvarying theme was George's own life story and the example it conveyed to his listeners. From humble beginnings and early struggles, hard work, ingenuity and perseverance had led him to national fame, wealth and a position of solid achievement. Basking in the cheers and applause, secure in the adulation of his hearers, he stood before them as someone already almost legendary. In his discourse, the locomotive was the central symbol. On its iron wheels, as everyone now realised, radical changes in national, indeed human, life were moving forward – and right here was the man who had spoken up for it and defended it as a lone champion when its abilities were derided and its potential pooh-poohed. Far more than that, his was the inventive spirit that had created those abilities and brought that potential into being. He had given the steam railway to the world. Nobody else, apart from Robert, received credit; not Booth, Seguin, Murray, nor Trevithick, and certainly not James. George was not setting out to give his hearers a history of the development of the locomotive, or even particularly to exalt his own fame: he was a man of 60-plus looking back on a remarkable life, entitled to present it as he chose, and he chose to give it a moral thrust, pushing less his own glory than his constant message: 'Persevere!' As Smiles observed, his own appearance and manner helped enormously – he looked the part and played it to perfection, without humbug or false modesty. His force and warmth of personality reached out and for a brief time his audiences could feel that they knew what it was like to *be* George Stephenson. Largely because of Smiles – who had his own 'self-help' agenda to promote – George's version of events became firmly fixed in people's minds, and to much of the world he is still the inventor of the steam locomotive and the father of railways.

John Hart reported from around this time: 'He had conceived a strong desire that his Autobiography should be written by me from his dictation, and he hinted at it several times, but never pressed it; whilst I, foolishly young, did not take up the idea with the warmth it deserved.'[5] How many were truly inspired by George is impossible to say, but his example and his message undoubtedly played their part in making the British working class, already 'the masses' and becoming ever more numerous, into a different sort of community to that which Karl Marx, still at that time living in Germany, would identify as a force for eventual revolution and the flowering of a communist society. Aspiration and the possibility of its fulfilment – with perseverance – helped to make Britain unconformable with the Marxist model. A new friend of George's from the early 1840s was another who had moved from the status of a low-paid working man to that of a wealthy capitalist. Once the Duke of Devonshire's gardener at nearby Chatsworth, Joseph Paxton was, by 1843, the duke's land agent and confidential business adviser, as well as a garden planner, architect and engineer. Interested in railway development, he was introduced by Stephenson to George Hudson and became a big investor in railways in his own right.

His interest in the practical arts and sciences led Robert to become a member of the Society for the Encouragement of the Arts, Manufacture and Commerce, to which he was elected on 20 April 1842, extending his early morning perusal of texts and journals to attendance at the Society's meetings and its public lectures given by eminent figures. But at home there was also sadness. Through another year of intensive activity on her husband's part, Fanny's condition was steadily worsening and by September it was clear that the end was near. From 29 September Robert remained at home, and on the 30th he wrote to Richard Creed: 'It is quite impossible to leave my house for I am engaged in the melancholy task of watching over the last moments of my dear wife',[6] and he was with her when she died, in the early hours of 4 October 1842. He wrote in his diary: 'My dear Fanny died this morning, at five o'clock. God grant that I may close my life as she has done, in the true faith, and in charity with all men. Her last moments were perfect calmness.'[7] On 11 October she was buried in Hampstead churchyard, and on the 12th he set his grief aside and resumed work, with this record in his diary: 'Home – Stockton and Darlington Bridge.'[8] On the 13th he was out on the site of a bridge over the Medway, on the Maidstone branch of the South Eastern Railway, where Bidder was in charge of the engineering works. It was business as usual. Fanny had often urged him to remarry after her death and, still only 39, he might have done so and become a father after all, but he did not

follow her advice and remained a widower. Jeaffreson records that he used often
to visit his wife's grave.

From Fanny's death one of the main differences between father and son
becomes more apparent: George was assured from within himself, while Robert
needed reassurance from outside. Within his make-up was a strong impulse
towards self-protection; perhaps innate, perhaps caused by the need in his youth to
justify himself to a demanding parent. It became an aspect of his personal charm
and appeal; he was not a natural loner, and always functioned best within a framework of supportive friendship. He managed this very well, first in his marriage,
with its integration into the Sanderson family network, which continued after
Fanny's death, and in business with George Bidder as friend and partner. Edward
Starbuck, Edward Cook and his solicitor Charles Parker were also anchor-points
in a web of relationships, men whose affection for him was reciprocated both in
feeling and in his generosity to them. Sustaining, loyal and relieving him of a great
deal of mundane activity, these networks were evidently more valuable to him
than a second marriage.

Whether through arguments about management, or because of the strains
of competition between the Bedlington and Forth Street locomotive works,
Michael Longridge's partners in Robert Stephenson & Co. bought out his
interest in the company at the end of 1842. The process was handled by Robert
Stephenson, who wrote to Edward Pease on 4 November: 'I am much obliged by
your concurrence in the proposal I made as to the mode of closing our connection with Longridge … I have heard from Longridge in reply to my offer and he
accepts it.'[9] Relations between Robert and Longridge had been cool for some
years, but the dissolution of this partnership did not mean a social break, and, if
anything, something of the old friendship resumed. At the same time, Stephenson
wanted to make Edward Starbuck a partner, but this was resisted by the Peases
and Richardson,[10] and Starbuck always remained a commission salesman. Apart
from his selling skill, he was very useful to Robert as a chaser and collector of roy-
alties due on Stephenson designs built abroad, and in noting any infringements
of contracts.

Early in 1843, Robert moved house to 15 Cambridge Square in the new
Bayswater district of London, a brisk walk through the London parks from his
Westminster office. But very shortly after he moved in, the house was seriously
damaged by a fire. His father was sleeping there that night and as: 'The first in the
house to sniff the smell of fire, he lost no time in taking care of himself. When
Robert Stephenson and his servants were in the act of flying from the house in
their night-clothes, the prudent father made his appearance in the hall, dressed
even to his white neckcloth, and with his carpet-bag packed and swinging in
his hand.'[11] Much, if not all, of Robert's presentation silver and his pictures must
have been lost. He also found himself, by the terms of his lease, obliged to pay
the rent on a house that took a year to be made habitable again, while he lived

in furnished lodgings. Jeaffreson notes this as an example of Robert's unworldliness in matters outside his professional activities. He had a capable housekeeper, Margaret Tomlinson from Newcastle, Bidder in the engineering practice and Edward Cook in Newcastle. Now, in late 1843, he made his brother-in-law, John Sanderson, into a kind of personal manager, whose job it was to deal with the mundanities and allow Robert to concentrate on what interested him, both in terms of work and relaxation. Sanderson and his wife Mary were already his close and confidential friends and, with the Bidders and a few others, provided a kind of family circle in which he was the central figure.

The Stephenson practice moved along Great George Street in June 1843 to larger premises at No 24, next door to the Institution of Civil Engineers. Bidder, Robert and George Robert Stephenson (from 1844) were the permanent occupants, though others came and went. John Hart, at least, used the house as a London base at times, but he was a close friend of George Robert.[12] The latter, aged 25 in 1844, was the only son of George's brother Robert and his wife Anne. His father had died in 1837, as manager of the Pendleton collieries in Lancashire. George Robert, a very capable but not original-minded engineer, was the only other grandson of 'Old Bob' to emerge as a professional man, though the great majority of his first cousins were girls. Bidder, on the ground floor, occasionally found himself having to engage in an impromptu wrestling bout with George Stephenson when the 'old gentleman' was in the office with nothing better to do. On at least one occasion the unfortunate Bidder was hurled about enough to break one or two chairs.[13]

Britain's trade was still in recession. Investment in railways had slumped, as had railway dividends. Only Hudson companies seemed immune. The York & North Midland continued to deliver 10 per cent to its shareholders, and Hudson, at the company's general meeting at the start of 1843, 'could not refrain from congratulating them as well as himself, upon the successful results of the last half-year, when all other railways – not even excepting the greatest of all railways, the London and Birmingham – had experienced a large diminution of income'.[14] Huge investments had been made in railways and rivalries were intense, both where lines competed for the same traffic and where companies vied for the right to open up a new route. In these games of hardball, played with scant concern for such rules as existed, and not for the timid or over-scrupulous, Hudson was in his element. He had just gained control of the North Midland company and was embarked on a strategy that would, by September, bring about an amalgamation of this railway with the Midland Counties and Birmingham and Derby Junction companies. The two latter, each offering a route between Derby and London, had been engaged in a ruinous fare-cutting campaign to attract passengers. Though he played no overt part in Hudson's rumbustious manoeuvres, which forced the two boards into submission by playing on the anxiety and greed of the shareholders, Robert Stephenson later described himself to a Parliamentary Committee as 'the chief instigator' of this campaign.[15] In September 1843 the Midland Railway was established – now a single company linked Birmingham, Rugby, Derby and York, with a branch to Nottingham. The effect was akin to that of a *coup d'état*, and its leader, already chairman of the companies linking York with Gateshead, became a national figure.

At the second shareholders' meeting of the Stanhope & Tyne, now officially the Pontop & South Shields Railway, on 26 August 1843, Robert Stephenson gave a very clear exposition of the negotiations with Hudson and the Newcastle & Darlington Junction Company over the Durham Junction Railway, of which the P & SS owned about half the shares. The owners of the Durham Junction had offered their line at cost to the N & DJ, but Hudson had said that it was far too much. There the matter rested, but Stephenson proposed that in the meantime the Durham Junction should be leased to the N & DJ; he wanted action 'for simplifying the means of communication' and considered it 'on many accounts desirable that the line from Newcastle to Darlington should be as nearly as possible in the hands and under the control of one company', to which his audience assented with a 'hear, hear, hear'.[16]

On 18 November 1843 a further announcement was made: 'The Pontop & South Shields railway will apply to Parliament to make alterations and to sell their line to the Newcastle & Darlington or some other Railway Company. At the same time the Durham Junction Railway is to apply to sell itself to the Newcastle & Darlington and the latter to build a station at Gateshead.'[17] A deal had been done and the Newcastle & Darlington Junction Railway company picked up the Durham Junction for £88,500 – less than the line had cost to build: 'An engineer and a promoter were to buy a whole railway between them, and put its shares into their pockets like a parcel of cheese,' wrote Richard Lambert[18] – but Stephenson had done his best for the Pontop & South Shields shareholders as a group and made no undue benefit himself.

While things were thus being tidied up south of the Tyne, Hudson and the Stephensons had a strategy ready for north of Newcastle. The planned Lancaster & Carlisle Railway, on the other side of the country, had not yet received its Act of Parliament, though Joseph Locke had surveyed its route, and they were keen to be first with a line to Edinburgh. Three companies were projected to build from Dunbar to Berwick, Berwick to Newcastle, and to build a high-level bridge over the Tyne between Newcastle and Gateshead. The desire of the Newcastle & Carlisle company to have the proposed Newcastle central station built on its line had helped to delay any action. Now, in 1843, matters went purposefully ahead. George Stephenson had always supported the plan for a high-level Tyne bridge and knew exactly where he wanted to place it. While George and Robert planned the line, which also required a long viaduct over the Tweed at Berwick, Hudson worked out the means by which these businesses would be financed. George had first surveyed a Newcastle–Edinburgh railway in 1838, selecting the coastal route via Berwick rather than a more direct line crossing the Cheviot Hills, and he returned to re-survey the same ground in the winter of 1843–44, while Robert was engaged on the completion of the Darlington to Gateshead line.

Also in that winter, Robert made a brief visit to Ireland to look at the Dublin & Kingstown Railway and its atmospheric traction system. He went at the behest of an important new client, the Chester & Holyhead Railway. This proposed trunk line, first surveyed by George Stephenson in 1838–39, was in the process of being fully surveyed by Alexander Ross, one of the Great George Street team, with Robert as consulting engineer, and its Bill was to be considered in the 1844 session of Parliament. It was backed by the London & Birmingham Railway, with its strong Liverpool contingent of investors, and, as the fast route to Ireland, government money was also anticipated. For Robert Stephenson, the country's leading locomotive designer and builder, to report on a rival form of traction was a sensitive matter. Both he and his father still bore the scars of the old campaigns in which their integrity and self-interest had been publicly questioned. But Robert's personal standing in the engineering and the wider public worlds was much higher than it, or George's, had been ten years earlier. Having seen the system for himself, he deputed two engineers, George Berkley and William Marshall, to undertake more exhaustive examinations before he produced and published a report in April 1844. There is little doubt that, like his father, he was a sceptic from the start, but his report was technically detailed and conscientiously balanced, accepting the actual and potential merits of the atmospheric system. He judged that it 'might be advantageously applied' on short lines in the vicinity of large towns, though not on urban lines like the Blackwall Railway (where he still believed rope haulage was the only way of running rapid-interval, stop-start traffic), and on steep gradients, but he spelled out the reasons why it was too inflexible and uneconomic for a long-distance trunk line. Copies of the report were sent to all foreign customers of the Forth Street Works, in case they might be tempted away from locomotive traction.[19] Brunel was enthusiastic about atmospheric traction, as was William Cubitt; Joseph Locke had no time for the notion. The Chester & Holyhead board accepted Stephenson's recommendation and, when the line got its Act in July 1844, it was for locomotive haulage, and he was appointed as engineer-in-chief with a team from Great George Street on the construction sites.

Edward Cook's letter-book for January–August 1844 shows Robert's constant involvement in the day-to-day business of the locomotive works, despite the claims of civil engineering projects; on 5 January Cook wrote to the Dutch engineer F.W. Conrad: 'Mr Stephenson has more to do at this moment than ever and perhaps as much as all the other Eng$^{rs}$ of England put together.'[20] That month Robert was in Holland and gave Conrad a 'present' and his daughter a dressing-case.[21] Export business remained very important, and for a time they considered opening a boiler plant in Germany to circumvent the heavy import duty on worked iron. Locomotives were also under construction for Paulin Talabot's Marseilles–Avignon Railway in France.

George Stephenson was in robust form when Thomas Sopwith met him in the Great George Street office on 7 May 1844: 'He is looking remarkably well, is very animated, and displays great kindness of manner in those directions where he has formed a favourable opinion. On the other hand he is said to be equally unbending under opposite circumstances.'[22] On 30 April the Yarmouth & Norwich Railway had opened for business with the customary celebrations. Seven weeks later there was a grander event. By the middle of June, the 39-mile (62.4km) Darlington to Gateshead line had been completed, and in his most grandiloquent manner George Hudson organised a triumphal opening on 18 June. A special train from Darlington carried him, the Stephensons (minus their luggage which got lost at Darlington[23]) and many other notables to the terminus at Gateshead. Greeted by bands, cannon-salutes and the civic dignitaries of Newcastle, they crossed the Tyne in a great parade to the Assembly Rooms, where a 7-hour banquet for more than 300 people awaited them. For George Stephenson in particular, this was a profoundly emotional moment. Back in his native Tyneside, with all the much-cherished and polished memories it evoked for him, he was the central figure of the festivities. Even if his ambition to oversee the growth of a national railway system had had to be abandoned, he could take pride in the fact that he and his son had responsibility for the building of almost every mile of track from Euston Square to the south bank of the Tyne. For the first time ever, the London morning papers could be read on the same day in Newcastle, brought by a 'flying train' to Gateshead. Truly the world was being compressed and speeded up. Twenty-seven healths were drunk in the course of the feast. Every speaker saluted George. Even Hudson was content to let himself be overshadowed on this occasion, claiming to be: '… only a tool in the hands of a genius – and probably a very pliant one at the hands of the Messrs Stephenson.' The Hon. H. T. Liddell, son of the man who had commissioned *Blücher* back in 1813, proposed the health of George Stephenson, as 'The Father of Railways' and builder of 'the first locomotive that ever went by its own spontaneous movement along iron rails'.[24] In his response, George did not contradict him. A more accurate encomium appeared at the end of the year, when the directors of the Liverpool & Manchester Railway commissioned a marble statue of their engineer to stand in the new St George's Hall in Liverpool: 'Father of the Improved Railway of Modern Times'.

In the course of 1844, money was becoming more readily available and investors were looking for opportunities at home because of the uncertain political situation in much of Europe. Interest rates, which had been at 6 per cent in 1839, had fallen to 2.5 per cent by September 1844. The stop-start progress of railway expansion in Britain was beginning to speed up, despite two major unresolved issues, in both of which the Stephensons and Isambard Brunel stood at opposite sides of the argument. The question of railway gauge, which had been latent ever since the completion of the Great Western's first main line in 1838, came to a head when the broad-gauge Bristol & Gloucester Railway, an ally of the GWR, and the standard-gauge Birmingham & Gloucester Railway, met at the latter place in 1844. The quiet cathedral city became a frontier town, with scenes verging on the chaotic when passengers and goods between Bristol and Birmingham had to be transferred from one system to the other. Transfers were hardly new when most journeys were still by coach and goods were regularly trans-shipped from cart to canal barge, but the inconvenience of changing trains was made more irritating by the fact that it was unnecessary and by the large number of people involved. Much of the fuss, however, was stirred up by propagandists of the standard gauge. Parliament took an interest in the controversy, and it seemed increasingly likely that it would have to give a lead on what railways' policy should be.

The other issue was that of the atmospheric railway versus locomotive haulage, which took a peculiar twist while George Stephenson and Hudson were preparing their plans for the railway between Newcastle and Berwick-upon-Tweed. Strong opposition was mounted by the Grey family, whose large estates on the coast would be traversed by Stephenson's planned route. Earl Grey, the former Whig premier, was 79, and the campaign was led by his heir, Lord Howick, who objected to the line surveyed by Stephenson, which he claimed interfered with his amenities. Hudson and Stephenson refused to make the expensive detour he demanded, and Howick then proposed that the railway to Edinburgh should take a completely different, inland route through the fells and over Carter Bar. George Stephenson had rejected this option in 1836 and was not about to change his mind. Howick then began to organise backing for his own railway over the inland route. It would be cheaper, its steep gradients worked by atmospheric traction, and it was to be a single line only, rather than the coastal line's double track. Apart from enlisting much local support, he succeeded in recruiting Isambard Brunel as the potential engineer of his 'Northumberland Railway'. When Brunel came to Newcastle late in 1844, George Stephenson – whether by accident or intention – met him there, and in the encounter: 'Stephenson ... good-naturedly shook him by the collar, and asked "What business he had north of the Tyne?"'[25] Behind the bluff bonhomie was real resentment. Brunel had never been well-thought of by George: he was the engineer who had dismissed the 'Stephenson' gauge out of hand; never having designed a successful locomotive himself, he was now backing the fad of atmospheric traction – and doing so for a company set up in opposition to George's, in George's own backyard. Even before Howick's proposal for a completely different railway, in writing to Michael Longridge, still at Bedlington, Stephenson showed the strength of his feelings:

I am rather astonished at Lord Howick's observations ... my senses are puzzled in judging how these people can set about making such paltry objections! It is compensation they want, nothing else … This species of objection is a genteel way of picking the subscribers' pockets: there cannot be a doubt but it is meant to do so.

I have never taken any part in politics, but I think I now will and become a Tory, and I shall buy a piece of land in Northumberland to oppose Lord Howick. I do not like this double-dealing; we shall not fear Lord Howick's opposition.

Is the great thoroughfare through England and Scotland to be turned aside injuriously for the frivolous remarks made by Lord Howick? No! The times are changed. The legislators must look to the comfort and conveniences of the Public. Are hundreds and thousands of people to be turned through a tunnel merely to please two or three individuals?[26]

Ironically, the contest pitted a Whig grandee, ostensibly on the side of industrial progress (though of the 'not-in-my-backyard' school), against someone whose name was synonymous with the expansion of industry. At 63, George Stephenson still had no coherent political philosophy, but certain clear ideas. His notion that legislators should look to the comfort and convenience of the public, rather than to the interest of the landed gentry, was shared by Liberal radicals like John Bright (who, as a Durham MP, had been a guest of honour at the opening of the Newcastle & Darlington Railway). When it came to railway planning, Stephenson was thoroughly *dirigiste*, but also utilitarian in a real sense, putting the good of the greatest number before the privilege of wealthy individuals. In his conception of changed times, and what they imply, he shows a very modern attitude for 1843. As an employer, though benevolent in a paternalistic way, he was more of his time. Trades unions had no place at Clay Cross; his manager, Charles Binns, declared that: 'Messrs George Stephenson & Co. have no intention of employing any man who is a member of such union';[27] although Stephenson himself had declared to the Mines Commission in 1842 that, 'I do not consider it desirable to employ children under twelve or thirteen years of age in coal mines',[28] two years later, of the 250 employees at the Snibston collieries, 50 were under 16, and some were as young as 10 or even 7.[29] Though the Mines Commission report shocked most people who had avoided knowing what happened in the pits, coal-owners were insistent on the point that, without the employment of children, they would not be able to operate profitably. George Stephenson had made a prototype coal-cutting machine, but, like others, it was not pursued because it was cheaper to use manual labour.[30]

South of York, too, there was railway warfare. Even as Hudson and the Stephensons were finishing their route from London, two sets of promoters combined on 17 May 1844 to sponsor a 'direct line' (an increasingly popular concept at this time) to run from London via Peterborough and Doncaster to York, a much shorter and potentially quicker route than the Hudson line, and which would certainly draw off a great deal of traffic, not least the lucrative mail contract.

To Hudson's great annoyance, the board of the Great North of England Railway expressed support for this proposal, just as they had done for Lord Howick's railway. With George Stephenson's encouragement, and with Robert's collaboration, Hudson set out to oppose the London & York (later Great Northern) proposal tooth and nail.

All these various causes came to a head in the course of 1845, a remarkable year for railways. In April, a House of Commons Select Committee (set up on a motion from Lord Howick, who was MP for Sunderland), having made a rapid and sketchy survey of the matter, pronounced in favour of atmospheric traction over locomotives for working railways. Since atmospheric railways were cheaper to build, being more lightly laid and with a single track only, this verdict was received with enthusiasm by railway promoters. It was also convenient for Howick with his Northumberland Railway Bill already in the pipeline. He had impugned Robert Stephenson's motives in a manner reminiscent of earlier disputes with the Croppers and others: '... his opinion would have had more weight with us had we not thought that we discovered very strong symptoms of that bias of his mind in favour of the locomotive system, which, considering how much its brilliant success is owing to the exertions of himself and his father, was naturally to be expected to exist.'[31] Robert had, in fact, been temperate in his comments on the atmospheric system, and Hudson made a strong riposte in his defence: 'The characters of the Messrs. Stephenson stand much too high in their native country to be affected by the observations of any individual who may attempt to lower them in public esteem. I would merely observe, that daily experience shows that Mr Robert Stephenson's view of the Atmospheric system is correct.'[32]

Determined to dislodge the obstacle of the Great North of England Railway, Hudson made an offer to its shareholders in May 1845, which, its chairman said, would be madness to resist.[33] Hudson proposed a five-year lease of the line on behalf of the companies he controlled at a guaranteed annual rate, with an outright purchase of all the shares at the end of it at a price of £250 for every £100 share. It was a very steep price, and 'felt by Mr Hudson to be so; but George Stephenson told him "it must be done," and with a reluctant hand he signed the contract'.[34] No source is given for this claim, which would make George's influence with Hudson powerful indeed. But the enriched Quakers of Darlington had their revenge for the events of 1841.

On the contentious ground north of Newcastle, Bills for both schemes were submitted to Parliament, with Robert Stephenson and Brunel as the prime witnesses on each side; Joseph Locke backed Stephenson, and Major-General Pasley, the government's Inspector of Railways, was among the supporters of Brunel and the plan for atmospheric traction. During the proceedings, Lord Howick called to see Robert Stephenson at Great George Street, but encountered George in the outer office, who cornered him in a chair and gave him a powerful lecture on the shortcomings of atmospheric traction, from which Howick was glad to make a retreat, to the parting words: 'Take my word for it, my Lord, it will never answer.'[35] With Hudson's powerful support, the coastal, locomotive-operated route was accepted and Lord Howick withdrew his Bill at the end of June. On 31 July 1845 the Newcastle & Berwick Railway was duly incorporated, most of its capital coming from shareholders in the Newcastle & Darlington Junction Railway. The Act for a high-level railway bridge over the Tyne at Newcastle was passed at the same time.

In March Robert was elected to the Society of Civil Engineers, the 'Smeatonians': a dining and discussion club for those at the top of the profession. The atmospheric controversy did not shake his mutual friendly regard with Isambard Brunel, to whom he wrote in the middle of the conflict: 'Hudson has left town and I shall not see him until his return after Easter. I will do all I can, but I fear Lord Howick and he have so misunderstood each other that a reconciliation will be very difficult – Temper has in this instance like many others stepped in, and to all appearances is riding rough shod over reason.'[36] Nor did the on-going gauge controversy, though it imposed greater strains. Both broad-gauge and narrow-gauge proposals were made for a railway to join Oxford, Worcester and Wolverhampton. Daniel Gooch of the Great Western (once a pupil at the Vulcan Foundry) had the bad luck to run into George Stephenson on the day in June 1845 that judgement was given in favour of the broad-gauge for this route. George, who had known the GWR's locomotive superintendent since he was an infant in frocks, berated him publicly for betraying 'the tradition in which I had been reared'. Gooch wrote later that he 'would never forget the passion George Stephenson got into when the decision of the Committee was announced'.[37] A Board of Trade inquiry into the gauge question resolved nothing, and following a Resolution proposed in the House of Commons by Richard Cobden on 25 June, the government set up a Gauge Commission in August 1845 to make recommendations about the gauge of future railways. Two of its three members, Professors Peter Barlow and G.B. Airy, a mathematician and an astronomer respectively, had also served on the Anglo-Scottish trunk line commission a few years earlier; the third was Lt Col Sir J.F.M. Smith, RE, Pasley's predecessor as Inspector of Railways.[38]

The commissioners' inquiries were intensive and lasted for several months. Their first witness was Robert Stephenson on 6 August 1845; this in itself was a testimony to his primacy among railway engineers. Although other people by now were designing and building railway locomotives, the focus was largely on Stephenson's products, and in particular on his recent 'long-boiler' engines. Things took an uncomfortable turn when Major-General Pasley, also a witness, appeared to suggest that these, when fitted with outside cylinders, had a dangerous tendency to 'yaw' at speed (45mph/72km/h or more) on the standard gauge. The first 'long-boilers' had been designed with an eye to reduced fuel consumption; but the fast trains of the Great Western had created a public familiarity with, and expectation of, speed. Stephenson's outside-cylinder engines had followed in 1843, with the intention of running faster, and the suggestion that the 'narrow gauge' was unsafe at speed had to be dealt with. Robert's replies were clear and measured. He stated that he would originally have preferred a gauge of 5ft (1,530mm), but that there was no difficulty in accommodating all the moving parts in a locomotive built to the standard gauge. In his considered view, gauge was irrelevant to the performance of locomotives and he repudiated the idea that narrower width was a contributory factor in accidents. On the question of 'yawing' he frankly admitted that there was a problem to be resolved:

> I cannot make up my mind about it. If you consider the action of the cylinder, it is perfectly rigid metal – engine and cylinder together. Now, when the steam presses upon the piston, it is at the same time pressing against the lid of the cylinder; the action and the reaction must be equal. Therefore, I do not believe it is the steam that causes the irregular action, but I believe it to be the mere weight of the pistons themselves, and therefore if we could contrive to balance the pistons by the weight upon the wheel, we should get rid of that very much ...[39]

Neither George nor Robert had been advocates of speed, believing the steam locomotive to be essentially a slow but steady machine. It was Brunel's vision and achievement of a fast railway that compelled Robert to respond. The Great Western engines were not mechanically superior, but their width and weight gave them greater stability at speed. On the standard gauge, the requirement for greater speed and hauling power had taken locomotive design into a new phase, in which it was not enough simply to make a bigger, longer boiler and fit bigger wheels. A 20-ton locomotive travelling at 40mph (64km/h) was a very different kind of machine to the 'Patentees'. Other engineers apart from Robert Stephenson were working on the question of balancing the heavy moving parts, including John Braithwaite of the ill-fated *Novelty*, who was now engineer to the Eastern Counties Railway. However, it was a French engineer, Louis le Chatelier, who wrote a technical treatise which showed how balance between the reciprocating parts and the turning wheels could be achieved.

George Bidder naturally backed up Robert Stephenson's answers, but so did a succession of other engineers with no connection to either Great George Street or Forth Street. Brunel, Gooch and their Great Western colleagues were almost alone in advocacy of the broad gauge. It was at Brunel's suggestion that a set of comparative performance tests between broad- and standard-gauge engines was made in December 1845 and January 1846. J.G.H. Warren condemned it as a last gambit, 'a counsel of despair' on the part of the broad gauge adherents.[40] Many people at the time considered it a pointless diversion, and the *Railway Chronicle* spluttered at 'the impropriety and utter worthlessness of this wretched series of experiments, ill-contrived, and calculated to embroil inextricably a question which they could not in any event tend to clear up'.[41] Hamilton Ellis wrote:

The 'Narrow' party agreed with bad grace, and insisted on their opponents batting first. Also, under the auspices of G.P. Bidder, they introduced all sorts of devices favourable and even exclusive to themselves, such as pre-warmed (nearly boiling) feedwater, 'flying starts' and stimulating fireboxes in ways not consistent with normal railway working. The Great Western, on the other hand, observed fairly normal working conditions … On one trip, the axles of the vehicles used, having been left overnight at Paddington, were found to have had the grease replaced by sand, the work of some anonymous hero in G.P. Bidder's secret service, though it would be unfair to assume that Bidder himself was party to such a trick.[42]

Bidder was in charge on the standard gauge side; Gooch for the broad gauge, and both bombarded the organisers with demands to take account of every small difference in wind and weather. To create absolutely equivalent conditions proved impossible. Piquantly, it was a contest between Stephenson designs: the Great Western's *Ixion* was simply an enlarged 'Patentee' type, and the two narrow-gauge engines were an inside-cylinder 'long-boiler' from the North Midland and a newly built locomotive from Forth Street known only as 'Engine A'. Embarrassingly for Bidder and his team, on 31 December the North Midland engine overturned with Professor Airy on board while travelling at almost 48mph (77km/h). No one was injured and the accident was ascribed to a broken rail chair. It may seem odd that the track of the Great North of England Railway between York and Darlington, on which the standard-gauge trials were made, should not have been fettled up for the tests, but the company was in financial trouble and looking for a buyer.[43] The Great Western locomotives were generally agreed to have outperformed their standard-gauge rivals, but the commissioners could not overlook the fact that standard-gauge engines were working satisfactorily on all railways other than the GWR and its associate companies. Uniformity of gauge was clearly desirable, and there was no case for altering the 4ft 8.5in (1,435mm) set by George Stephenson. In January 1846 the commissioners recommended that this gauge 'be used in all public Railways now under construction or hereafter to be constructed in Great Britain'.

# 14

# BIG BRIDGES

'I have astonished the old Gentleman with a Melon and he has requested Mr Sanderson to preserve some of the seeds as he fears he has lost all that he had, I suppose from this his crop is a failure.'[1] A letter to Georgey Bidder from her husband in April 1845 notes George Stephenson's horticultural concerns. The events of that year prevented George from spending as much time in his hothouses as he might have liked. From his unique perspective, he was also looking out at a railway scene which could scarcely have been imagined even five years previously.

William Gladstone, as president of the Board of Trade, had tried and failed to regulate the expansion and management of railways: his Railway Act of July 1844 was a greatly watered-down version of his original Bill (this was a victory achieved by George Hudson). An official Advisory Board on railway construction, set up in 1845, had only a brief existence, being dissolved in July of the same year because of pressure from the railway companies. The economy was in good shape; interest rates were low and banks were relaxed about lending money. Railway companies were – mostly – paying good dividends. Towns and agricultural districts that had rail connections were seen to benefit from increased trade and income, and other towns and districts wanted a railway too, afraid that without one they would fall into decline. In these ways the scene was set for expansion of railway construction. What happened was a chaotic scramble, as the 'Railway Mania', its extent far transcending the boom of 1836, took possession of the population of Britain.

In its beginning it was linked to a genuine desire to build railways, supplemented by the fear that if a particular town did not get its railway connection *now*, it would be too late. Many people subscribed for shares and soon the market began to balloon – railway shares were the thing to invest in, so good that they could be sold at a profit even before the railway got its Act. Increasingly, it was not due to any interest in a particular line, or even in a reliable investment that would pay a steady dividend, but a highly infectious get-rich-quick fever. Among the new railways proposed in 1845 and 1846 were perfectly sensible schemes with proper planning and a sound financial basis. But as the fever spread, hundreds of schemes were rushed up quickly, often in competition for the same routes. At the bottom of it all was the alluring notion of safe speculation: that it was possible to commit to buy shares in a new railway and sell them again, at a profit, before having to find the money for the original purchase, or having paid only a small deposit.

In the course of 1845, Parliament authorised the construction of 2,170 miles of railway in Great Britain, effectively doubling the existing extent. In September 1845, 457 new railway schemes were registered,[2] and late in the year, with new proposals already swamping the Private Bills office, the government set a deadline of 30 November for Bills to be considered in the 1846 session. Sir Robert Peel, as Prime Minister, announced in January 1846 that plans for 815 new railways had been lodged with Parliament, at a prospective cost of £350 million. Britain's national expenditure at the time was less than a sixth of that amount,[3] and the entire national income was only around £200 million. In all, the plans would have meant the building of 21,000 miles (33,600km) of track, though many were duplications of one another, others preposterous and others fraudulent proposals. Some were broad gauge, some were standard gauge and some proposed to use atmospheric traction. Parliament, the only arbiter of the avalanche of schemes, was a prime source of corruption:

> Peers and members of the Lower House were avowedly engaged as traffickers in the railway market ... Members attached to 'the railway interest' voted for each other's projects ... there was in those years scarcely a person, either in the House of Lords or in the House of Commons, who was not, personally or through his connections, anxious that a bill should be obtained for some particular new line.[4]

The mania raged on, until on 16 October 1845 the Bank of England raised the interest rate by 0.5 of 1 per cent (to 3.5 per cent). At this touch on the brake, a degree of sanity was introduced, but following the splurge came the hangover. Sobriety led to a new panic when everyone was anxious or desperate to sell and nobody wanted to buy. For any shrewd investor there were good pickings now, with the shares of 'good' companies also forced down, but the instincts of the herd ruled in most cases.

For railway engineers, this tremendous surge of greedy self-delusion among the public, fomented by most of the newspapers and pervading society at every level, represented a combination of golden opportunity and ethical dilemmas, and of course a heavy overload of work. Such top professional men as Stephenson, Brunel, Locke, Rennie, Vignoles, Rastrick and a handful of others were in especial demand, since their names attached to a route survey, as 'Consulting Engineers', would lend great authority and credibility. An engineer could earn 100 guineas a day as an expert witness before the Parliamentary Committees. Retaining fees were also offered, though these were often no more than bribes:

> One morning Robert Stephenson and Mr Bidder received, through the post, cheques from various companies, amounting to more than £1,000. No previous overtures had been made by the directorates who forwarded these fees ... Both Robert Stephenson and his friend lost no time in returning these cheques, with intimations that their evidence was not a power to be bought and sold.[5]

For George Stephenson, so lately acclaimed as 'Father of Railways', it was a depressing spectacle, but he, who had always believed in the orderly development of the system, could scarcely be held responsible for what the British public had brought on itself, and he looked on with resigned detachment. Seeing it coming in 1844, he had warned against the dangers of railway speculation at a dinner given by the directors of the Leeds & Bradford Railway in October of that year – indulging, as ever, in reminiscence, he also told the gathering that he might have been the first person ever to sell a railway share at a premium: being the owner of three shares in the Stockton & Darlington in 1825, he had sold one to an insistent friend at a premium of £1 13s.[6] In 1846 George's name was down as engineer for only one main line, the Buxton, Macclesfield, Congleton & Crewe Railway, in which he had a personal interest as a coal-owner. Noting this, Smiles comments on others '... appearing as consulting engineers for upwards of thirty lines each!'[7] What he does not say is that Robert Stephenson was one of them: Robert might have returned unsolicited cheques, but he lent his name to thirty-four applications for Bills.[8] The thought of a free-for-all in railway development was of course repugnant to him, and the events of the mania years influenced his subsequent attitudes and speeches as a public figure; however even though he accepted only worthwhile proposals, his participation helped to swell the bubble. Bidder was just as busy, writing cheerfully to his wife on 16 July: 'I have agreed to give evidence in the Broad gauge, the York, the West Cornwall, Wakefield & Goole & in the Portsmouth ... In short, I seem to be getting more in request daily.'[9] Brunel was unusual in refusing association with any scheme in which he was not fully involved. George Hudson, whose enrichment of himself and his shareholders from dealing in railway stocks had been a significant, if involuntary, contributory cause of the mania, warned the public against wild speculation[10] and undoubtedly realised that the mania's inevitable collapse would gravely damage his own affairs.

As people counted the cost of the mania, a ghost from the past flickered momentarily. William James, briefly George Stephenson's ebullient friend and partner in locomotive sales, had died in 1837, in very reduced circumstances. A move was made to get up a posthumous testimonial to him, with a public subscription of money that could be put to the help of his surviving children. The text of the testimonial gave due credit to James' role in the history of railway development. When approached by William Henry, James' eldest son (himself an inventor and engineer, though financially unsuccessful), Robert Stephenson agreed to join the Subscription Committee and his name, at the head of the list, was followed by those of other leading railway engineers, including Locke, Brunel and Vignoles, but not the most eminent of them all. George Stephenson has been reproached for cold-shouldering James in death as in life. Acknowledgement of James' vision might have slightly occluded his own reputation, but only to a tiny extent. On the other hand, James had made no secret of his belief that George had schemed against him. We do not know the circumstances of George's refusal to sign, and he may have disliked the manner in which he was invited to subscribe. Though officially declared at the Mansion House, London, on 24 June 1846, the testimonial was a failure, and William Henry James died eventually in a Croydon almshouse.[11]

In their discussions of how to circumvent, if not actually prevent, the London & York direct project, Hudson and the Stephensons must have pondered long over the opportunities offered in East Anglia. George had been chairman of the Yarmouth & Norwich Railway until it merged with the Norwich & Brandon to form the Norfolk Railway in 1845. The Eastern Counties Railway had finally got to Norwich, but was paying its shareholders a dividend of only 1 per cent. Whether Robert was the architect of Hudson's plan, or even agreed with it, is impossible to find out, but Hudson's intrusion into the region made his and George's earlier effort at railway strategy there seem a very minor exercise.

Taking advantage of the extreme dissatisfaction of the Eastern Counties shareholders, and at the time when the mania was mounting inexorably, Hudson, with the same feigned reluctance as new Speakers of the House of Commons once adopted, allowed himself to be elected first a director and very soon afterwards chairman of the Eastern Counties Railway. Assuring a receptive audience of shareholders that they were at last on to a good thing, he unveiled the prospect of a north–south railway extending from their present London to Cambridge line, through Lincoln and Doncaster, eventually joining with the York & North Midland to make a new trunk route that would link London, York, Newcastle and Edinburgh. When Hudson spoke of such matters, a perpetual stream of shining guineas always seemed tantalisingly close, and the scheme was received with delirious rapture by his listeners, but Robert Stephenson must have had some misgivings. He had close links with the London & Birmingham Railway, to which he was still consulting engineer, and whose line between Euston Square and Rugby, part of the existing London–York route, was bound to lose traffic to that now proposed by Hudson. At successive general meetings of the numerous companies which he chaired, Hudson's initiatives, with the lids being whisked off new schemes like a succession of sumptuous new culinary delights, were beginning to acquire an air of desperation. Liverpool financial interests were strong in the Eastern Counties Railway, and the 'Liverpool Party', who had never liked Hudson, were disgruntled to see him impose himself on its affairs. A shareholder in one of Hudson's companies might be wilfully purblind as long as the dividends came in, but a man of Robert Stephenson's knowledge and perceptiveness, from his semi-detached position and with his own range of contacts in Liverpool, Birmingham and London railway circles as well as the north-east, could hardly fail to see that the Railway King could not keep up his dazzling sleight-of-hand indefinitely.

The team of engineers based at Great George Street was busy to the point of exhaustion.[12] Quite apart from burning the past-midnight oil to prepare plans for new schemes, under Robert's auspices his associates were building the Chester to Holyhead line, the Trent Valley line between Rugby and Stafford, the Newcastle & Berwick Railway, numerous lines in East Anglia, and others, including the Londonderry & Enniskillen Railway in Ireland. Here the resident engineer was Alexander Ross, who was accused by some shareholders of colluding with the contractor to inflate the construction costs. Appealed to by the protesters, Stephenson investigated the claims and made a stout defence of his engineer.

Despite intense pressure on their time from British commitments and potential clients, Robert Stephenson and George Bidder also struggled to maintain their European interests both as consulting engineers and in actual supervision of new construction. George Stephenson's plan for retirement was put on hold in 1845, during which he made two visits to Belgium, in March and August, examining proposed lines in the Borinage coal-mining district and in West Flanders. On the March visit he also went to Brussels, where he was feted at a grand dinner given by Belgian engineers and, on the next day, had a private audience with King Leopold, where coal and railways were the subject of conversation. Sopwith, who had made a survey of the terrain of the Sambre & Meuse Railway, went with him in March, and Edward Starbuck accompanied him on both trips, as aide-de-camp and interpreter. The Belgian government was inviting tenders for nine concessions, totalling over 500 miles (800km) of railways, and the Stephensons were acting as consulting engineers to four British investment groups.

Between George's two Belgian trips, and three days before Robert Stephenson gave evidence to the gauge commissioners, Elizabeth Stephenson died at Tapton House on 3 August. She had been married to George for twenty-five years, a self-effacing presence in his public life, but a warm and positive one in private; as Hart says,[13] and the Bidder letters also occasionally show, 'homely, good, and kind'. Robert was always fond of her, and a letter from Bidder to his wife on 16 July 1844 noted: 'Mr R. Stephenson has gone to Derby and intends I believe to invite Mrs Stephenson to London to spend a few days with him – How pleased the Old Lady will be …'[14] Having known many homes with George Stephenson, from Paradise Row to Tapton House, she was buried at Holy Trinity church in Chesterfield.

His wife's death did not alter George's lifestyle. He continued to live at Tapton House, extending his hothouses, managing his mines and lime-works, and involving himself in railway matters. Betty had been a regular churchgoer, with a taste for Methodism and chapel sermons, but though he sometimes accompanied her, George remained a sceptic in religion to the end of his life. Summerside, a lay preacher, occasionally dared to have a go at the 'old man' about the afterlife, but George came back with: 'Ah! Summerside, none of them ever come back to tell us'[15] and when asked to contribute to foreign missions: 'Ah! Summerside, I will send the locomotive as the great Missionary over the world.'[16]

Always taking the way things worked as a point of departure for his studies, George had bought a powerful microscope, and one of the hazards faced by his guests was the demand for a small sample of their blood for examination. Intrigued by the motion of the cells, he came to his own conclusion that 'they were respectively charged with electricity, positive at one end and negative at the other' – quite an advanced piece of science for the mid-1840s. In January 1845, as a guest of Sir Robert Peel, the politician whom he most admired and Prime Minister at the time, he brought his microscope and his needle, but as George advanced on his host, keen to see how swiftly the cells of a politician's blood moved, Peel protested squeamishness, 'sensitively shrunk back, and at length the experiment, at least as far as he was concerned, was abandoned'.[17]

In September 1845 he accompanied his friend Sir Joshua Walmsley to Spain. Walmsley was representing a group of British investors interested in backing a railway from Madrid to Bilbao, with Hudson to be its chairman,[18] and the mission was both to estimate the costs and traffic potential, and to negotiate with the Spanish government on what terms the line might be built. On the way through France, George crossed the Massif Central to Nîmes, where he met an old friend, Paulin Talabot, who as a young graduate of the *École Polytechnique* had made several visits to English railways in the 1820s. Later to be the first managing director of the Paris–Lyon–Méditerranée Railway, he kept up a lifelong friendship with both Stephensons. Talabot's coal railway from Alès to Nîmes had been built very much on Stephensonian lines, but Stephenson was so impressed by the volcanic landscape around Le Puy that the French engineer had some difficulty in drawing his attention to *des objets plus prosaïques*.[19]

In Spain, surveys had already been made, but George conducted his own reconnaissance, working his way inland from Bilbao to Madrid across mountainous country. It was a strenuous ten days or so, with nights sometimes spent on the floors of shepherds' huts. On Stephenson's advice, Walmsley asked for substantial concessions to be given to the railway promoters, including free land for the track and duty-free imports of materials and machinery. The Spanish government was slow to answer and the visitors had to leave without an agreement. On the return journey, they paused to look at some railway works in France, where George took pleasure in showing the navvies how to load a barrow properly, and then challenged them to a hammer-throwing contest in which, of course, he was the victor. By then they knew who this athletic elderly man was, and cheered him to the echo.[20] Altogether, the trip had taken five weeks, and by the end of it Stephenson was becoming ill. On the steam packet from Le Havre to Southampton, he developed pleurisy and was bled (the treatment of the period) by a doctor before the ship left. Walmsley got him to the Arundel Hotel in Haymarket, London, and looked after him 'as his nurse and sole attendant' for six weeks.[21] Smiles says that George gratefully remembered this, but in a letter to Longridge, sent from Tapton as he convalesced, on 22 November 1845, Sir Joshua is not mentioned:

> I am now at home and quite well, my recovery has been most extraordinary, the attack I had was pleurisy: I think it was first occasioned by taking unwholesome food at Bordeaux from that place I travelled night and day to Paris: I took very ill there, but still perservered in getting to England: on arriving at Havre I was obliged to have a Doctor who took 20 oz blood from me on board the Steam boat: I was then very weak but still wished to get on to England: the boat sailed at 5 o'clock P.M. I got to London the next day about half past 2 o'clock & there got the best advise; they got two Physicians to me; they put me to bed and cupped me on the right side; how much blood they took by cupping I cannot tell, they then put a blister on my side and gave me Calomel every four hours from Saturday night to Tuesday night: I then became so weak that they durst not give me any more, on the Wednesday morning I was considerably relieved from pain and could then eat a little – my rapid recovery from then has been astonishing. I am now quite as well as I ever was in my life, but I am advised to keep quite for a while.

To this account of his illness – a reminder that in his time medical treatment might still sometimes be more life-threatening than the disease – he added a postscript:

> I have had a most extraordinary journey in Spain. I crossed the Pyrenees 5 times; and rode on horseback 50 miles amongst the mountains seeking out the lowest pass – we had our carriage drawn up by bullocks on to the mountain passes where a carriage had never been before – we passed just under the snow range, I shall give you an account of my travels when I see you: we travelled 3000 miles in 33 days: stopped 4 days in Madrid, 2 days at the summits of the mountain passes – I was kindly received in every Town where I was known.[22]

Though his recovery shows the strength of Stephenson's constitution, after 1845 he spent much more time at Tapton House. But he was still involved in railway affairs, which meant, in part, Hudson's affairs. In October 1845, while George was in the Spanish mountains, Hudson bought the Durham & Sunderland

Railway for twice its (low) current share value and also arranged for the Pontop & South Shields Railway to be worked by the Newcastle & Darlington. At that time Robert Stephenson was also selling off the Pontop mine leases. Hudson had already bought the Brandling Junction Railway in October 1844 and transferred it to the Newcastle & Darlington company, and these new purchases were also on its behalf so that it now controlled all the permanent way between Darlington and Gateshead. Outright purchase of the Pontop line was only a matter of time, and George Stephenson presumably shared his son's sense of relief that the saga was coming to a happy end. But, that autumn, something happened which made a break between the elder Stephenson and the Railway King. In the same letter to Longridge as quoted from above, George wrote:

> Hudson has become too great a man for me now. I am not at all satisfied at the way the Newcastle and Berwick line has been carried on, and I do not intend to take any more active part in it – I have made Hudson a rich man but he will very soon care for nobody, except he can make money by them. I make these observations in confidence to you.

In fact, while lying ill in London he had made identical comments to Joshua Walmsley, predicting 'his speedy downfall. No-one appreciated more than he did the pluck, the strong will, and business capacity of Mr Hudson ... but that no eminence can long be held by unprincipled means, was a tenet too firmly held by the "old man" to allow him to be swayed or blinded by the Railway King's position'.[23] What had happened? Construction of the Newcastle & Berwick line was going ahead with Robert as engineer-in-chief and control delegated to Thomas Harrison, who had also been resident engineer on the Newcastle & Darlington line. George's worries were not about that, but about Hudson's plans for amalgamating the railways between Newcastle and Berwick into a single York, Newcastle & Berwick Railway company. As a broad strategy this made perfectly good sense and would have appealed to George's desire for a unified system, but Hudson's motive was also financial – he saw an opportunity for making a lot of money out of the amalgamation. Always happy to cut his friends into the profits, it would seem that Hudson revealed more of his intentions to Stephenson than was wise. George had no objection to anyone's self-enrichment through railway management or railway investment, so long as it was a natural consequence of the railway's successful operation. When it was a matter of artificially manipulating a share price, he balked.

After ten years' running with Hudson like a pair of engines double-heading a train, he found that the other was ignoring too many signals. A less high-minded reason may also have played its part in his change of attitude, however. At various times since the mid-1830s there had been suggestions for some kind of national testimonial to George Stephenson. In August 1844 John Moss had proposed to the Liverpool & Manchester board that they sponsor a marble statue of George, and in September the board ordered the printing of a prospectus for a testimonial to be funded by railway companies rather than private individuals, committing itself to £500.[24] Now, at shareholders' meetings in June and July 1845, Hudson expanded this into the idea of a grand testimonial to George Stephenson, with a public subscription to raise funds for the presentation of silver plate and the erection of a statue on the High Level bridge to be built in Newcastle. While the proposal was well received, a group of Hudson's courtiers also proposed a testimonial to the Railway King himself – indeed it was rumoured by his enemies that it was done at Hudson's prompting, with himself drafting the text of the appeal and supplying a list of those who should cough up most handsomely,[25] including George Stephenson.

While George's languished, Hudson's testimonial was pushed ahead. Hudson was still the idol of his shareholders, however, and though the disenchanted Stephenson expressed his views in private – including, one supposes, to Robert – he made no public comment. To do so, of course, would have been risky in a number of ways, among which the most obnoxious to George would have been a general collapse of public confidence in the Hudson companies, and a possible stoppage of construction of the Berwick line and the High Level bridge at Newcastle. Even if Hudson was no longer potentially useful, he could not be repudiated, though there was a certain dryness in George's remark, in a speech at the opening of the York & Scarborough Railway in July, that at the time of their first meeting, 'Mr Hudson was not to be led into a rapid movement with respect to railway speculations; he then looked very carefully at these undertakings ...'[26] Anyway, George still liked the other's bravura and show, and when Hudson, elected in 1846 for a third term as Lord Mayor of York, gave a banquet of Babylonian opulence at the city's guildhall on 17 December, 'a high festival in honour of the railway system' according to the *Railway Times*,[27] a prime figure among the aristocrats, prelates, civic dignitaries and high officials present was George Stephenson.

Robert Stephenson remained very much in Hudson's counsels, but he kept his independence, helped, as before, by his involvement across a wide range of railways. He did not entirely seal himself off from Hudson's insistent generosity. When the Railway King bought up the Brandling Junction Railway in the autumn of 1844, he made the purchase in his own name before selling it on to the Newcastle & Darlington Junction Railway. To do this, he arranged for the company to raise £550,000 of new capital, in £25 shares. Twenty thousand shares were allocated to shareholders in the N & DJ company in proportion to their existing holdings. On the stock market they instantly appreciated in value by £20, offering an immediate 80 per cent profit. The happy shareholders voted for 2,000 of the shares to be given to the directors and Hudson received 1,600 of them, while Robert Stephenson (not a director) accepted a gift of twenty-five.

At this time, Smiles records that Robert's offices in London were 'crowded with persons of various conditions seeking interviews, presenting very much the appearance of the levee of a minister of state. The burly figure of Mr Hudson, the "Railway King", surrounded by an admiring group of followers, was often to be seen there ...'[28] Robert Stephenson's prestige and diplomatic skills were of particular value to Hudson in his struggle to prevent a rival railway from London to York. In 1845 Stephenson's diplomacy detached the cross-country Manchester & Leeds Railway from an alliance with the London & York company, making a deal with Hudson's group instead. In the autumn of 1845 Robert also went abroad to visit Italy,[29] presumably to inspect work on the Leopold Railway linking Florence, Pisa and Leghorn, on which he was engineer-in-chief with two resident engineers.

The Italian project, following the wide valley of the Arno, presented no special problems, but the Chester & Holyhead Railway met a variety of difficulties as it cut through or skirted the headlands of the North Wales coast. Ledges had to be cut in steep slopes plunging to the sea, and just beyond Penmaenmawr tunnel a seawall was partly washed away in October 1846, replaced by an elevated section right on the shoreline, where the waves at high tide might come right under the piers. Apart from the sea below, rock-slips from above were also a danger, and avalanche walls were needed. Generally, though, tasks that had seemed daunting in the building of the London & Birmingham line had become virtually routine for railway builders by now.

Robert's attention was focused very much on bridges at this time, since there was an outstanding double challenge on the Holyhead line – a new solution was needed for the bridging of the Conwy estuary and the Menai Strait. For his turnpike road to Holyhead, Thomas Telford had designed two fine suspension bridges; the lofty Menai span, opened in 1826, in particular being considered as a wonder of the modern world. Suspension bridges of that era were not suitable for locomotive-hauled trains, however. Only one had been tried: over the River Tees on the Middlesbrough extension of the Stockton & Darlington Railway. Designed and constructed by Samuel Brown (later Sir Samuel), the pioneer of chain cables, ever since the opening of the line in 1830 it had had to be shored up to prevent excessive undulation, and in 1842 Robert Stephenson had been called on to build a replacement. When George Stephenson surveyed the Holyhead line, he proposed, as Vignoles had done before him, that the railway track should occupy half of the Menai suspension bridge's deck, and that trains should be drawn across by horses in sections, with a fresh locomotive to take over on the far side.

Though this plan formed part of the railway's original Act in 1844, it was scuppered by the owners of the bridge, the Commissioners of Woods and Forests, who refused to allow half of it to be monopolised by the railway. The railway company had to resign itself to building its own bridge and Robert Stephenson was instructed to draw up plans for submission to Parliament in a supplementary

Bill. Clearly, this viaduct would not cross the Strait in a single span, and Robert identified a suitable site to the west of Telford's bridge, where the Britannia Rock, conveniently located in the middle of the seaway and exposed at low tide, provided a solid foundation for a pier. His first plan was for a structure of two great cast-iron arches, but the Admiralty, which had to approve the design, got three engineers including John Rennie and J.M. Rendel, who was building the port at Holyhead, to inspect the site and the drawings. Following this, the plan was rejected in March 1845; the Admiralty insisted upon a completely clear passage over the Strait, with not less than 100ft (33m) headroom throughout. This put an arched bridge out of the question.

Anticipating the verdict, Stephenson was already considering an alternative design. Somehow, he had to devise a means of providing a self-supporting level deck across two main spans of 460ft (140m), at a height of 100ft (33m) above the channel. Nothing like this had ever been attempted before. Only a few weeks after the Admiralty's 'no', a Bill for a bridge of a totally new type, in which the rails would be carried within huge wrought-iron tubes, was presented to Parliament in May 1845. A key witness at the hearing was Major-General Pasley, the Inspector-General of Railways. He had already privately approved the general design, though he insisted that suspension chains must be a feature, and on this basis the Act for the new bridge was passed on 30 June. A similar bridge, at a much lower level, was to span the Conwy estuary.

Work on the easier stages was already under way. In a little over a year, the first section of the line out of Chester would be completed, but no one could tell when the two main water-gaps would be bridged, or even if Stephenson's scheme was feasible. He had first conceived a horizontal lattice structure, with cross-braces between each side, and additionally strengthened by lateral chains suspended from towers at the middle and each end, forming what he described as 'a roadway surrounded on all sides by strongly trussed framework'.[30] For the building material, wood was considered and rejected because of its perishability and fire risk. Cast iron, though immensely strong, was too brittle for the great lengths needed. So Stephenson came to consider wrought iron. While cast iron was formed in moulds to make rods and girders, wrought iron required a longer smelting process, which reduced its carbon content and minimised flaws and impurities. It was then hammered or rolled into thin sheets or plates. Compared to cast iron it had much greater tensile strength, but it had never been used in anything other than small structures – with one important exception. The hulls of iron ships were now being built using wrought-iron plates. But its properties and behaviour in the application Stephenson was considering were quite unknown.

Robert Stephenson's situation was not a comfortable one for a man liable to anxiety and self-doubt. For the next five years, referred to by himself as 'years of increasing and intense anxiety',[31] through many other projects and personal crises, the Menai Bridge remained a permanent presence in his mind, and, just as its rising pillars might be shrouded in sea-mists, so his project was wreathed in clouds of worry, arguments, calculations and controversy. These years drew heavily on his resilience and resourcefulness, and his underlying will and determination, quite apart from his managerial skills and his gifts as an engineer. There was a hint of resemblance to the position in 1829, when, with a rapidly approaching dead-line, boldness and initiative had triumphed. But that had been a short period, and his father had been very much part of the project and its head; this was a lengthy process, and he was the chief, the man with ultimate responsibility. George might offer encouragement, but twenty years of technical progress had elapsed since the Gaunless had been bridged and the elder Stephenson would have been able to give only general advice, in which the concept of 'persevere' doubtless figured.

What Robert needed was expertise in wrought iron, and he described how chance put him in touch with a man who had more practical experience with this material and its application than anyone else in the world.

> My late revered father, having always taken a deep interest in the various pro-posals which had been considered for carrying a railway across the Menai Strait, requested me to fully explain to him the views which had led me to suggest the use of a tube, and also the nature of the calculation I had made in reference to it. It was during this personal conference that Mr William Fairbairn accidentally called upon me, to whom I also explained the circumstances of the structure I had proposed. He at once acquiesced in their truth, and expressed confidence in the feasibility of the project, giving me at the same time some facts relative to the remarkable strength of iron steam ships, and invited me to his works at Millwall ...[32]

This was none other than his father's friend from the old days at Willington Quay, who had become a very successful iron founder and shipbuilder, with establishments at Manchester and at Blackwall on the Thames. Robert does not give any reason for Fairbairn's apparently fortuitous call at that conceptual moment. Fairbairn wrote: 'I was consulted by him.' There were no drawings or calculations: 'The matter was placed unreservedly in my hands.'[33] In many ways Fairbairn resembled George Stephenson – he had acquired his knowledge, wealth and position in the world by dint of hard work, intelligence and a strong entrepreneurial strain. Strongly believing in practical experience and getting on with the job in hand, he too was a vigorous and even combative personality. Aged 56 in 1845, he was much more likely to have dandled the infant Bobby on his lap than Richard Trevithick, and was delighted to renew old friend-ship and to become involved with such a grandiose scheme, which also held out the promise of a far wider future for wrought iron. In turn, he obtained Stephenson's consent to draw in Eaton Hodgkinson, FRS, at that time director of the Manchester Mechanics' Institute and the country's leading expert in the

behaviour of iron under stress. A series of experiments began at Millwall and
in Manchester to test the strength of wrought-iron plates under compression
and tension. At this time it may be remembered that the Gauge Commission
was about to sit, the Newcastle & Berwick Bill was in the final stages and the
railway mania was in full furore, with Robert Stephenson and George Bidder
hurrying from committee room to committee room at Westminster to give
evidence for promoters or opposers of projected lines, as well as attending to
contracts in Europe.

On the Newcastle & Berwick Railway, the long viaduct over the Tweed at
Berwick was planned as a conventional stone structure of twenty-eight arches.
Nineteen abortive designs, at various sites, for crossing the Tyne at Newcastle by
high-level or low-level bridges, and building a joint station, had already come
and gone,[34] with the first serious proposal for a high-level bridge coming from
the luckless Thomas Storey in 1836.[35] Robert's location was that also chosen by
his father, close by the twelfth-century keep, which gave the city its name, and
cutting through the Castle Garth (George was no respecter of antiquities when
they were in the way of railway lines). A two-deck bridge had first been proposed
by John Dobson, who had produced his own design in 1843, and the concept
was strongly backed by Hudson, who saw road tolls as a useful contribution to
funds. The High Level bridge at Newcastle was also of a tried design, of cast-iron
bowstring girders combined with horizontal wrought-iron members, supported
on five tall sandstone piers. Three rail tracks were to be on top, with a roadway
suspended beneath flanked by two footpaths.

Visits to the Tyne bridge site would also have made it possible for Robert
Stephenson to visit the Forth Street Works, but in the late 1840s he does not
seem to have spent much time on locomotive design. A change was made in
the owning partnership in the summer of 1845 when William Hutchinson,
who had started working for George Stephenson in 1821 and had long been
works manager, asked for 'a small share in the works' and was admitted a partner.
The balance sheet at the end of the year showed the partners as Robert and
George Stephenson, Thomas Richardson, Joseph Pease and William Hutchinson,
each with a capital holding of £6,600. This was also the year in which Edward
Bury's decade-long contract with the London & Birmingham Railway expired.
His fleet of lightweight engines had long been struggling to maintain a serv-
ice, and from early in the year the company had been in touch with Robert
Stephenson & Co. about new locomotives. At this time, locomotive builders, like
every other department of railway activity, were overwhelmed with orders, and
Starbuck had trouble finding spare capacity anywhere; in April he could only
offer six 'Patentees' to be delivered by a Manchester builder between March and
July 1846.[36] On 30 October 1845 Richard Creed, the L & B Company Secretary,
wrote to Robert Stephenson:

I am desired to say that our Company are prepared to deal with you for a supply of Engines to an extent that would probably make it worth your while to devote your establishment to the execution of our orders exclusively, and with a view to more prompt delivery of them than might under other circumstances be thought convenient ...[37]

This neat attempt to jump the queue did not find favour with the partners. Prices and profits were going up in line with demand. An extension factory was leased at Forth Bank West, and by May 1846 the company had 224 locomotives on order and a workforce of 850.[38] Early in 1846 Edward Cook died, and Robert Stephenson no longer had the personal and confidential link with Newcastle affairs that 'Uncle Edward' had provided. The factory was now run by a triumvirate of Hutchinson, W.H. Budden, the new Secretary, and William Weallens, head of the drawing office. George Stephenson had not given up his own involvement with the Newcastle works. In 1846 he took out a patent, along with William Howe, the most talented of the skilled workers at Forth Street,[39] whom he had lifted to become engineer at the Clay Cross collieries, for a three-cylinder locomotive. The pioneering three-cylinder arrangement was made, not in the interest of additional power, but with the intention of achieving greater stability at speed. Very soon after the gauge tests, it shows that the Stephensons were sensitive to the 'yawing' problem and were working to find a solution. This was not it, however: only two examples were built,[40] though one was recorded as having run 41 miles in 40 minutes on the London & Birmingham main line on 7 May 1847, with a maximum speed of 64mph (102.4km/h). With George's final locomotive design, perhaps a comment from Jack Simmons, doyen of twentieth-century transport historians, should be noted:

There is ... something strange about the way he seemed to give up railways around 1845. Something appears to have snapped in him. It was the height of the mania, yet he kept out of it. Something made him withdraw and I think whatever it was happened around 1842–3. Perhaps he was worried by Hudson's success, jealous of Robert even, or perhaps he just physically suddenly aged and decided to retire? I've never seen anyone suggest an explanation.[41]

The Hudson connection apart, perhaps there is less to explain than Simmons suggested. George had good reasons for avoiding the mania and the extent of his 'withdrawal' can be overstated. He was, moreover, in his mid-60s and greatly relishing the opportunities offered him by wealth and fame; he had no lack of things to do. In the heroic years of the 1820s there had been one great rope to pull – with that task accomplished there were now dozens of lesser strands of development being drawn out by experimenters in various countries. Locomotive design had become a matter of detail improvements around the basic 'Stephenson cycle' of fire turning water into steam, which operated the pistons and, in its exhaust blast, sucked more heat through the boiler tubes, and in doing so also made the fire burn hotter to make more steam. In the 1830s he had struggled in vain to control or moderate the rate and form of railway expansion, and took a few bruisings as a result. By now, he had attained a degree of detachment and public perception of him was gradually shifting towards the status of a grand old man and visionary figure, 'The Hengist of Railways' as the *Railway Times* called him.[42]

The extent of his own railway commitments did not deter Robert Stephenson from taking on other responsibilities. Within the Society of Arts, prompted by Henry Cole, one of its leading figures, a plan was being hatched for a grand exhibition in London which would celebrate the extraordinary technical progress of the previous decades. Prince Albert, consort of Queen Victoria, was an enthusiastic supporter. A special committee was formed in the summer of 1845 to drive the proposal along, and Robert Stephenson not only joined the committee, but offered a loan of £1,000 to support its activities. In December 1845 he became one of the first shareholders in the Electric Telegraph Company, the first commercial exploitation of telegraphy. Through George Bidder's enthusiasm, Robert had taken an interest in telegraphy since early experiments with the London & Birmingham Railway, and by 1844 the system was greatly improved. The Yarmouth & Norwich Railway used the electric telegraph system for train control on its single track line. The Electric Telegraph Company was set up by Bidder as a sideline to an already hectic professional life, and he continued to work for the Stephenson practice, though by now as an almost equal partner. The 'Electric', as they called it, grew quickly in size and profitability, initially on railway contracts, and Stephenson was a director and, in 1858, briefly chairman.

For the more enthusiastic or gullible investors of the mania, 1846 was the time to lick burned fingers and contemplate shattered hopes. The railways whose Bills had been passed were now calling on their subscribers to pay up. By the end of that year the three courts of common law had issued 24,000 more writs against defaulting subscribers than in the previous year.[43] Many found it impossible to raise the money and there was a spate of bankruptcies. Others found that they had paid a premium price for shares in hyped-up railway projects whose promoters had mysteriously vanished with the money. Overall, a substantial amount of the nation's wealth had changed hands and there were thousands of *nouveaux pauvres*, with many cases of hardship and even suicide, but the economy withstood the shock.

Railway shares, so recently considered as the key to instant wealth, were viewed with deep suspicion, especially those of new companies. Existing railway companies maintained their trade, paid their dividends and saw their share prices reach a reasonable level. New companies often had a tremendous struggle to claim their subscription capital, but the result of the mania was a substantial increase of railway mileage, which expanded the foundations for further exploitation of coal and iron reserves, and for the continuing growth of the manufacturing industry.

George Hudson, who had been elected as Conservative MP for Sunderland in
August 1845 when Howick went to the Lords, and whose railways, when all was
said and done, were real railways doing real business – even if not at a 10 per cent
dividend level – emerged from the smoke and mirrors of the mania period still
doing his tightrope walk, spinning ever more plates as he went. Shrewder investors – not personal or political enemies – were becoming uneasy. The novelist
Charlotte Brontë wrote to her friend Miss Wooler on 30 January: 'The York and
North Midland, is, as you say, a very good line, yet I confess to you I should wish
for my part to be wise in time. I cannot think that even the best lines will continue for many years at their present premiums and I have been most anxious for
us to sell our shares ere it be too late and to secure the proceeds in some safer, if
for the present less profitable, investment.'[44] Charlotte did not join with Ann and
Emily, who each sent £1 towards Hudson's testimonial; she also did not sell her
shares in time to avoid a loss.

Bad news for Hudson in 1846 was the merger of the London & Birmingham
and Grand Junction Railways, with others, into the London & North Western
Railway, controlling a railway route from Euston Square to Carlisle and in friendly
alliance with the new Caledonian Railway north of the border. It was not bad
news for Robert Stephenson, who retained his position as consulting engineer to
the mighty newcomer's southern division, the former London & Birmingham;
however, as rivalry for the route to Scotland sharpened up, his presence in both
camps might have become embarrassing. It was a test of his diplomatic skills, but
these were well-honed, and Robert was, by nature, a negotiator, a pacifier and a
facilitator. Helped by his own prestige and the magic of his name, he was a dis-
creet and effective middleman. In addition, the Euston party and George Hudson
were at one in their detestation of the 'Johnny-come-lately' Great Northern
Railway. In September 1844 Joseph Locke had been forced by pressure from the
Grand Junction Railway, on which he was consulting engineer, to abandon his
agreement to become engineer-in-chief on the Great Northern.

# 15

# INVITATION
# TO EGYPT

Robert was, of course, engineer-in-chief on trunk lines for the West Coast party and for Hudson, and work was going ahead well on both. On 9 February 1846 he reported to the Chester & Holyhead board on progress with the tubular bridge design, observing that 'I saw the importance of avoiding the influence of any preconceived views of my own ... For this purpose I have availed myself of the assistance of Mr Fairbairn and Mr Hodgkinson';[1] and Fairbairn and Hodgkinson, already wrangling about their contributions to the work, made their own separate reports. Oval or round tubes, suggested as a possibility by Stephenson, had been rejected in favour of a rectangular shape, but beyond the little group of planners there was a good deal of scepticism about the overall design. 'Many doubted the efficiency of riveting to unite such a mass of plates; some foretold the most fatal oscillation and vibration from passing trains, sufficient even to destroy the sides of the structure; while others asserted the insufficiency of the lateral strength to resist the wind.'[2]

Arrangements were made for construction of a one-sixth scale model tube section in wrought iron, at Fairbairn's yard in Millwall, between April and July. As tests and calculations went on, more friction arose between Stephenson's duo of experts, with the ultra-cautious Hodgkinson expressing reservations about the pushful and sanguine Fairbairn's methods and results. Believing that the work should establish the standards for all future wrought-iron girder construction and application, as he had already done for cast iron, Hodgkinson would have liked to have been in control of the tests, but Fairbairn, seeing him merely as an adviser to the process, was determined to do things in his own way. Meanwhile, the directors were anxious to see signs of progress on the site, and by mid-March the designs for the masonry work were completed by Francis Thompson, who had first been employed by Robert for the buildings of the North Midland Railway in Derby in 1839, and whom he retained as architect to the Chester & Holyhead Railway.

Stephenson would certainly have thoroughly discussed, and ultimately approved, the plans for the 'Egyptian pylon' towers. Thompson also designed the castellated stonework of the Conwy bridge. When the first stone of the Menai Bridge was laid on 10 April 1846, it still had not been decided whether chains would be used or not, either to help in installing the tubes, or as permanent fixtures. This explains the height of the towers, to allow for them just in case, though the aesthetic factor cannot be ignored either. Fairbairn, in his February paper, had said chains were unnecessary; Hodgkinson disagreed; Stephenson had not formed a definite view. Fairbairn's status in the project was underlined by a board minute of 13 May 1846, which noted that he was 'to superintend the construction and erection of the Conway and Britannia Bridges in conjunction with Mr Stephenson'.[3] Design is not mentioned. By summer, both experts were pursuing their own tests with little or no mutual communication, a frustrating situation for Robert. He was relying heavily on his resident engineer on the Menai site, Edwin Clark, a mathematician and surveyor, but with no engineering experience,[4] whom he had met in 1845. Clark, then aged 31, was clever and cheerful, acting as Stephenson's 'eyes and ears',[5] rather than the executor of his designs, not least helping to sustain his chief's confidence in the ultimate resolution of the immense difficulties they faced.

For the installation of the iron tubes, each of the four main ones weighing 1,550 tons, Robert Stephenson had considered using the suspension chains, erecting the tubes section-by-section along them. As the iron plates had been ordered in July 1846, and contracts had to be negotiated for erection, a decision on this procedure could not be indefinitely delayed. William Fairbairn believed himself to be the originator of the method of erecting the tubes on land, floating them into place between the piers and hoisting them by the use of hydraulic jacks. Edwin Clark also claimed credit for proposing this hoisting method.[6] In a letter to Stephenson on 15 July 1846, Fairbairn proposed floating and jacking the first of the Conwy tubes. This would make the suspension chains quite redundant, since, as Fairbairn wrote: 'I have made the tube of such strength, and intend putting it together upon such a principle, as will ensure its carrying a dead weight, equally distributed over its bottom surface, of 4000 tons. With a bridge of such powers, *what have we to fear?*'[7] Stephenson replied on the 20th that he and Alexander Ross had already discussed the idea of floating the tubes and rejected it for both bridges because of the strong tidal currents. In the same letter, he wrote:

> I should be delighted to get away, but there are two or three Boards of Directors threaten me with actions for damages, if I absent myself from London now that they have got their Committees in the Lords appointed … as to getting down to Manchester this week, I am afraid it is wholly impracticable … I dare not leave in spite of the Doctor, who is insisting on my leaving Town for some days, in order to get some *quietness.*[8]

His dependence on Fairbairn, and Clark, was evident, and even though he considered the employment of dependable lieutenants to execute his ideas as an essential part of the engineer-in-chief's role, his absences encouraged Fairbairn to consider himself as the real mastermind of the project. By October, Stephenson had agreed to floating the Conwy tubes and recommended this to the board, which gave its approval. The decision was encouraged by the willingness of the Conwy contractor, William Evans, to take responsibility for managing the whole exercise of assembling the two tubes, one for each railway track, at the bridge site, and raising them by jacks. The height of the deck above high water was 17ft, much less than the 105ft of the Menai Bridge.

Despite Fairbairn's confidence and all the preparatory activity, it was not clear beyond dispute whether the tubes would actually do the job, and even though 15,000 tons of wrought-iron plates had been ordered in July 1846, his Millwall yard was still testing the large-scale model. Edwin Clark was still coping with the tricky business of liaison with the expert consultants and the collation of Fairbairn's results of destruction-testing on the model tube with Hodgkinson's separate test results and figures. Both claimed credit for the final design of the top section of the tubes, a layer of cells which provided the necessary resistance to buckling. Clark treated his own efforts to reconcile the prima donnas with good humour, writing to Stephenson on 18 October 1846: 'I have seen Fairbairn and Hodgkinson and commenced my mediative commission with little success so far. They continue to hate each other most enthusiastically.'[9]

Robert Stephenson crossed to Norway from Leith with George Bidder on a chartered steam vessel late in August 1846; Jeaffreson describes it as 'a trip of pleasure' and Stephenson clearly needed a rest, but Bidder's letters leave no doubt that, though they thoroughly enjoyed the outing, railway business was behind it.[10] The North Sea crossing was bracing: 'On Friday morning we commenced our Morning Baths in a huge Tub which being filled with Salt water we plunge into it as soon as we wake and then one of the men pours several buckets of Salt water on our heads, then we return to the Cabin and complete our *toilets*.'[11] Of the three-day voyage Bidder wrote to his wife: '... constant exposure to Sea Air and complete relief from business has produced the utmost effect both on Mr Stn and myself.'[12]

John Crowe, the British consul, who had called on Stephenson in London earlier that year,[13] met them at Christiania (now Oslo) and introduced them to the viceroy (Norway being subject to the kingdom of Sweden at that time), and everyone was pleased to see them, 'for they appear to hail our visit as the advent of Railways on which they rely for great improvements in every way'.[14] Stephenson and Bidder were taken along the line of a proposed railway from Christiania to Lake Mjøsa, the beginning of the Norwegian Trunk Railway, and promised to produce estimates and plans, which they duly did, though actual building operations would not begin until early 1851. In early September they sailed back to Newcastle.

Through August and September, George Stephenson was off on a long European tour with Joseph Paxton and two other Derbyshire friends, passing through Belgium on Stephenson-built railways, and making their gradual way through Germany to Munich, where they were presented to King Maximilian of Bavaria. Despite George's earlier encounters with Leopold of Belgium, Paxton noted: 'Mr Stephenson was in a pretty stew, as he said these great folks did not quite fit in with his previous habits.'[15] But the king expressed himself as 'very happy to know such distinguished men'.[16] The party travelled as far as Linz in Austria before returning to England; at one point Stephenson and Paxton enjoyed a discussion on electricity while caught in a thunderstorm. It was George's last journey abroad.

Soon after his return from Norway, Robert became involved with a project that was to demand his occasional attention and involve him in increasing controversy over the next thirteen years. It also introduced him to a country which had interested him since his boyhood visit to London and the exhibition of the Egyptian tomb. The scheme was for a ship canal across the Isthmus of Suez between the Mediterranean and Red Seas. Egypt was a dependency of the sprawling Ottoman Empire, ruled by a viceroy on behalf of the Sultan in Constantinople with a substantial though imprecise degree of self-government. To France, which controlled most of the North African coast, the notion of detaching Egypt from Turkish suzerainty was seductive. Ever since Napoleon Bonaparte's military occupation in 1798–1801, with its teams of cultural and scientific investigators, Egypt had been a preserve of French scholars and scientists, and sporadic, inconclusive initiatives for a canal had been going on for forty years.

In 1830 a railway was proposed and, in expansive mood, George Stephenson had anticipated being its engineer, writing to Longridge: '… you and I are to go and set the Railway out from the meditirranean to the Red Sea. & we shall send the young men to execute it …'[17] A railway had been started and abandoned in 1835, without the Stephensons being involved. In France the *Société d'Etudes du Canal de Suez* was formed in October 1846, with the aim of establishing some basic facts relating to a canal link, before forming a company to take on construction. At the time it was thought that there could be a 34ft (10m) difference in sea level between the Mediterranean and the Red Seas, and an accurate survey was essential. French engineers of the *École Polytechnique*, the *Saint-Simoniens*, were the moving spirits, but they were fully aware of the international implications of the scheme, and it was through his French friend, Paulin Talabot, that Robert was drawn in. An Austrian, Luigi Negrelli, Imperial Inspector of Public Works, was also invited to participate. On 30 November it was agreed that Talabot, Negrelli and Stephenson would each send a team of surveyors to the isthmus: the British to examine the Red Sea coast, the French to trace an inland route and the Austrians to survey the Mediterranean coast.[18]

Stephenson's participation was on a personal basis. He would, of course, have known that the isthmus was a hot-spot in every way. Five empires, Britain, France,

Austria, Russia and Turkey, converged here in terms of power, influence and strategy. Britain had a vital strategic interest in the isthmus, which formed the only land break on the 'short' sea route to India, avoiding the Cape of Good Hope. Cutting across diplomatic terrain that was much shakier than the isthmus itself, the idea of a canal had always been resisted by the British government. If Robert did not know that already, he would not remain in ignorance for much longer.

Closer to home, his experience of Ireland was relatively limited. The Londonderry & Enniskillen affair had left a bad taste and he and Alexander Ross had both withdrawn from the project. Now he found himself drawn into discussions on the catastrophic events following the failure of the Irish potato crop in 1845, and the recurrence of the same devastating blight in 1846. Of Ireland's population of 8 million, more than a quarter were facing, not merely destitution, but actual starvation. Among the British politicians who took an active interest was Lord George Bentinck, hitherto something of a playboy. Like many members of Parliament in that decade, his support for particular causes cut across ill-defined party lines, and his backing of Protectionism against free trade made him an ally of George Hudson's and brought him into contact with Robert Stephenson, also an ardent Protectionist. Bentinck's solution was to provide work for the poor so that they could buy food, and the work he had in mind was railway construction. Numerous railways in Ireland had been authorised by Parliament, but the country's economic plight had dried up funds and brought construction to a halt. He consulted Hudson and Stephenson, and at Stephenson's suggestion, George Bidder's younger brother Samuel visited Ireland in 1846 to examine the state of railway construction and road building. Bidder reported that no railways were under construction and that road gangs were doing more harm than good to the roads. In his view it would be better to feed the people and leave the roads alone.

In October 1846 the first section of the Chester & Holyhead Railway was opened, on easy terrain, crossing the Dee and following the south bank of its estuary to Flint, the only notable feature being the iron bridge over the Dee. Concern about the stability of the foundations had led to the replacement of an arched stone design by iron girders. Large numbers of similar bridges had been designed and built by the Stephenson practice, though this was the longest of its type. By this time, the design of the tubes for the wrought-iron bridges had been largely finalised, though at Hodgkinson's behest changes continued to be made until February 1847. He and Fairbairn were still at odds with each other, but Stephenson kept on good terms with both. On 18 October 1846 Fairbairn took out a patent for a hollow-beam wrought-iron girder in various forms. According to his later account, it was 'a joint affair between Mr Robert Stephenson and myself. It was in my name as the inventor, but he paid half the expense, and was entitled to one ½ the profits'.[19] George Stephenson, too, was drawn in. Evidently Fairbairn had been remembering their old wrestling bouts, and on 5 January 1847 George wrote a playful letter to him from Tapton House:

My dear Sir,

It will give me great pleasure to accept your kind invitation to Manchester when you return from Ireland ... Now for the challenge to wrestle. Had you not known that I had given up that species of sport, you durst not have made the expressions in the letter you have done. Altho' you are a much taller and stronger-looking man than myself, I am quite sure I could have smiled in your face while you were lying on your back! I know your wife would not like to see this, therefore let me have no more boasting, or you might get the worst of it. Notwithstanding your challenge, I remain yours faithfully,

Geo. Stephenson[20]

The elder Stephenson was still in cheerful mood when he wrote to J.T.W. Bell, a Newcastle surveyor who wished to make some form of dedication to him and to know what letters followed his name, on 27 February:

I have no flourishes to my name, either before it or after, and I think it will be as well if you merely say Geo. Stephenson. It is true that I am a Belgian Knight, but I do not wish to have any use made of it.

I have had the honour of Knighthood in my own country made to me several times, but would not have it; I have been invited to become a Fellow of the Royal Society; and also of the Civil Engineers society, but I objected to these empty additions to my name. I have however now consented to become President to I believe a highly respectable mechanics' Institute at Birmingham.[21]

The last sentence refers to what is now the Institution of Mechanical Engineers. On 7 October 1846 a group of engineers, mostly though not all connected with railways, had convened at the Queen's Hotel in Birmingham to form a society which would 'enable Mechanics and Engineers engaged in the different Manufactories, Railways and other Establishments in the Kingdom, to meet and correspond, and by mutual interchange of the ideas respecting improvements in the various branches of Mechanical Science to increase their knowledge, and give an impulse to inventions likely to be useful to the world'. Although the Institution of Civil Engineers included within its ambit all aspects of the rapidly expanding activities of non-military engineers, the new body had several reasons for an independent existence. It was less exclusive in its admission policy; it was based (until 1893) in Birmingham and so more accessible to the new industrial districts, and it acknowledged that mechanical engineering was in itself now a wide and diversifying field of knowledge and innovation.

Its first chairman, and honorary secretary, was J.E. McConnell, locomotive superintendent on the southern division of the London & North Western Railway. George Stephenson attended its first meeting and was elected president. Both its utilitarian emphasis and provincial location would have appealed to him.

Robert was one of the seventy founder members;[22] at this time he was also on the Council of the Institution of Civil Engineers and would be elected a vice-president in the following year. In April 1847 he was elected a vice-president of the Society for Encouragement of the Arts (it received its royal charter that year as the RSA), and was also elected president of the 'Smeatonians'. His father still held aloof from metropolitan institutions, and his reference to 'the Civil Engineers society' has often been pondered over. No evidence exists that he was ever formally invited (or applied) to become a member of the ICE, though it is very likely that some members discussed the possibility with him in conversation. By now his stature was such that, as he rightly said, almost any honour would be an empty addition for the 'Father of Railways'.

Robert Stephenson's stint as a railway chairman came to an end on 1 January 1847, when the Pontop & South Shields Railway was formally taken over by the Newcastle & Darlington Junction company. Transfer of this unwanted responsibility to his friend Hudson's broad shoulders must have been a relief. During that spring he was also considering the Egyptian survey expedition agreed with Talabot and Negrelli. The Austrian was already in Egypt with a team examining the Mediterranean coastline. Either Stephenson planned to have Bidder with him, or to delegate the responsibility; a letter to Bidder from his friend Sir Henry Jardine on 2 April remarks: 'I envy you the prospect of your going to Egypt. It would be a grand affair to make a junction between the Mediterranean and the Red Sea.'[23] In the meantime, Britain's official attitude to a canal had been firmly reiterated. On 8 February Lord Palmerston, Foreign Secretary, had sent an instruction to the Hon. Charles Murray, consul-general at Cairo, to '... lose no opportunity of enforcing on the Pasha and his ministers the costliness, if not the impracticability, of such a project ... the persons who press upon the Pasha such a chimerical scheme do so evidently for the purpose of diverting him from the railway which would be perfectly practicable and comparatively cheap'.[24] Given the strategic importance of the scheme, it would be most unlikely that Stephenson did not discuss it with British government officials.

In March he received a letter from Thomas Waghorn, who had established the coaching route across the isthmus in 1835–37, imploring him to detach himself from the canal project and promote a railway instead. Vehemently anti-canal and claiming that 'I speak and write from a greater practical knowledge of this Desert than any other European',[25] Waghorn also sent a copy of his letter to Lord Palmerston. Whether this, and official opinion, played any part in Stephenson's decision is impossible to establish, but in the end he sent no survey team, and merely dispatched some Admiralty charts of the Suez coast to Talabot.[26]

Other matters arose to demand his attention. At this time he was making more frequent visits to the Conwy and Menai construction sites, and negotiations on contracts for assembly of the tubes were almost complete. On 24 May 1847 he crossed the Dee bridge just outside Chester on his way to Bangor. Later that day, a load of 18 tons of crushed stone was spread on its wooden decking to protect it from possible fire caused by dropped cinders. Apart from construction traffic for the Holyhead line, it was also used by the Shrewsbury & Chester Railway, a separate concern to the Chester & Holyhead, which used the short length between Chester and Saltney for its service to Wrexham and Shrewsbury. That same evening, as a train from Chester to Ruabon was crossing, one of the bridge's iron girders broke into three pieces. The locomotive and its tender gained the opposite bank, but the carriages were tipped into the river. Five passengers were killed and sixteen were injured, some seriously. For Robert Stephenson, always haunted by a sense of walking on eggshells, his waking nightmare had come true.

# 16

# AN ENGINEER
# AT BAY

The spread of railways had brought an enormous expansion of bridge-building. Between 1825 and 1856 more bridges were built in England than in its entire previous history. Many were constructed of wood in the initial years, to be replaced by more permanent materials later; stone and then brick were most common. But railway engineers, always looking for economy and speed of construction, were attracted to the idea of iron bridges from the start. Among their potential advantages were less time and manpower in building; they were easier to build on a skew site which crossed a gap at other than a right angle; and a bridge of iron beams could be of uniform height, while an arched over-bridge forced road- or water-borne traffic to compete for the centre of the way. Disadvantages included greater initial cost, particularly as the price of iron rose steeply in the mid-1840s; a shorter life-span than stone; and the need to be painted to keep them from rusting. Some engineers, notably Joseph Locke and Brunel, were reluctant to use cast-iron bridges,[1] but Robert Stephenson, like his father, had used them from the beginning of his career as a civil engineer.

The technology of the time could produce beams or girders only in cast iron. But they were of limited value because their lack of tensile strength made them unusable on anything other than a short span, enough only to cross a small stream or road. The form of the beam was developed by Eaton Hodgkinson, an L-shape with its lower flange much thicker and broader than the upper one. In order to bridge wider gaps, two systems were developed to relieve the tensile strain on cast iron by supplementing it with wrought iron. One was the bowstring girder – a bow-shaped strip of riveted cast-iron plates, anchored at each end to the bridge piers and from which vertical wrought-iron tie-rods were attached to a horizontal cast-iron beam. On such a bridge, the wrought iron took the tensile strain and the cast-iron beams took the compressive strain. Robert Stephenson was first to construct bowstring girder bridges, on the London & Birmingham Railway, over the Regent's Canal in London and the

Grand Junction Canal at Weedon, though credit for the design is attributed to his young engineer Charles Fox.

The other method has been ascribed to George Parker Bidder, who seems to have been first to use it in bridging the River Lea near Tottenham on the Northern & Eastern Railway in 1839, but Stephenson, as engineer-in-chief, is likely to have also been involved in what was a new technical departure, and he took full responsibility at the Chester inquest. Up to three lengths of cast-iron beam were bolted together to form a single long girder, and at each end and at the joins, curved flanges or 'shoulders' of cast iron were bolted on the upper edge to strengthen the joint. On such a three-piece girder, each end-piece had a downwards-sloping tie-rod of wrought iron bolted in position across its vertical surface, joined to a central rod across the lower face of the middle piece, thus forming a flat-bottomed valley shape. The tie-rods were intended to relieve the girder of tensile stress. This was the 'trussed compound girder', much less expensive to make than the bowstring girder. The Chester bridge was the longest yet to use it, with three trussed compound girder spans of 98ft (30m) length resting on the abutments and two piers set in the river. General Pasley had certified it as suitable for traffic in October 1846.

Railway accidents were not unusual, but their causes were normally human error, defective couplings or lack of brakes. For an iron bridge to fail was unprecedented. The fact that it was one of Robert Stephenson's bridges made the matter sensational, not only because of his great reputation, but because it was speedily realised that there were numerous other bridges of this type on Stephenson-built railways. Robert himself was quickly on the scene. Reporting a week later to the Chester & Holyhead directors, he said that he had closely examined the bridge only a few hours before the disaster when on his way to Bangor and had seen nothing amiss. After the accident, he 'could arrive at no other conclusion than that the fracture of the girder arose, not from inability to support the weight, but from a violent blow given by the tender, which he conceived to have got off the rails, probably from the fracture of one of the wheels, while crossing the bridge'.[2] He repeated this hypothesis as a witness at the coroner's inquest, which took place in Chester in early June.

Local feeling was strong and generally hostile to the railway company. The inquest was naturally of great interest to members of the engineering profession as well as to the survivors of the accident, the relatives of those killed and the wider public. Among those who squeezed in was Francis Conder, who ten years before had worked for Stephenson on the London & Birmingham Railway, and, still a keen observer, noted that Stephenson:

> … betrayed the most intense vexation and inquietude. It was evident that the sympathy of the man with the loss and pain caused by this occurrence was the emotion most present to his mind, but the disquiet, as is so often the case, betrayed itself by an unfortunate disturbance of the usual manner of the great Engineer. Pale and haggard, he looked more like a culprit than a man of

science ... His manner was abrupt and dictatorial, betraying extreme irritation at the remarks of the jurors; and on more than one occasion he attempted, on the score of professional knowledge, to put down with some contempt the questions of Sir E. Walker [the jury foreman], who, nettled in his turn, affected to treat Mr Stephenson as a culprit on his defence.[3]

The coroner, Mr Hostage, kept an even-handed control, and though Conder suggests Stephenson was unpleasantly aware that the jury might return a verdict of manslaughter, even murder, it is unlikely that this ever seemed a real possibility. Three fellow engineers, Locke, Vignoles and Thomas Gooch, joined Stephenson in refusing to believe that the girder had failed. Locke 'corroborated Mr Stephenson's testimony; he said the bridge did not fall from any pressure downwards, but from a side blow; this the examination of the fragments proved to him'; perhaps unfortunately he went on to say 'he considered all forms of iron bridges objectionable'.[4]

Sir Charles Pasley stated that he now viewed all trussed compound girder bridges as inherently unsafe, while Robert Stephenson 'flatly contradicted nearly all Sir Charles Pasley's statements',[5] and insisted that a derailment of the train must have caused the fracture. The only eyewitness of the accident, a boy of 15, had, however, seen no sign of anything untoward until the girder broke. The train had belonged to the Shrewsbury & Chester Railway, whose engineer, Henry Robertson, contested Stephenson's wheel breakage theory with scientific evidence of the breaking strain of cast iron, adding: 'The opinions of Mr Stephenson and Mr Locke, founded on the alleged facts as to the paint on the tender, the broken carriage wheel, and the snips in the [rail] chairs, appear to fall to the ground, as they must have been misinformed in these particulars, which can all be disproved.'[6]

Two independent engineers, James Walker and Captain Simmons of the Royal Engineers, were commissioned by the Board of Trade to make a rapid inquiry, and their report, published on 15 June, before the inquest was at an end, rejected Stephenson's explanation. Though they accepted that the bridge was of sufficient strength if the cast-iron and wrought-iron elements of the design acted together as required, they emphasised the great difficulty of ensuring that this would actually happen. Their report was well short of calling the design defective. This, and the coroner's exclusion of the possibility of negligence on Stephenson's part in his summing-up, helped to ensure that the jury returned a prompt verdict of accidental death, to the great relief of Robert Stephenson and his employers, although it continued: 'We are further unanimously of opinion, that the aforementioned girder did not break from any lateral blow of the engine, tender, carriage or van ... but from its being made of a strength insufficient to bear the pressure of quick trains passing over it.' They considered all such bridges unsafe, 'in proportion to the span; still, all unsafe',[7] and recommended a government inquiry into the safety of bridges of this type.

These days of early summer 1847 were fraught ones for Robert Stephenson. At a time when he was engaged in constructing the most ambitious iron bridge yet

to be built, his competence as a designer and builder of iron bridges was thrown
into the public arena to be chewed over, not only by a coroner's jury, but by
every saloon-bar expert in the land. Did he genuinely believe in the derailment
theory, or could he not admit, even to himself, that the girder design was fatally
flawed? Did he stick to it through the inquest because the Chester & Holyhead
solicitor, Timothy Tyrrell, insisted he do so, to avoid what would otherwise have
been a huge liability for damages consequent on the deaths and injuries? Rolt
claimed that he was torn 'between personal honesty and loyalty to the Company
he was serving. He had been prepared to make an honest admission but had been
vehemently dissuaded from doing so by the Solicitor to the Chester & Holyhead
Company …'[8] Rolt gives no source, but was probably improving on Conder's
account, which clearly implies that the railway company's whole intention was
to evade responsibility rather than to help establish the truth. Conder says nothing about Stephenson's personal motives. Stephenson was in a terrible double
bind. For him to have used a lethally faulty design was unthinkable – but perjury
was both dishonourable and a crime. The only possible solution was obstinate
belief, against the evidence, the independent inquiry and the jury, that somehow
a derailment had occurred. He never accepted that the trussed compound girder
was a flawed design. In the debate over his 'long-boiler' locomotives he had been
willing to admit that there was an unresolved problem, but here he was adamant.[9]
Too much was at stake – not only in terms of public confidence, but in terms
of Robert's own self-perception. Apart from the verdict itself, which blamed no
one, there were two things he could be grateful for in the Chester ordeal. One
was that no 'enemies' emerged, as they had troubled him in the days of building
the Leicester & Swannington – his professional peers might have stood aside or
criticised since, after all, he was much more identified with iron bridges than
anyone else, but they stood by him. The other was his employment of Fairbairn
and Hodgkinson – their months of testing, stressing and calculation left no doubts
about the strength and safety of the tubular bridges.

Fairbairn had written to Stephenson on 24 March to say: 'I have now completed, or nearly completed, the whole of the drawings for the framework, girders,
&c. for lifting the tubes. The arrangement of the hydraulic apparatus, chains, etc.,
is also complete; and as soon as we have copied the drawings, the whole shall be
laid before you.'[10] Late in May or early in June, however, Fairbairn had offered,
or threatened, to resign his position following a dispute with the board over the
terms on which he had transferred his own part-contract for tube fabrication to
another company, and Stephenson had to find time to placate him.[11]

General Pasley had hardly figured better than Robert Stephenson as a witness at the inquest, and the Board of Trade lost no time in asking all railways to
make a return of information on iron bridges on their lines and their means of
construction, following up with a general repudiation of responsibility for any
iron structures which might have been inspected and passed as fit for use by its

officers. In August a Royal Commission on the Application of Iron to Railway Structures was set up to investigate the whole question of iron bridge design and construction. The compound trussed girder would be abandoned. Stephenson and his team had already worked out urgent alterations to bridges of this type on the almost-ready Trent Valley line and on other railways. Hasty instructions for reinforcement went to the Leopold Railway in Tuscany, where three viaducts over the Arno were all on the Chester model.

While Robert Stephenson was in the toils of the bridge inquest, news came of the death of Eleanor Stephenson, 'Aunt Nelly', in June. By then she was a widow and had been ill for some time. A succinct letter from George to his nephew Robert (son of John) in Newcastle indicates the firmness of his hand as head of the family:

Dear Robert,
    I wish to give you instructions with respect to your Aunt Ellen.
    I wish Mrs Willis to see that your Aunt is properly looked after as I am informed by Mr Hardcastle she is very poorly. In case she shall die and I cannot be there Mrs Willis will, I think, be best to take the management and afterwards I wish the Daughter [Margaret Liddle, then aged 19] to stop with Mrs Willis. She can take a larger house so as to contain the furniture. I know nothing what learning the girl has got. I wish her to learn to write, read and keep accounts, and Mrs Willis will learn her something of cooking. These are my views until further consideration – you can show this to Mrs Willis. When I get to Newcastle again I shall talk to her on the subject. I am, yours truly, Geo. Stephenson.[12]

In 1847 his elder brother James also died, at Snibston, leaving three daughters. Of the six children born to Bob and Mabel, only George and Ann,[13] in America, now survived.

This was a time of intense political controversy. In May 1846 Sir Robert Peel as Prime Minister had succeeded in abolishing the Corn Laws, with the support of the opposition Whigs and amidst furious resistance from most of his own Tory party. George Stephenson, who had a high regard for Peel, would have been pleased by the move towards free trade, but his son supported the Protectionist views expressed with passion, if not conviction, by the young Benjamin Disraeli. From late 1846, Robert Stephenson was moving closer to active political involvement. He supplied much of the information for a scheme proposed by Lord George Bentinck to the House of Commons in February 1847, seeking government investment in railway construction in Ireland, both to provide work for the destitute and to form a basis for the same kind of commercial growth that had been generated by railways in England. Bentinck's proposal met a frosty reception from the Peel government and the House voted to shelve it for six months, which was equivalent to a rejection.

Among its supporters was George Hudson MP, who had collaborated in its preparation and was more than pleased at the notion of himself becoming Fat

Controller of a government-funded Irish railway system. Hudson had once intended to make himself the member for Whitby, where he was immensely popular, but even he could only represent one constituency. Aaron Chapman, the sitting member for Whitby, a Tory, was due to retire, and the local committee made the mistake of selecting a new candidate, John Chapman, without consulting Hudson. This man, of Peelite sympathies, was persuaded to withdraw, but then Hudson's own preference, Major William Beresford, having accepted nomination, dropped Whitby for North Essex.[14] Only then did Stephenson enter the frame, undoubtedly at Hudson's instigation.

On 14 June, during the Chester inquest proceedings, Robert accepted the nomination. Hudson managed the whole thing. The Liberal press, in the form of *The Yorkshireman*, might call Stephenson 'the nominee of Mr Hudson'[15] and suggest he was no better than a carpet-bagger, but this was mild political abuse for the 1840s. In Hudson's view, Whitby's Conservatives were there to do his bidding, and the seat was his to dispose of; naturally he wished to install a friend there. Robert Stephenson was sound on Protection, the major issue of the day, and in most other respects a Tory of deeper conviction than Hudson himself. Hudson might have preferred George, but George was not interested. He had already declined an invitation to stand for South Shields.[16] For Robert, there was no obvious overriding motive to submit himself to the electorate. He had certain firm opinions, most of which coincided with those of the anti-Peel wing of the Conservative Party; he was rich and could afford it (MPs had no salary or publicly funded expense accounts); he was prominent in railway and engineering circles, with views on how government should behave towards railway companies; he was 44 and intended to reduce his involvement in business over the next few years; he was a northerner and Whitby was in the north; not least, perhaps, Hudson could be immensely persuasive. In those days, also, Parliament sat for less time and transacted much less business than in later years.

By happy coincidence, the Whitby and Pickering Railway was just being converted to steam traction and the new station beside Whitby harbour was opened by Robert Stephenson on 1 July. A local bard celebrated the event:

When Whitby's wealth in yearly steps decay'd,
By Stephenson's bold scheme her fall was stay'd;
And now, behold! The works his labours won
On wings of steam bring here his far-famed son,
Whose powerful genius and inventive mind
Shall show us all how best our local wealth to find.[17]

Parliament was dissolved on 23 July and the General Election took place on the 30th. Robert, with his father at his side, appeared in Whitby on the 27th to promote his candidature. Although Hudson was working to build up its reputation as a holiday resort, Whitby's business was with shipping, ship-building and fishing, fields in which Stephenson had not shown any previous interest. He had been briefed, however, and spoke out firmly against the proposed Navigation Laws (which would remove certain privileges from old maritime boroughs like Whitby) in a speech to supporters, adding a few platitudes about 'wooden walls' and 'brave British sailors'. In the event, he was unopposed and, after being paraded with pink banners from the Angel Hotel to the hustings, was duly acclaimed as the member for Whitby. George Hudson arrived in time to give a speech of congratulation.

Robert's election showed: 'Victory without opposition, and triumph without anger ... Politicians of all grades united in demonstrating joy and gladness.'[18] His acceptance speech pledged himself as 'dwelling most forcibly, as a member of the Church of England, on the vast importance of that Establishment in con-nexion with the State, and his determination to support all measures calculated to increase its efficiency and usefulness ...', as well as to oppose any changes to the Navigation Laws. Fourteen hundred people gathered at a celebratory ball, under banners proclaiming 'Stephenson and Protection to All Clases', 'Stephenson the Sailors' Friend' and similar slogans, and dancing from 7 o'clock, opened by the Hon. Member with Mrs Wakefield Chapman, was 'kept up with great spirit until four next morning'.[19] He was one of 102 MPs with close railway connections; Joseph Locke was another, elected Liberal MP for Honiton, where he had made his home. But the divided Conservatives were thrust out of government and a Liberal administration under Lord John Russell took over direction of the affairs of Queen Victoria's realm.

Most of the railway pioneers and promoters, if they had political inclinations, were Whigs – or Liberals, as they were beginning to be called. The great Tory preserve was land, not industry. And the coming of the railways had revolution-ised national life in a way that no other technical innovation had ever done; for one of its prime creators to call himself a Tory might seem rather as if St Paul had continued to claim himself an old-fashioned Pharisee while exhorting the Greek cities to turn Christian. By 1847, railways were well advanced in bringing rapid national distribution of everything from newspapers and the mail to coal and fresh fish; their schedules standardised time across the country; they speeded up life in all manner of ways; they democratised transport, with lord and labourer conveyed in the same manner and at the same speed (if not the same comfort) – all with results whose ramifications rippled outwards endlessly. But neither of the Stephensons had set out to make this revolution; seeking to 'get on' in the traditional manner, they had lit a fuse which had found unsuspected reserves of powder. Neither of them shrank from or wanted to disown the broad social impact of their work, but Robert Stephenson did not see it extend beyond a material effect, as 'agents of benevolence and ameliorators of the condition of the human race',[20] and ignored or discounted its deeper political and social signifi-cance. He never published a political credo, but like his younger contemporary John Ruskin (a dedicated railway-hater all his life), who also claimed to be 'a

violent Tory of the old school',[21] he believed in the rightness of order and social
hierarchy in a stable society founded on honest workmanship, religion and obedience. He did not believe in emancipation of the working class, though he did
believe in individual members of the working class emancipating themselves, as
his father had done – if they could. Seen in this way, his professed Toryism is perfectly genuine, though he and Ruskin would have disagreed on many details. On
several occasions as an MP Stephenson would vote the opposite way to his Tory
colleagues and his own expressed views were not always consistent.

Robert Stephenson MP returned to London, to the affairs of the Great George
Street practice, the aftermath of the Chester accident, and the Britannia Bridge.
Just before the Dee bridge collapse, negotiations on the building of the great
tubes had been completed, and the contractors had begun the construction of
massive wooden platforms 1,000ft (300m) long by the Menai shore, to which the
iron plates would be delivered and on which the tubes would be assembled. On
10 August the first rivets were hammered in to join the plates together: the start
of a process that would go on by day and night with scarcely an interruption until
March 1850. At least one aspect of that process might raise modern eyebrows. The
furnaces had to be on ground level, not on the wooden platforms:

> One furnace will supply several sets of rivets, and, not being portable, the rivets
> have frequently to be passed a considerable distance to the work. This is done
> by throwing them with a pair of pincers; and no spectacle has afforded so much
> amusement to spectators as the unerring process by which boys only eleven or
> twelve years of age toss them to their destination. And when a large number of
> riveters were engaged on the tops of the tubes by night, the constant succession
> of these red-hot meteors, ascending in graceful curves to a height of 30 or 40
> feet, formed an interesting sight.[22]

Each riveting team was formed of two riveters, one holder-up and two boys,
one to work the bellows at the furnace, the other to pass on the heated rivets.
Brunel had been invited down by Stephenson to see the start of the riveting:
'I shall be delighted if you will go with me & give me the aid of your thoughts
about these tubes both as to riveting & hoisting.'[23] At Conwy, things were somewhat further advanced, but there was relatively little for Stephenson to do now
until it was time to begin floating the first tubes into position, an operation
that he dreaded. In the previous winter he had written to Edwin Clark that,
'... my doubts and fears will never end until the floating is actually finished.
Turn it what way you like it is a frightful operation.'[24] William Fairbairn may
have not understood the degree to which Robert's expressions of foreboding
concealed his determination to see matters through. On 16 August he wrote to
Stephenson offering to take charge of the floating and raising of the tubes. The
answer was dated a week later:

I was surprised at your letter this morning, asking if I wished you to take charge
of the floating and lifting. I consider you as acting with me in every department
of the proceedings, and I shall regret if anything has been done which has conveyed to you the idea that I was not desirous of having the full benefit of your
assistance in every particular.[25]

Characteristically polite, the letter defines Fairbairn's role as 'assistance' – Stephenson
was in charge. But Fairbairn was increasingly viewing himself as 'owner' of the
project and Stephenson as a kind of absentee landlord.

Around the time of the opening of the new Parliament, Robert moved house,
taking a lease on 34 Gloucester Square, a short distance from Cambridge Square,
but a grander house. Parliament's time was largely taken up by the miseries of
Ireland, for which it had no solution, and by the onset of economic recession,
for which the government managed a slight palliative by an increase in the
circulation of banknotes and the formation of a select Committee of Inquiry
into Commercial Distress, of which George Hudson was a member. Robert
Stephenson did not speak in the House of Commons during this session. After
the boom, new railway business had largely dried up, and even Hudson was calling a halt to fresh schemes. In August he had achieved his ambition of uniting the
Newcastle & Berwick Railway with the York & Newcastle in a single company,
managing, in the process, to concoct a share-fix that considerably enriched those
in the know. These did not include Robert Stephenson, but he did join with
Hudson in another commercial venture that winter, when both became partners
in R.W. Swinburne & Co., which had a glassworks at South Shields.

The intention seems to have been that the factory should supply glass for station
roofs and buildings and perhaps also for the windows of new carriages. The five
other partners were three members of the Swinburne family, George Cockburn
Warden and Nicholas Wood, once Robert's prentice-master, now a prominent
figure in north-east industry. An indication of the way in which father and son
still shared their plans may be had from the fact that George Stephenson went to
inspect the Swinburne works before the partnership deeds were signed. His visit
was remembered more than fifty years afterwards, by the junior partner, George
Warden, because of a prophetic utterance about electricity. Despite his dismissal
of atmospheric traction as 'gimcrack' (by this time it was generally accepted as a
lost cause[26]), Stephenson was not a blinkered adherent of steam power. Touring
the works, he paused to explain to Warden that:

several of our operations were the results or effects of electricity – and then
proceeded to give me quite a lecture on the subject ... He said finally, to
quote as nearly as I can remember his own words – 'I have the credit for
being the inventor of the locomotive. It is true that I have done something to

improve the action of steam for that purpose, but I tell you, young man, I shall not live to see it, but you may, when electricity will be the great motive power of the world.'[27]

In September 1847 a large and well-equipped French survey team set off for the Isthmus of Suez, in order to explore the feasibility of a canal and trace a possible route. The Austrian group had already produced a report stating that a canal was possible and that there were no insuperable technical difficulties.[28] Through 1847 the British government, via its consul-general in Cairo, was pushing the proposal for a railway and denigrating the canal plan.[29] The Egyptian viceroy felt himself increasingly squeezed between British and French agents, both claiming to be advising him in his best interests, and also had to allow for the policy of the Imperial government in Istanbul, whose officials were always ready to suspect Egyptian schemes of secession. The British, with secure naval control of the long sea-route, did not want a ship canal which would be advantageous to the French and Austrians, and perhaps the Russians too, in terms of access to Far East trade and colonial opportunities. But that could not be openly stated.

Within this nexus of imperial pressures, where did Robert Stephenson stand? Unfortunately, no direct evidence survives to tell us. At the end of 1847 he went to Vienna for a meeting with Talabot (who had not accompanied the French party) and Negrelli, at which they apparently all agreed to go together to Egypt in March 1848. This expedition, however, never materialised,[30] probably because of the civil and political disturbances of that year throughout Europe, and the Vienna meeting marked the end of Stephenson's participation in the Canal Study Group.

While Robert coped with these diverse commitments, his father pursued the more leisurely tenor of his own ways. In December 1847, at Frederick Swanwick's house near Chesterfield, he met an American celebrity, the writer Ralph Waldo Emerson, who was on a lecture-tour of Britain. Emerson was keenly interested in the impact of industrial development on England's antique traditions and established ways. On the last day of his first trip to England, in October 1829, he had gone to Millfield yard, by Crown Street Station in Liverpool, to look at and ponder on *Rocket*, the mute black-and-yellow harbinger of a new era. Now he met the man most responsible for it – 'Dined at Mr Swanwick's, Chesterfield, with George Stephenson, inventor of the railroad car' – and noted a mantra for the new age: '¼lb coke will carry one ton one mile.'[31] Writing to his sister, Emerson described George as 'One of the most remarkable men I have seen in England',[32] and a few weeks later he mused: 'Stephenson executed the idea of the age in iron. Who will do it in the social problem?'[33] George invited him to come to Tapton House and see Chatsworth, but the invitation was not to be taken up.

When in relaxed mode, and diverted from reminiscing about his youthful struggles, George was convivial company, if sometimes rather unsettling with his blood-seeking needle. With a large staff inside and out, he was not deprived of human company at home. Spending more time at Tapton in 1847 than in previous years, he was active in planting and landscaping his hilly property, and in growing fruit and vegetables. The cultivation of pineapples was a passion to him, and among the outbuildings was a pinery at 140ft (42.6m) long and a pine-house of 68ft (20.7m). Altogether he had ten glass forcing-houses, heated by hot water pipes, in which he also grew melons and grapes. These were not just for eating, but very much for show: with a competitive spirit as strong as ever, he exhibited at agricultural produce shows round about – 'his grapes took first prize at Rotherham, at a competition open to all England'.[34] Cucumbers were another preoccupation. Objecting to their habit of growing in a curved shape, he tried hard to straighten them, eventually coming up with the solution of growing them in glass cylinders. Bringing a fully-grown straight specimen into the house, he showed it to some visitors with the remark: 'I think I have bothered them noo!'

As he had forecast, he became a farmer, with a particular interest in stock-breeding. He also returned to keeping rabbits. In a sense, his wheel had turned full circle, back to the pursuits of his boyhood, but conducted on a very different scale and at a higher level of knowledge. There was nothing childish about these activities – at last he was giving himself time to do things which had always pleased him, but which in the years of work had hardly been possible. In one major way he diverged from the conventional country-gent image – he abhorred blood sports. Staying at Walmsley's country house, where shooting was the main pastime, 'at dinner Stephenson broke out into earnest remonstrances at the inhumanity he had witnessed that day, at the cruelty exhibited in the wholesale destruction of the wild creatures that enjoyed life so harmlessly'.[35] But in his late sixties he still enjoyed the occasional 'quiet wrestle on the lawn ... In the evening he would sometimes indulge his visitors by reciting the old pastoral of "Damon and Phyllis" or singing his favourite song of "John Anderson My Jo".[36] Old Newcastle friends might be fed with a nostalgic 'crowdie' as well as the more refined products of the kitchen and the stocks of the wine cellar.

Indoors, his house was run by a housekeeper, Ellen Gregory, daughter of a farmer from near Bakewell. While it does not seem that Robert, who had been a widower for longer than his father, ever considered remarriage, or even had a romantic relationship, George was more susceptible. He became fond of Ellen, as a hasty letter written from the Great George Street offices in 1847 makes clear:

My dear Ellen,
I have just got Mr Marks letter which gave me much pain to hear your illness. I do hope you are better. I shall not leave here until Tuesday morning so that I can have a letter before [I leave] London to say how you are, [If] you cannot write in time for me here I shall hear from you at Tapton if you are no better I will come down to you but do not let me come unless you feel very ill.

I shall be home to be with you by the end of the week.

I am my dear Ellen

Your loving friend

Geo. Stephenson[37]

On 11 January 1848 George married Ellen Gregory at Shrewsbury, her family's home town. The news of his intention had been received by Robert as an
unpleasant shock. Thomas Summerside, who knew George well and saw him
often around this time, was also taken by surprise. Robert's reaction and subsequent behaviour suggest strongly that he did not like Ellen Gregory. It would be
tempting to suppose that this was the natural reaction of the heir to a fortune
who sees his elderly father marry a comparatively young woman. George rewrote
his will in 1848 to take account of his remarriage. Robert had also been very fond
of Betty and perhaps expected his father to venerate her memory rather than
replace her in his bed. It was a Trollopean situation (the novelist was just embarking on his writing career at this time): the country house, the vastly rich old man,
the designing housekeeper and the affronted heir, but fortunately also the reconciling friend. The biography of Sir Joshua Walmsley relates:

> The marriage was contracted without Mr Robert Stephenson's knowledge,
> and it caused some ill feeling between him and his father. Sir Joshua Walmsley
> became the mediator between the two. Letters before us, of a nature too deli-
> cate and private for publication, show the tact and zeal with which he pursued
> his self-imposed task. No-one understood better than he the strong affection
> that bound the father and the son. Mr Robert Stephenson has said that he never
> had but two loves in his life, his wife and his father; and this only son was the
> chief pride of the old man's heart.[38]

Walmsley's emollience succeeded in reconciling father and son.[39] George
Stephenson, who had come to the bridge sites often before, was in the party that
watched as the first of the great tubes at Conwy was floated into position, on the
high spring tide of 6 March. Someone else who wanted to see the operation was
Robert's friend from his Colombia days, R.S. Illingworth, to whom he wrote
a hasty note: 'We are in such a state of confusion and uncertainty that I could
hardly wish you here – I shall be so anxious and engaged that you would be left
alone without any one to explain matters. – If however you still wish to see the
operations I have ordered an Engine to be at Chester under the direction of Mr
Lee the Resident Engineer to bring you to this place.'[40] The wooden platform on
which the tube had been assembled was right at the water's edge, and the staging
had been cut away until the ends of the tube rested on two stone piers which
had been built to support it. The beach was excavated so that a line of six pontoons could be got under the tube, and, on a spring tide, raise it above the level

of the stone supports. The first stage was a marine operation more than anything else, and, almost certainly at Isambard Brunel's suggestion, Robert entrusted its management to a practical naval man, Captain Christopher Claxton, RN, an old friend and colleague of Brunel's from Bristol, among whose exploits had been the re-floating of the iron steamship *Great Britain* after it had lain aground in Dundrum Bay for nine months from September 1846. Brunel was also present, joining Robert who had been on the site since 20 February.

A great capstan on each bank drew in or paid out chains to hold the line of pontoons and swing them with their load into position between the piers of the bridge. Justifying Robert's apprehensions, there was a hitch. At the Conwy side, the pontoons moved out of line, were caught against a rock, and the position could not be retrieved before the tide began to ebb. Four tense days of manoeuvrings ensued, with a succession of heart-stopping moments, including one when a falling tide swept some pontoons out from below and threatened to bear the whole assemblage seawards. Edwin Clark, who had lost a joint of his big toe in a capstan accident on the 3rd, wrote that 'Mr Stephenson quite despaired of the possibility of effecting his object',[41] but on the 11th the rogue pontoons were again secured in place and the tube was successfully nudged the last few feet into its final position.

The hoisting could now begin, and by 1 May the north tube was in place, passed its inspection and was ready for use. Trains could now run as far as Bangor, where, on 17 May, a celebratory dinner was held. Neither of Stephenson's external advisors was present. Hodgkinson had already withdrawn from the project and Fairbairn had written on the previous day with a half-proposal of resignation, as '... the object for which I was engaged has now been attained', though he added that he would be guided by Stephenson's view.[42] On the 19th he wrote to say he had decided to resign. Between the two letters, Robert Stephenson had made a speech at the Bangor dinner which has been described as 'marked by surprising self-adulation and with niggardly acknowledgement to Fairbairn and Hodgkinson'.[43] It might have been made by George Stephenson rather than by his usually modest son: '... Parliamentary powers were granted for the construction of a bridge over the Britannia rock, with such conditions attached to it as rendered it all but, if not absolutely, impossible. It was then, to use a common expression, that I felt myself "fairly driven into a corner".' Describing his decision to use the tubular structure, Robert went on: 'It was then I called in the aid of two gentlemen, eminent, both of them, in their profession, Mr Fairbairn and Mr Hodgkinson ... well qualified to aid me in my research. They heartily went into it ...'[44]

This was as much credit as his collaborators were to get. Such an attitude was untypical of Robert Stephenson, but it seems clear that Fairbairn was already asserting his claim to be both prime designer and supervisor of construction. Never having for a moment intended that the Britannia should be considered Fairbairn's bridge, the engineer-in-chief was showing his teeth, determined to assert his ownership of and responsibility for the great, and not yet completed, scheme that he felt he had borne, and still was bearing, on his own shoulders. From then on Fairbairn took no further part in operations on either bridge, and embarked on the writing of a book detailing his side of the argument, depreciating Stephenson's role.

George may have mellowed, but age did not abate his capacity to be offended. When, in April 1848, a letter signed 'Nemo' appeared in the *Darlington & Stockton Times* in praise of Edward Pease as 'the man who first conceived the idea, and so successfully carried out the completion of the first Railway for the conveyance of goods and passengers …', he dispatched a letter to Nicholas Wood, on 20 April:

> My Dear Sir
> I am going to give you a little work to do if you like to, but if you don't then send back the enclosed observations.
> I am disgusted with a letter that has been published in the Stockton & Darlington [sic] Times of April 8th which I enclose you.
> You are the only man living except myself who knows the origin of Locomotive engines on the Stockton & Darlington Railway, I shall be obliged if you will take the matter up and shew that Edward Pease is not the man of science who has done so much for the country.
> I am My Dear Sir
> Yours faithfully
>
> Geo. Stephenson
>
> P.S. When are you going to bring Mrs Wood and your daughter here? I have grapes now ripe and the Duke of Devonshire has not.[45]

Wood, of whom Robert quipped that 'His motto is, "Nothing for nothing for nobody"',[46] took no action. George, who had presented Pease with a gold watch inscribed 'Esteem and gratitude: from George Stephenson to Edward Pease',[47] was not going to lose an ounce of credit for his own achievements, even to his old friend. In July he stayed with George Hudson at his palatial house, Newby Park, in company with two dukes, an assortment of lords and the American ambassador, and attended the Royal Agricultural Society's show in York, at which the Railway King was Master of Ceremonies. Later that month he went to Birmingham and presented a paper to a meeting of his 'Mechanics' Institute' on the subject of the fallacies of the rotary engine. Returning to Tapton House he became ill and a new attack of pleurisy was diagnosed. This time, his once powerful constitution could not withstand both the disease and the best efforts of medical science. It became apparent that he was gravely ill and Robert was sent for, arriving with his friend Edward Starbuck. On Saturday 12 August 1848, just before midday, George Stephenson died, with Robert at his side.

17

# 'THE TUBES FILLED MY HEAD'

It was an unexpected death, and for Robert, with no reason to anticipate it and minimal time to prepare for it in any way, a severe shock. Though he had kept his own household since his marriage, his father was by far the dominant personality in his life. The years between 1808 and 1820, when his character had formed and been influenced by his father, were the backbone of his own being. Like everyone who loses a parent, he was assailed by memories, regrets, small guilts, recollections of lost opportunities and unfulfillable wishes. He also had the solitary task of the only child, to make arrangements for his father's funeral and sort out his affairs. Much had to be done quickly as George's friends were many, living across a vast swathe of the country, and had to be informed. He wrote to Sir Joshua Walmsley: 'I am desirous that the funeral should be as private as possible, attended only by a few of his pupils, who like myself have been dependent upon him for our professional success.'[1] George Bidder arrived on the 13th to lend support, writing to his wife on the 14th:

> Robert I am sorry to say is not so well this morning he can get very little sleep and this day old recollections come over him and depresses him very much besides which he had a long and very trying conversation with Mrs Stephenson last evening – He relies very much on me for keeping his mind occupied and relieved and I hardly expect to get away until after the funeral on Thursday ...[2]

The funeral[3] was in Chesterfield, an old market town whose new prosperity as an industrial centre had been sparked off by George Stephenson. To coal and steam it owed its growth, as well as the coating of soot which was already darkening its half-timbered façades. The town did its best for him: shops closed and some 300 people, including the corporation and mace-bearers, walked in front of the hearse and seven coaches of mourners to Trinity church. Among the few who came from any distance was old Edward Pease, with John Dixon:

Went in the forenoon to Tapton House, late G. Stephenson's residence, and received from Robert a welcome reception; had a serious friendly conference with him, under a feeling expressed to him of my belief that it was a kindness to him his father was taken, his habits were approaching to inebriety; his end was one that seemed painfully to feel no ground, almost, for hope. I fear he died an unbeliever – the attendance at his funeral appeared to me to be a right step due to my association with him and his son ... In the church I sat a spectacle with my hat on, and not comforted with the funeral service.[4]

Of George Stephenson it might have been said that his true monument spanned the nation. Already, there were few places of consequence not on the railway. But Joshua Walmsley was perhaps not the only one to say, 'He ought to be lying in Westminster Abbey, and not in a country churchyard'.[5] Robert's desire for a private funeral may have owed something to his father's rejection of public honours, but was also a mark of the intense personal significance to him of his father's life and death. In later speeches and at almost every event where he himself was honoured, he would refer to 'my late revered Father' and emphasise George's influence on his life and career.

If Robert had had worries about Ellen's settlement, they could be forgotten. George's first will, of 3 January 1839,[6] had made him sole heir and executor of the entire estate, with only two additional bequests. Elizabeth Stephenson, 'my dear wife', was allowed use of all or part of the household furnishings and plate in her lifetime, as long as she did not remarry, and was left £200 a year on the same condition. To his brother James, George left £50 a year for life. By the time George made his final, and much more detailed will[7] on 4 April 1848, James was dead. It specifically leaves to Robert the silver tankard presented for the Geordie lamp in 1817, and to Ellen all her 'paraphernalia and wearing apparel', the silver plate, jewels and trinkets, pictures, linen, china and 'such portions of the furniture as she may select for her absolute use', and a lump sum of £1,000. An annuity of £800 a year was payable as long as she did not remarry or try to mortgage it: in either event it was to drop to £100. George Robert Stephenson was to receive £1,000, as were his other nephews, Robert and James. Smaller bequests went to Ann Nixon, by now a widow, and to his nieces. Everything else was Robert's. Ellen's annuity of £800 was a very comfortable income for 1848, but still bore little relation to the size of George's estate. Summerside noted:

> ... she found it too little, and even contemplated coming to Matlock to reside, in order to economise. She wrote to her step-son, Mr Robert Stephenson, for an increase in her allowance, but he significantly referred her to his solicitor. I only name the above by way of contrast, knowing that it will strike all the intimate friends of his second wife who would know that her watchword was frugality, economy and saving.[8]

Ellen emerges from his description as little better than a failed gold-digger. She formed no part of Robert Stephenson's family circle. One feels that, from the tenor of George's only surviving letter to her and from a report of Starbuck's that she had nursed her husband devotedly, she deserved rather better treatment. She did not remarry, and died at Beauchamp near Shrewsbury in 1865. Incidentally, the annuity enjoyed by Robert's godfather, Robert Gray,[9] is not mentioned in George Stephenson's will – presumably he had been provided for earlier.

Far off in Boston, New England, Ralph Waldo Emerson learned the news: 'George Stephenson died the other day in England ... the father of railroads, – and not an engineer on all our tracks heeded the fact, or perhaps knew his name. There should have been a concert of locomotives, and a dirge performed by the whistles of a thousand engines.'[10]

Three days after his father's funeral, Robert Stephenson was back in North Wales, where Fairbairn's withdrawal had made his supervision of the bridge works all the more necessary. On 20 August he narrowly escaped becoming the most celebrated victim of a fatal railway accident since *Rocket* ran over Huskisson. With Hedworth Lee, engineer of the eastern portion of the line, he was sitting in a first-class carriage at Conwy station while it was being pushed over points from one track to another. Before the manoeuvre was complete, a train entered the station and hit the corner of the carriage, throwing it off the rails and smashing the framework and glass fittings. Robert climbed down from the wreck, but collapsed on the ground, from shock rather than injury. Lee was unhurt. On the following day, Stephenson was back at work. There was a great deal to do, but it was taxing his strength; on 15 September he wrote to Moorsom postponing a visit to Wales: 'I must have a few days quiet.'[11]

From 27 September to 5 October, he and Bidder sailed in a chartered yacht (by coincidence, its name was *Fanny*) round the English and Welsh coasts from Southampton to Conwy. As with their trip to Norway, they thoroughly enjoyed the cruise, despite being tossed about in a severe storm off the Devon coast. 'We live very simply and regularly sometimes nearly altogether on fish',[12] reported Bidder to Georgey. After the stresses of the previous months, the break was welcome. At Conwy, Robert watched the second tube being floated on 12 October, an operation which went far more smoothly than the first time. By 8 December the second tube was raised to rail level and the completed bridge was passed by the inspector on 2 January 1849.

On the Newcastle & Berwick Railway, bridge construction was also well advanced. The west coast route to Scotland, via Lancaster and Carlisle to Glasgow, had been open since February 1848, and Hudson was eager to get the complete east coast service in action. The citizens of Newcastle had watched the five piers of the Tyne bridge, the tallest being 146ft (44.5m), rise above the masts of the shipping. With the capable Thomas Harrison in charge, this bridge did not contribute to Robert Stephenson's anxieties. Compared to earlier building, construction was made faster and easier by the use of steam-powered machinery, and this was the first major bridge for which supporting piles were driven by Nasmyth's steam hammer. By July 1848 the piers were at full height and assembly of the ironwork could begin, supported by a temporary timber framework, which also made it possible for trains to cross. By the end of August, through services between Newcastle and the south began. With the completion of a temporary wooden rail-deck over the rising arches of the Tweed bridge, by 10 October the east coast route could at last run trains between London and Edinburgh.

As these great structures went up, the reputation of George Hudson, the man who had made them possible, was on the brink of disaster. The only hope for his financially over-extended empire was further expansion and more investment, but the government's refusal to subsidise Irish railway construction, and the post-mania climate of caution, hemmed him in. His difficulties were compounded by the slump in trade and commercial confidence caused by the political and social turmoil across Europe in 1848. The Eastern Counties Railway had totally failed to fulfil his golden promises, and on the day of George Stephenson's funeral he faced an angry and disillusioned shareholders' meeting, whose only quiet moment came when he asked for a short silence in memory of his friend. Even on the York & North Midland, the dividend had slipped to 6 per cent. Share prices in all railway companies were falling, but the Hudson lines, with no cash reserves and heavy liabilities, were especially vulnerable.

On 28 December 1848 Edward Pease recorded in his diary: 'Pecuniarily, I have cause to admire how an effort to serve a worthy youth, Robert, the son of George Stephenson, by a loan of £500, at first without expectation of much remuneration, has turned to my great advantage. During the course of the year I have received £7000 from the concern at Forth Street.'[13] By that time there were just over 5,000 miles of railway in Britain and new railways were being built in many other countries. British locomotive design had reached a plateau, with essentially similar machines being produced by all the manufacturers, who by now included some railway companies like the London & North Western, which built its first locomotive at Crewe in 1845. Robert Stephenson was still signing working drawings at Forth Street in 1846,[14] but seems to have withdrawn from active involvement in design from around 1848, though retaining his interest in locomotives, which he always claimed to have been the first love of his professional career. In 1849 he participated in a discussion at the Institution of Civil Engineers about the need for locomotives to have a low centre of gravity, a fashionable thesis which he – and Brunel – did not accept. Perhaps because railway work was in abeyance, but more likely because other large-scale public works were now being embarked upon, he took on the role of consulting engineer to the Nene Valley Drainage and Navigation Improvement Commissioners in September 1848, along with Bidder and George Robert Stephenson.

As his father's heir, Robert became a partner in the Snibston collieries, with Walmsley and Joseph Sandars the younger, and the largest shareholder in George Stephenson & Co.; a company established by George to exploit the Clay Cross coal and Crich limestone, and which by now was also engaged in iron smelting. He succeeded his father as its chairman on 15 November 1848. This was a large-scale and dynamic business in its own right, and the need for its directors to watch and develop its affairs perhaps played a part in his comparative neglect of the Forth Street Works. By now he was again in need of a restful break, and wrote on 7 November: 'I am leaving on the 16th for Florence I shall probably be absent for 2 months.'[15] Florence was one of the traditional wintering places for Britons escaping their climate. While in Italy, Robert inspected the completed Leopold Railway in Tuscany, with its redesigned girder bridges, writing when he returned on 10 February to Edwin Clark to complain about Fairbairn's 'astounding egotism'.[16] Captain Claxton too was annoying him, by acting as if everything at the Britannia site depended on himself: 'Really Claxton is getting too bad,'[17] he wrote on the 28th.

Even as the final links in the east coast route were being made permanent, everyone with an interest in the nation's business life was shaken to learn that Committees of Inquiry had been set up by the York, Newcastle & Berwick, and the Eastern Counties Railways into the conduct of their chairman, George Hudson. At the half-yearly meeting of shareholders in the York, Newcastle & Berwick, on 20 February, it had been revealed that the company had bought shares in the Great North of England Railway from Hudson at a higher price than had ever been quoted on the Stock Exchange. In Hudsonian terms it was a modest transaction, and the price paid was an error made by the company secretary,[18] but with this detail, a knife-point was inserted into the complex knotting of the Railway King's financial dealings, which proceeded to unravel a whole series of dubious actions, going back through several years and the many amalgamations and share issues that had been devised and carried through by the dynamic Hudson. In many cases they were transparent deals which could have been challenged at the time if shareholders had been less greedy and complacent, and directors less cowed and sycophantic. The committees reported in April 1849, having found that Hudson had manipulated the accounts of both companies to pay dividends that their actual revenue from traffic could not support. In the case of the now bankrupt Eastern Counties Railway, Chairman Hudson, during his period of control, had authorised dividends of £545,714, almost double the amount that should have been paid.

With this news the share values of the northern companies took a dive. Already the York & North Midland's customary 10 per cent had been reduced to 6; now, in the first half of 1849, it paid no dividend at all. Commenting on the scandal, *The Times* blamed the system at least as much as the man: '... without rule, without order, without even a definite morality ... He had to do everything out of his

own head, and among lesser problems to discover the ethics of railway speculation and management ... Mr Hudson's position was not only new to himself, but
absolutely a new thing in the world altogether.'[19] On 19 April Hudson resigned
as chairman of the Midland Railway; on 4 May he gave up the chair of the York,
Newcastle & Berwick, and on the 17th he also withdrew from the board of the
York & North Midland, which he had nourished for so long and whose shareholders now reviled him as much as they had once fawned over him.

Continuing inquiries threw up further evidence of the way Hudson had played
fast and loose with the financial orthodoxies. Charles Greville, clerk to the Privy
Council and secret diarist, noted that: '... most people rejoice at the degradation
of a purse-proud, vulgar upstart, who had nothing to recommend him but his
ill-gotten wealth. But the people who ought to feel most degraded are those who
were foolish or mean enough to subscribe to the "Hudson Testimonial", and all
the greedy, needy, fine people who paid abject court to him in order to obtain
slices of his good things.'[20] A more thoughtful reflection came later from William
Ewart Gladstone, who had crossed swords with Hudson over his Railway Bill in
1844, and had been worsted on that occasion: 'It is a great mistake to look back on
him as a speculator. He was a man of great discernment, possessing a great deal of
courage and rich enterprise ... a very bold, and not at all an unwise, projector.'[21]

Considering that some people – the railway historian John Francis has already
been quoted – saw Hudson as pushed by George Stephenson, and that Robert
was one of his closest business associates, their reputations suffered not at all in
the Railway King's fall from grace. George had backed off from business involvement and Robert had always taken care to be Hudson's adviser, not his tool. They
had been very useful to one another, but in a complementary way. George, if still
living, would have been entitled to say 'I told you so' to his confidants. Robert
Stephenson has left no comment on Hudson's downfall, though he was prompt
to slap in some very large charges for his services to former Hudson companies, including one in excess of £40,000 to the Midland Railway.[22] Unlike some
others who had worked closely with Hudson, he made a smooth connection
with the new managements. But he was not among the many who dropped the
link (Hudson never admitted to any wrongdoing and informed the sympathetic
inhabitants of Sunderland on 1 May 1849 that 'a sense of conscientious rectitude
sustains me'[23]). With Hudson a partner in the Clay Cross and the Swinburne
Glassworks companies, he would have found it difficult to cut off relations. Also,
Robert had never had anything but benefits from his association with Hudson.
The slump in value of his little parcel of free shares would not have troubled him.
They also remained political allies at Westminster, though Robert still had not
yet found a topic on which to address the House. There is no reason to suppose
that he did not also feel genuine sympathy for Hudson's fate, even if the other
man's lifestyle and personality did not appeal to him, and he certainly was of help
to Hudson on at least one future occasion. Hudson's biographer could not have

claimed of Robert as he did of his father, that Hudson was 'George Stephenson's closest friend from their first meeting to the end of his life'.[24] But that is an overstatement. Sir Joshua Walmsley was certainly closer to George, as was Charles Binns, his former secretary, who became manager of the Clay Cross Works. But it is debatable whether George Stephenson had any really intimate friends. A man who has made himself a living legend is not easy to get close to.

In December 1848 Robert Stephenson was proposed as a Fellow of England's premier scientific body, The Royal Society, and elected on 7 June 1849. Compared to Brunel (elected 1830) and Locke (1838), he had to wait a long time for this acknowledgement of distinction. It seems unlikely that his sponsors, who included Sir John Rennie, Brunel, Major-General Pasley and Eaton Hodgkinson, would not have proposed him earlier, unless he had discouraged the idea. Sir Joseph Banks had died in 1820, Sir Humphry Davy in 1829, but it is very probable that Robert was unwilling to have his name put forward during his father's lifetime.

As the spring of 1849 blended into early summer, one issue was dominating Robert's life to the point of obsession. The first of the great Menai tubes was nearing completion, and 19 June was set as the date for floating it into position between the piers. Although both Conwy tubes had been successfully installed, things had almost gone badly wrong with the first one, and he described himself as 'nervous and extremely anxious'.[25] The side tubes were being built out without any particular difficulty on massive wooden trestling, but the tide race in the channel, and the 100ft hoist needed for the four main tubes, haunted his imagination. Intensive planning for months past had included the making of a large-scale model of the scene with 'real water'.[26] The knowledge that William Fairbairn was preparing his own account of his involvement with the tubular bridges did not help his equanimity, and he had already arranged that Edwin Clark would write an 'official' account of the projects. Without the confident, encouraging presences of Fairbairn or his father, he vented his anxieties on Clark, continually thinking of what might go wrong. Later he wrote to Thomas Gooch about this time:

> Often at night I would lie tossing about, seeking sleep in vain. The tubes filled my head, I went to bed with them and got up with them. In the grey of the morning, when I looked across the square, it seemed an immense distance across to the houses on the opposite side. It was nearly the same length as the span of my tubular bridge.[27]

Captain Claxton was again in charge of the nautical side of the operation; Brunel was again there to lend moral support; and Joseph Locke also came, at Stephenson's invitation. George Bidder too was a member of this formidable supporting party. Numerous reporters were present and, from a distance, a huge crowd of sightseers waited – triumph or disaster, it would be an intensely public spectacle.

Eight new pontoon floats had been constructed, six of wood and two of iron.
The iron ones, at each end, were fitted with pumps, making them partly submersible so that they could be manoeuvred under the 1,500-ton tube, and then
pumped out to rise and bear its weight. Powerful capstans had been mounted
on both shores to draw the pontoons into position by hawsers. Timing was vital:
the movement had to begin just before high tide, and the tube had to be in place
within an hour and a half, before the ebb tide, flowing at 6mph, could sweep the
pontoons away. On the 19th, one of the capstans broke and the flotation had to
be put off until the following day. Another capstan problem delayed things to the
evening tide of the 20th, but finally, all seemed well:

> Captain Claxton was easily distinguished by his speaking-trumpet, and there
> were also men to hold the letters which indicated the various capstans, so that no
> mistake could occur as to which capstans should be worked; and flags, red, blue,
> and white, signalled which particular movement should be made with each.[28]

Silence was broken only by Claxton's commands as the massive assemblage slowly
swung out into the channel and then, on the calm surface, crept towards the gap
between the piers. Abruptly there was potential disaster – coils of thick rope over-
rode each other on a capstan on the Anglesey side; the capstan took the full strain
and was yanked out of its housing. Only quick thinking by Charles Rolfe, the
engineer in charge, saved the situation; he 'with great presence of mind, called the
visitors on shore to his assistance, and, handing out the spare coil of the 12-inch
line into the field at the back of the capstan, it was carried with great rapidity up
the field, and a crowd of people, men, women, and children, holding on to this
huge cable, arrested the progress of the tube'.[29] Ledges had been built at the bases
of the towers, and as the pontoons were drawn into line between them, the ends
of the tube protruded over these, then came gently to rest on them as the pontoons were again partially submerged. Cannons had been brought to the shore,
and the boom of blank artillery fire saluted the success as the pontoons were
hauled away to await the completion of the next tube.

After all the tensions, the smoking of innumerable cigars, the efforts to inter-
pret the ponderous movements, the moment of horrid alarm, it was time for
relieved laughter, congratulations and back-slapping. But for someone of Robert
Stephenson's temperament, it merely heralded a new set of worries connected
with the lifting process. On 31 July he wrote to Edwin Clark that, 'New operations where experience does not aid are cruelly uncertain and even humiliating'.[30]
The next step was to install the hydraulic presses and their steam-powered pumps.
Patented by Joseph Bramah in 1796, the hydraulic press was extremely strong
and these, with their 40 horsepower engines, were among the biggest yet made,
capable of a 6ft (2m) lift. Mounted in the hollow sections of the towers, where the
tubes would eventually be fitted, they drew the tubes upwards by means of heavy

chains. The Admiralty having conceded partial obstruction of the channel for a few weeks, the tube lay on its ledges until 10 August, when lifting began. Regular inspections revealed no sagging of the structure. Vertical niches as wide as the tubes had been left in the faces of the piers, and these were built up as the tube ascended. At Stephenson's insistence, teams of workmen did not wait for each 6ft hoist to be completed, but fitted wood-block packing under the rising tube as soon as enough space was left. His foresight was justified when the bottom of one of the pump cylinders 'burst out' on 17 August and 1½ tons of iron fell 80ft on to the top of the tube, killing a workman and making a large dent in the surface. A lift of 2.5ft (76.5cm) had just been made, but the tube fell less than 1ft (30cm) because of the packing. Repairs to the tube and the pump were put in hand, and Stephenson gave an account of the accident to the British Association for the Advancement of Science at Birmingham in September. Lifting resumed on 1 October and the first tube was successfully placed at full height on the 13th and fully installed by 9 November.

# 18

# CALOMEL

With the occlusion of George Hudson, the completion of Newcastle's High Level bridge on 15 August 1849 was not the scene of an orgy of self-congratulation such as the Railway King would have arranged. The formal opening, by Queen Victoria, took place on 28 September. At that time the road deck was still being built, and the new central station was also still under construction. Patronage by the Queen, travelling south from Scotland, was a useful accolade for the east coast route.

Sometime that summer Robert was visited in London by Samuel Smiles, a personable, rather earnest individual of 37; a railway manager who occupied his spare time with journalism and had dreams of becoming a full-time author. Born in Haddington, Scotland, and trained originally as a doctor, Smiles had heard George Stephenson speak in Leeds, and later wrote a short account of George's life for a popular magazine, *Eliza Cook's Journal*, relying mostly on gossip from his railway contacts.[1] Realising that his article was grossly inaccurate, and also that there was a real story to be told, he called on Robert and proposed himself as the author of a full biography, with the result described in the Introduction. Smiles' day job then became more demanding and no more was heard from him for four years.

At the time, Robert was not at all well; his health had been seriously undermined by mental stress and physical fatigue over a long period. Since the Chester accident in May 1847 he had had to cope with a succession of crises, and by the summer of 1849 he was again referring to himself as 'under doctor's orders'. In October he went to the Lake District for two weeks of rest, and on the 13th wrote to Clark to say that he had been 'low and suffering a good deal from acute pain in the region of the heart';[2] although he continued to demand a daily report from Clark, he engaged in some intensive botanising, noting a few days later that 'I have been rallying myself by cryptogamic studies ... this country is rich indeed in ferns', and a long letter discusses the features of the black spleenwort and related plants.[3]

Unsurprisingly, he ignored a crude attempt at a commercial challenge, the last gesture of an old rivalry that went back to the Rainhill trials, from John W. Hackworth, son and partner of Timothy. At their small Soho Works in Shildon, the Hackworths had just completed *Sans Pareil II* – their most advanced engine – and the younger threw down the gauntlet in an open letter to Stephenson published in several railway journals:

> Understanding that you have now … a locomotive engine which is said to be the best production that ever issued from Forth Street Works, I come forward to tell you publicly that I am prepared to contest with you, and prove to whom the superiority in the construction and manufacture of locomotive engines now belongs … Relying on your honour as a gentleman, I hold this open for a fortnight after the date of publication.[4]

Early in December the second Britannia tube was floated (Bidder reported laconically to his wife, 'Tube all right', on 4 December). Publication of William Fairbairn's book caused Stephenson anger and distress. Copies were sent by Fairbairn to his own friends in the engineering world, including James Nasmyth, who wrote to thank him on 15 December, saying that he intended to let it 'run about telling the truth in many a quarter where the truth ought to be known'.[5] Fairbairn claimed the credit for the conception and design of the self-supporting rectangular iron tube, and for the method of raising it by flotation and hydraulic jacks, backing his account by publishing many of the letters exchanged between himself and Stephenson. With Edwin Clark still busy on the Menai site, Stephenson had to bide his time before his own version of events could be published, but his feelings can be judged from an explosive letter to Clark, written from London on 29 December: 'What infernal devil can have invented the story that you alluded to in the Bangor paper – I am not conscious of having deprived anyone of the merit which was their due – I wish I had never seen the Holyhead Railway.'[6]

Robert had acquired a taste for sailing, and in 1847 was among the founders, and first vice-commodore, of the Royal Welsh Yacht Club at Caernarfon. In 1849 he indulged himself, in a manner he could easily afford, by ordering the construction of a large iron-hulled yacht, to be rigged as a topsail schooner. Its designer and builder, the Glaswegian John Scott Russell, had just taken over Fairbairn's shipyard on the lower Thames, as Robinson & Russell, and Stephenson's yacht, *Titania*, was one of the first commissions; the name, that of the fairy queen in *A Midsummer Night's Dream*, was her owner's choice.

Seven days into 1850, the century's half-way mark, the second main tube of the Britannia Bridge was raised to rail level, and work began immediately on joining the four lengths of the northern tube to make a single vast hollow beam 1,500ft (457m) long, reaching across the Strait, supported on the piers. A continuous beam was not part of the original plan, which envisaged independent spans, but the final version, in which the shorter end sections helped to balance the whole beam, gave even greater strength and rigidity. This work was completed on 4 March, and Robert Stephenson was on hand to drive in the last of some 2 million rivets. Rails were already in place and on the following day he drove the leader of three locomotives pulling the first train to cross the bridge, passing under entrances above which an uncompromisingly large inscription gave one form of his response to Fairbairn: 'Erected Anno Domini MDCCCL. Robert Stephenson, Engineer.' Later that day a monster train of 503 tons, formed of some 38 carriages carrying 700 people and 45 coal wagons, was drawn across. At a dinner held in Holyhead that evening, Stephenson's speech did not mention his collaborators at all, but made contemptuous references, not to Fairbairn, but to those who had derided the whole idea of the tubular girder bridge: '… individuals who to the worst features of ignorance and error, superadded malignity of motive and of will.'[7] Evidently, through these stressful years he had borne the additional burden of feeling that hostile watchers were waiting and hoping for his downfall.

The single line was opened to normal rail traffic on 18 March. Now that one tube was complete, the whole structure of the Menai Bridge could be appreciated, especially when seen from the north, and it was generally considered to be an entirely worthy neighbour to Telford's masterpiece. Professor William Pole felt it was a pity there was no central span, 'architectural beauty requiring an opening in the centre and not a pier', though he accepted that the central rock gave Stephenson no choice.[8] Although at one stage the railway company's funds had run dangerously low – the final cost of the bridge, at £674,000, was over three times his original estimate – Stephenson had succeeded in having the purely decorative feature of four couchant lions, each carved from eleven massive limestone blocks by John Thomas, the leading monumental sculptor, placed in position. Two at each end, they conveyed an impression of 'Repose and Dignity' according to one observer.[9] The two bridges attracted many sightseers and special rail excursions were laid on. Perhaps the least fortunate were the train passengers, who caught only a glimpse of sea and space at each end before entering the tube, and then they might as well have been in a tunnel.

The other form Stephenson's response took to Fairbairn's claims was Edwin Clark's book, published in the course of the year, a more imposing production than its rival, in two large volumes with a third containing plans and plates. In his Foreword, Stephenson is a little more generous to Fairbairn, expressing indebtedness for his zeal and confidence and 'the sound practical information which he brought to bear on the subject, and the assistance he rendered me as we progressed'.[10] But in relation to the extent of Fairbairn's contribution, it was thin praise. While the concept of the tubular wrought-iron girder rests with Stephenson, it was his collaborator who saw it through into practicable form. For several years, opinion among engineers was sharply divided on whose bridge it really was. In December 1852 Stephenson wrote a sharp letter to Andrew Ure, whose *Dictionary* suggested he had stolen Fairbairn's credit: 'I am not so fond of strife & notoriety as to think it either necessary or becoming to seize every opportunity of reiterating claims which once fairly asserted must rest on their truth and justice …'[11]

In Fairbairn's eyes, and in those of modern experts, Robert did not help his case by suggesting that a bridge he had designed to take the Cambridge Road over the Lea Navigation at Ware, predating Fairbairn's work, had anticipated the cellular roof structure which preserved the rigidity of the tubes: it was not built in that form and the only drawing he furnished does not show a comparable design.[12] If Fairbairn had been less concerned with claiming overall credit for design and execution, Stephenson might have conceded him more. As it was, Fairbairn was leaving him with little more than having had a vague idea of what to do and the good sense to have commissioned Fairbairn to do it for him. To the public at large, in any case, it was always Stephenson's bridge. He and Fairbairn were never reconciled, though he remained on friendly terms with Eaton Hodgkinson, even acting as a witness at his second marriage in October 1853.

Another festive dinner in the almost completed Newcastle central station on 16 January 1850, with Robert Stephenson and the architect John Dobson as joint guests of honour, had marked the final completion of the High Level bridge. Stephenson was clearly feeling much better after his recuperative visit to the Lakes, and with trains running regularly across the Menai Bridge, he could feel more relaxed about the continuing work on its southern main tubes. By May *Titania* had been launched and fitted out, and he was able to try it out, in testing conditions. Under the master, Loving Corke, of an old Cowes family, he had to maintain at least the nucleus of a permanent crew, since he and his guests were supercargoes who did not actually work the ship. Robert was an enthusiastic sailor who did not mind a bit of rough weather and liked to go fast; he was also an involved owner who had mastered the technical jargon of boats and boatmen:

> I have had a sail or two in the Titania in very rough weather … She behaved very well, beating any thing we could find to run south. She is certainly very fast say 11 knots or 12½ miles an hour, and as a necessary consequence *rather wet*. I suppose it is in navigation as in mechanics, you cannot have both power and speed at the same time. We must therefore not expect velocity & dryness in a vessel in a boistrous sea & beating to windward.[13]

On 10 May he was proposed for membership of the socially exclusive Royal Yacht Squadron by the Marquess of Conyngham and F.P. Delme Radcliffe, but his application was anonymously blackballed on 12 July.[14] Stephenson's sponsors were sufficiently affronted by the rebuff to re-enter his name on the same day, with the Earl of Wilton, Commodore of the Squadron, as his proposer, and his seconder the Earl of Ilchester. Nineteen other 'heavyweight' members added their names, and he was elected on 12 August, the first member not to have an 'upper-class' background. It was a tribute to his prestige, but also to the manner in which he had inconspicuously moved into an upper-class lifestyle, grander than the old household on Highgate Hill.

After the travails of 1849, Robert enjoyed a much happier year in 1850, though it was not without stressful moments. The date set for the Great Exhibition, May 1851, was looming close, with much still to do. From January 1850, direction of the undertaking had been assumed by a Royal Commission, and though the Royal Society of Arts had been struggling to cope with the scale of the project, the handover caused some strains. Stephenson had joined the RSA's Executive Committee for the Great Exhibition late in 1849, and from 5 January 1850 was its chairman, and also a member of the Building Committee. The new commissioners were keen for him to join their number, but Robert stood up for the RSA's continuing involvement with the scheme. In a clumsy preparatory move, the commissioners replaced Stephenson as chairman, without consultation or indication of their further intentions. Angered by this, he submitted his resignation from the Executive Committee on 8 February, along with Henry Cole and Charles Wentworth Dilke, the two most dynamic members. After some hasty diplomacy, he withdrew his resignation, and was appointed a member of the Royal Commission on 12 February. One of his responsibilities was to agree a settlement with the disgruntled brothers James and George Munday. These entrepreneurs had contracted with the RSA to finance the project, build the exhibition hall and manage the Exhibition, in return for a percentage of the profits, but now that the government was involved, this method of financing was not considered appropriate. In May 1850 the Mundays indicated that they would accept Stephenson as an arbitrator, and fourteen months later a settlement was agreed.

Funding remained a major problem until a guarantee fund was opened in July, with Samuel Morton Peto guaranteeing £50,000, and Robert Stephenson £3,000. The Building Committee, which also included Brunel (from January 1850, probably at Stephenson's instigation) and Sir William Cubitt, had a much more immediate problem, as it had not yet approved a plan for a suitable building in Hyde Park, though it had rejected 245 proposals. Brunel produced a design for a long brick edifice with a huge cast-iron dome, and this seemed likely to go ahead. By virtue of a chance meeting on Derby station, on 20 or 21 June, Stephenson was the first of the Exhibition commissioners to see Joseph Paxton's design for a vast iron-and-glass building to house the show. His father's friend had taken his own design for hothouses at Chatsworth and greatly enlarged it, and though the entry date for submissions had passed, and his plan did not meet the specifications set down, he intended to put it forward. As they travelled together to London, Robert examined the plans and reacted enthusiastically: 'Wonderful! Worthy of the magnificence of Chatsworth! A thousand times better than anything that has been brought before us! What a pity they were not prepared earlier!'[15] and he agreed to show them to the Commission.

Unfortunately, Robert was not present on 29 June when the plans were considered, and '… at first the idea was rather pooh-poohed',[16] but Stephenson continued to be supportive and Paxton later said to his wife that Robert had 'helped him like a brother'.[17] The determined Paxton got engravings of his plans published in the *Illustrated London News* on 12 July and public acclaim was immediate. Peto, whose huge financial guarantee gave him considerable clout with the commissioners, also backed the plans; the Building Committee reconsidered its verdict and Paxton's design was accepted.[18] The Exhibition provided the subject for what is believed to be Robert Stephenson's maiden speech to Parliament, made on 4 July, almost three years after his election. Curiously for a self-professed Tory of deepest dye, he spoke in opposition to a much crustier reactionary, Colonel Sibthorp of Lincoln, the one-time virulent adversary of steam locomotion. Sibthorp deplored the very idea of the Great Exhibition and its intrusion on the green space of Hyde Park. In a well-received speech, Stephenson gave a reasoned defence of the chosen site, stressed the urgency of the matter and said that for the House to oppose the Exhibition would disgrace Britain before the nations. Sibthorp's motion was defeated by a majority of 120.[19]

It was a summer of culminations and congratulations. In a fraught day on the Menai Strait, on 10 June 1850, again with Brunel on hand and Captain Claxton with his megaphone, Robert supervised the floating of the third main tube of the Britannia Bridge. Everything almost went horribly wrong. The breeze was strong, and some capstan lines fouled, allowing the pontoon-line to swing away and the tube 'struck the buttress beneath the Britannia tower with a fearful crash. To add to the difficulties of this position, the current forced the tube into the timber; and as the tide was rapidly rising, the piles began to be drawn out of the ground, in which case the tube must have passed by the tower and have been irretrievably lost.'[20] Heroic sawing and axe-work by the carpenters cut the engaged timbers away, fresh ropes were attached and catastrophe was narrowly averted. The fourth tube was floated on 5 July without mishap.

On 30 July the York, Newcastle & Berwick Railway organised a great celebratory banquet inside the newly completed station building at Newcastle, with Robert Stephenson as guest of honour. Under the triple arched roofs of the first curved iron-aisled building in architectural history, 400 other guests assembled, including such old friends as Nicholas Wood and Michael Longridge, and the proceedings were chaired by the Hon. Henry Liddell, son of Lord Ravensworth. As backdrops, large-scale cartoons of the High Level, Tweed and Menai bridges were mounted. Speeches and toasts abounded. In sharp contrast to his Chester & Holyhead speech, Stephenson was at his most self-deprecatory, with the inevitable reference to 'him who is no more', and in the course of proposing a toast to Thomas Harrison, he described how the role of the engineer-in-chief had evolved from the days when, as:

... intelligence was not so widely diffused as at present, an engineer like Smeaton or Brindley not only had to conceive the design, but had to invent the machine and carry out every detail of the conception: but since then a change has taken place, and no change is more complete. The principal engineer now only has to say, 'Let this be done!' and it is speedily accomplished, such is the immense capital and the resources of mind which are immediately brought into play. I have myself, within the last ten or twelve years, done little more than exercise a general superintendence and there are many other persons here to whom the works referred to by the Chairman ought to be almost entirely attributed. I have had little or nothing to do with many of them beyond giving my name, and exercising a gentle control in some of the principal works.[21]

He added that in the Newcastle district especially, he had been 'most fortunate in being associated with Mr Thomas Harrison. Beyond drawing the outline, I have no right to claim any credit for the works above which we sit.' It was, of course, good and proper form for the chief to give credit to his principal assistant, showing a generous and expected modesty. Stephenson displayed this handsomely at Newcastle, but the game had to be played both ways, and in April 1846 Harrison had written of the High Level bridge design: 'The plans have been prepared under my direction: the designs are not mine, but my friend Mr Robert Stephenson's.'[22] There can be no doubt that the general design of the High Level bridge is owed chiefly to Robert Stephenson. Of all his civil engineering works, this was the one closest to home, in a figurative as well as a purely literal sense. Newcastle was fast becoming a city, bent on improvement, with a range of handsome public buildings in its central area, including the station building. The bridge was Robert's contribution to the evolving appearance and expansion of the place where he had grown up, an old town transformed into a dramatic new cityscape. It was a Stephenson monument, a powerful and dramatic statement of achievement; and of course, doubly useful with its two decks. While the memory of George Stephenson was warmly recalled, another name was not mentioned. The new chairman of the railway was George Leeman, a long-standing adversary of George Hudson. Revelations about Hudson's financial operations were still appearing, and though the new Sunderland Dock had been opened with Hudsonian éclat only two months before, he was not invited to the Newcastle festivity.

On 16 August the final main span of the Britannia Bridge reached rail level, and on the 29th, the Royal Border Bridge was formally opened by Queen Victoria. It had been made known to Robert Stephenson that the queen was willing to bestow a knighthood upon him, but he declined the honour. Jeaffreson states: 'He had reached a point of life and fame when such rank could afford him neither pleasure nor profit.'[23] This comment is surely questionable. Robert Stephenson was already MP, FRS, a member of the Royal Yacht Squadron, a vice-president of the Institution of Civil Engineers and president of the Institution of Mechanical

Engineers, a vice-president of the Royal Society of Arts, wearer of the Cross of St Olaf and of the Belgian Order of Leopold to boot. He had just been awarded an honorary MA by Durham University. Why should he refuse an honour from the hand of his own sovereign – especially given his own deeply conservative political views? Knighthood would also confer distinction on the still-new profession of mechanical engineer, and on his own practice in Great George Street. His father, by his own statement, had also more than once turned down a knighthood. Somehow the action seems more in character for George, who never had any letters to put after his name other than his self-supplied CE, than for Robert. It may be that Robert, whose often-reiterated veneration of his father's memory went well beyond any requirements of conventional piety, had decided that he would not assume a title which his father had declined. Foreign honours, it could be said, were in the line of business; an accolade at home was a personal thing. The magazine *Punch* suggested that a knighthood was wholly inadequate: 'We are glad he sent back the insulting offer.'[24] As he left no word on his unwillingness to become Sir Robert, any answer is guesswork.

Robert Stephenson was a multi-layered character and, like his trussed compound girders, combined weaknesses and strengths. To some people he could behave in a distant and chilly way, or at least be perceived by them to do so. Petulance and irritability have also been ascribed to him. There were moments in his active engineering career when his reaction to criticism or adversity seemed to verge on the hysterical. But everyone suffers from moods and moments and in his work he was faced with demands and stresses which few people ever meet. Perhaps his greatest task in life was to find his own sense of identity and his own voice, having grown up with the magnetic and monolithic personality, the block-like certainty, of his father. George had been a fond and, to some degree, an indulgent parent, and, to a boy with a mechanical and scientific bent, also an exciting and sometimes exhilarating one. But if Robert was his pride and joy, the demand was also there, from 'Mind the buiks' to 'Robert must wark'. The son might assume, on his father's part, standards of expectation which each success merely set higher. Robert could remember the time when his father had been Geordie the Killingworth brakesman – he had been 9 when the High Pit engine was fixed. He had been bound to the rocket whose trajectory rose until, instead of 'Geordie', there was George Stephenson Esq. of Tapton House, 'The Hengist of Railways'. Now George was dead, and the perky boy who had carried the miners' picks was a man of 47, who had had to find out how to make his own light within the bright glare and deep shadows created by George's confrontation with the world.

For Robert, to have that massive presence in his life was like living on a satellite of Jupiter: the view of the universe could only have a skewed perspective. In his older teenage years he showed considerable resource, and some necessary deviousness, in getting away from that tremendous gravitational pull and declaring his own independence. A succession of lesser mentor figures helped him, wittingly or otherwise, to achieve this: Wood, Losh, James, Longridge, Thomas Richardson, with William James perhaps most important, as the first to get him into a setting where being 'Geordie's boy' did not count. On Robert's return from Colombia, he quickly established himself at a distance from his father, and was the prime architect of the relationship that ensued: two lives, two households, but an intimate bond of affection and experience; a shared vision of what had to be achieved; collective views of what was necessary to achieve it; mutual support and advice, and a trust in technical matters that went all the way back to the bangs and flashes of the safety lamp experiments.

For a few years, from 1829 to 1835, there was virtually a single entity as far as the engineering aspect of their lives was concerned, a George-and-Robert. This ended with Robert's appointment to the London & Birmingham Railway, but by no means halted their consultations and collaboration, and George was fully integrated into the London chambers. Right up to George's death, Robert would talk to him about 'our' projects,[25] and it seems that in a real sense he felt his achievements were his father's as well as his own, as if George was working through him. Now there was no one he had to prove himself to. The dynamic between the 'real Robert' and the 'ideal Robert' was no longer under tension.

From around 1849, another factor increasingly dominated his life – his own health. Spells of illness of varying duration and degrees of severity, but always returning, interrupted his enjoyment of work and relaxation. At such times he was tetchy and bad-tempered, showing 'a passing peevishness and irritability on trivial provocations'.[26] Stephenson's physical ailment was a chronic form of nephritis, a malfunctioning of the kidneys whose effects included the presence of albumin in the urine, back pains, nausea, feverishness and accumulation of fluid in the body. Swelling became first noticeable in the face and ankles, but 'the genitals may swell alarmingly' and hypertension is also a feature of the condition.[27] The connection of these symptoms and signs with kidney disease was a recent one, made by Dr Richard Bright in 1827, and the condition was then known as Bright's Disease. Although it was also linked to alcoholism and venereal infection, these were far from being the only causes or aggravators, and it is possible that Robert's exposure to bad air in his time as a mining apprentice played a part; as also the chills and draughts of his London & Birmingham years.

Medical science at the time was still based on purgatives and blood-letting, in the belief that the causes of disease could be flushed out: 'The evils that are attendant upon an inattention to the due unloading of the bowels, almost surpasses the common belief.'[28] The great specific was calomel, or mercury chloride, an insoluble, almost tasteless yellow-white powder that was prescribed for a whole range of ailments, and was so common that the medical chapter of Mrs Beeton's *Household Hints*, first published in 1859, included instructions for pills and powders containing it. As Francis Conder had noticed, Stephenson had been taking calomel ever since his time on the London & Birmingham Railway. It was not a narcotic, but its use was normally combined with opium, in order to relieve the 'morbid irritability, accompanied with restlessness and want of sleep' induced by mercury.[29] A letter of December 1857 from Stephenson to Brunel mentions: '... this morning I had a rather severe bilious attack which the Doctor says a strong charge of Calomel will work off perhaps tonight ...'[30]

Unfortunately for Stephenson and other sufferers – including, by a curious parallel, Brunel himself (who was one of Dr Bright's patients until the latter died in 1858) – calomel was of no benefit whatsoever and the accumulation of mercury helped to generally weaken the system. The 'dropsical' side-effects of nephritis account for Robert's somewhat flushed and puffy complexion in his later years. Constant cigar-smoking – 'he smoked cigar after cigar, shall I say furiously?'[31] – had its own impact on his nervous system and lungs. Some doctors were already warning against tobacco and Stephenson, with leading medical men among his friends, might have known about Dr John Lizar's *Cigars: the Uses and Abuses of Tobacco*, published in Edinburgh in 1856, which listed many unpleasant possibilities as consequences of smoking – but he smoked to the last. Diet may also have worsened rather than helped matters: 'efforts to maintain the strength and improve the quality of the blood by strong nourishment' were part of the treatment.[32] Smiles noted: 'He was habitually careless with his health and perhaps he indulged in narcotics to a prejudicial extent. Hence he often became "hipped" and sometimes ill.'[33] Perhaps a degree of fatalism entered into his attitude, since it must have been clear that Dr Frederick Bird, who attended him in London, was unable to effect any sort of cure or to prevent the attacks from becoming more frequent and severe.

But Jeaffreson also shades matters more darkly: '... few suspected how he struggled against melancholy, and how he looked forwards to death. The quiet of his house, when it was without guests, he could not endure. Often he walked about the lonely rooms, and sat down to yield to sorrow which in the presence of others he courageously suppressed.'[34] This was a very different Robert Stephenson to the convivial host and diner. Depression could be an effect of the calomel treatment, but 'Melancholy' at that time was still a medical term, since replaced by melancholia, and still seen to have its ancient link to black bile, one of the 'four humours'. It was an illness rather than a mood, and he had to struggle against its attacks just as he had to cope with bouts of physical illness. Characteristics of the condition include an exaggerated sense of anxiety and a lowering of self-esteem. How much of a part his religious beliefs had in sustaining his spirits is hard to say. Unlike his father, he was always a convinced Christian and a committed Anglican, but religion is scarcely referred to in his surviving letters except for growls against Catholic emancipation, and he had no close association with any church. He did insist on reading divine service on Sunday mornings to the crew and guests on *Titania*, but this was conventional practice in the navy-influenced Royal Yacht Squadron. It seems that religion was a background influence in his life, reflecting a deep concern for order and structure, rather than a spiritual yearning or a sense of divine presence.

# 19

# TITANIA AND AMERICA

A few days after the opening of the Royal Border Bridge, Stephenson and Bidder set off for Switzerland. Only two short railways had so far been built in that country, and the 2-year-old federal government was keen to expand the system. Individual cantons controlled railways within their borders and there was controversy as to gauge and which inter-cantonal routes should be authorised. With little experience to draw on at home, the Swiss consul-general in London was given a list of British engineers to approach:

Stephenson, Robt., MP, London
Brunel
Vignoles
Cubbit
Parker Bidder  George St., 24, Westminster.[1]

When contacted in the early part of 1850, Robert Stephenson had been reluctant to get involved personally and proposed that George Bidder should act as the Swiss government's consultant. Foreseeing problems of inter-cantonal rivalry, the Swiss were keen to have Stephenson's fame and authority applied, and it was agreed that Henry Swinburne would be sent out to make a detailed investigation and that Stephenson would come over at the end of the parliamentary session (his rate, incidentally, was still 7 guineas a day).[2] The two engineers, both sailing enthusiasts, were the only passengers not to be seasick on a stormy passage to Ostend,[3] where they were joined by Sanderson and his wife; the journey became a tour in a special train provided by the Belgian Railways' chief mechanic, Henry Cabry, one of George Stephenson's 'young men', who had gone out to Belgium with a locomotive from Forth Street in 1836 and stayed on.[4]

Pausing to inspect stationary engines on the steeply graded Liège–Aachen Railway, they went on into German territory. Having admired the scenery as they travelled up the Rhine by steamboat from Cologne to Mainz, they took a train to Wiesbaden and gazed with the disapproval of respectable Victorians at the gaming rooms of its casino. Bidder gave his wife details of one of their meals in Offenburg: 'as a specimen of a good German Inn dinner I'll give you a list of our dishes – Soup, Trout Cutlets, Boiled Beef Potatoes (admirable), Cauliflower, Sour Krout, Carrots sliced, Omelet, Roast Hare, Partridge, Roast fowl, stewed Pigeons – Coffee Custards a beautiful pudden with a delicious sauce – to drink 1 Bottle Rudesheimer – 1 d° Sellar (a red wine) 1 d° sparkling Hock Kirschwasser – Coffee Noir and then to Bed …'[5]

On arrival in Basel on 11 September serious activity was added to the sightseeing, examining possible routes between Basel and Zurich, and up the Rhine valley as far as Chur. Bidder and the Sandersons returned to England on 24 September, but Stephenson remained in Switzerland until mid-October, planning railway lines in Ticino Canton, and studying potential trans-Alpine routes into Italy with Henry Swinburne. Both men signed the final report for the Swiss government on the 12th. Robert was pleased with what he had accomplished in Switzerland, believing that he had laid the foundations for a sensible railway system which provided the necessary communication without duplication. It was, however, a very conservative set of proposals. Even in 1850 he was pessimistic about the ability of locomotives to mount gradients with trains, and the plans provided for rope-haulage on inclines and locomotive use only on relatively level track. He found no way for a railway to cross the Alps towards Lake Maggiore or Lake Como, though any engineer who proposed the necessary gradients and long tunnels at that time might have been considered a wild man. Although his findings were accepted by the Swiss Confederation, within two or three years they would be overtaken by railway technology developed within Central Europe.

After more than five years' work, on 19 October 1850, the Britannia Bridge's final span was opened. Successful completion of the project established the tubular wrought-iron bridge as the state-of-the-art solution to long-span railway viaducts (for a few years), and the Stephenson practice was already building another example over the River Aire.[6] Details of the design, passed on through Clark's and Fairbairn's books and the Proceedings of the Institution of Civil Engineers, were noted by engineers all over the world. With that long saga finished, Stephenson made his first visit to Egypt, in *Titania*, between November 1850 and February 1851. Primarily for pleasure, according to Jeaffreson,[7] it also had a professional aspect, and Michael Borthwick and Henry Swinburne, two of his engineers, came with him. Even in the far-off United States, his purpose was reported in *Scientific American* in January 1851: 'Mr. Robert Stephenson, the celebrated engineer, has gone up the Nile to inspect the barrage of that river, which has been carried on for many years, and take a survey of the Suez desert, with regard to the practicability of a railroad between Cairo and Suez. Any suggestion from Mr. Stephenson would be very valuable in this country.'[8] Nearly two years before, in March 1849, a British commercial delegation headed by a former Lord Mayor of London (and P&O director), Sir John Pirie, had come to Egypt to discuss the construction of an Alexandria–Cairo–Suez railway with the new viceroy, Abbas Pasha, who had taken over from the senile Mehmet Ali. Abbas chose to favour the British rather than the French, who had dominated his grandfather, and was receptive to the idea of the railway.

Swinburne's diary of the expedition is preserved at the Institution of Civil Engineers. The three men shared their boat with an interpreter, a native guide, a Nubian servant, Mr Thomas, *Titania's* cook, and a quarrelsome Egyptian crew. Poor Swinburne suffered from severe diarrhoea in the early days and feared cholera. Stephenson was a considerate master: 'Chief administered opium and Calomel';[9] on 1 December the sufferer was still 'making up bread rolls into each of which ¼ grain of Calomel was carefully introduced'. By the 15th he was well enough to set about his main task, an English translation of Talabot's substantial report on the Isthmus of Suez. The barrage, at the head of the Nile Delta, was another French project, designed to control flooding and irrigation. Welcomed by the French, they inspected the works and then moved on upriver, poled by the crew when the wind failed, pausing for 'geological and coprolitic strolls' as well as archaeological excursions at Thebes and Luxor, where they were greatly impressed by the architecture, but apparently uninterested in its historical background. By 10 December they reached Aswan and rode to Philae, where they rambled on Elephantine Island.

Returning northwards, one day they tried crocodile cutlets: '... very unequal as to tender and tough – part being very fibrous and part excellent.'[10] Christmas Day was passed, 'reading, smoking, writing ... and not the least important part of our Anglo-Saxon religious observances – eating and drinking to the honour of the day ... Dominoes and whist aided the digestion.' Back in Cairo, on 27 December, letters were waiting: 'Mr Stephenson's news was of a cheerful tone, a heavy outstanding bill has been paid at last. Chester & Holyhead shares have risen to nearly double their price when I left England in June last, when I was forced to be a seller against my better judgement, consolatory for Mr S, however who bought some soon after that.' While in Cairo, Swinburne persuaded his two seniors to try the Turkish baths, which sadly, 'I was surprised to find inferior in every way to those of Siout, and not least so in the want of decency on the part of the chief *scourer* & manipulator who greatly shocked my friends and rather disconcerted my more hardened self'. At dinner on 29 December they '... heard something of the Chief's interview with the Pasha [viceroy] who seems to be a very civil and approachable man tho averse to being victimised by Engineers'.

On 30 December Swinburne 'after dinner coached the Chief through the last three chapters of the *Isthme de Suez*', and next day they travelled to Suez itself, where they spent a week on the coast, looking for the best access valley for a railway line from Cairo.[11] On the 8th they set off inland, mostly on camelback, noting vestiges of the ancient (280BC) canal, and reaching the southern edge of the basin of the Great Bitter Lakes. From here they turned westward back to Cairo, arriving on 13 January; then on the 16th they rode north again, on packhorses now: 'Our proper course lies along the Shoubra road to the Barrage where it crosses to the west side of the Rosetta Branch & down by the skirts of the Lybian hills to Lake Mareotis – the most probable line for a railway between Alexandria and Cairo of which there is some talk among the English party here just now.'[12] By the 22nd they were in Alexandria, and on the 23rd, 'Chief paid Pasha a long visit, Dined with him'. On 25 January *Titania* set sail again from Alexandria on the long journey home.

Plainly Stephenson had abandoned the idea of a canal in favour of a railway even before perusal of Talabot's report enabled him to state that a canal was 'quite out of the question'.[13] But Talabot did not dismiss a canal, proposing a lengthy alternative route which involved an aqueduct over the Nile to reach the sea at Alexandria. Negrelli remained convinced of the practicability of a direct canal and actively promoted the idea. In Stephenson's discussions with Abbas Pasha, the subject was a railway, and 'the result of the intercourse between the Pasha and the engineer, was a commission from His Highness the late Said [sic] Pasha to connect Cairo and Suez by a line of railway'[14] in April 1851. Middle Eastern potentates did not grant railway commissions out of conversations with touring engineers, and Stephenson's contract was the fruit of a campaign lasting several years, driven by the British consuls in Alexandria and Cairo. In a placatory gesture to the French, the railway project was reduced to an initial stage of Alexandria–Cairo,[15] and arrangements were made in London to facilitate loans to the Egyptian government to cover the building costs. Announcement of the railway contract dismayed the Canal Study Group: 'They alleged it was being built as a result of British influence as a way of stopping the building of a canal … Stephenson was accused of acting against the interests of the Study Group of which he was a member.'[16]

Robert Stephenson's surviving letters are mostly of a business or professional character, but a rare surfacing of feeling comes in a letter written from Malta on his way back to Egypt, to Edward Starbuck. After describing a 'terrific gale' in which they had to lie to, and 'ate cold viands off the cabin floor', he fulminates against the Vatican's re-establishment of a Catholic hierarchy in England under Cardinal Nicholas Wiseman: 'I am now very anxious about getting home as I feel I am deserting my colours in the matter of papal interference. It is a sad exhibition of the worldly character of religion – a battle as to the mere *form* in which the creator shall be worshipped – the true spirit of Christianity is never allowed to appear.' As far as Stephenson was concerned, the Pope 'must be a fool as I always considered him. Wiseman, on the other hand, is a cunning, designing, talented rogue'.[17] It was the orthodox line of outraged Protestantism at the time, but evidently he felt deeply on the issue. It does not seem that he made any public statement on the matter, however.

Egypt remained the eastern limit of Robert Stephenson's travels, but he undertook consultative work for early railway projects in India, beginning in 1845 when
he was appointed consulting engineer to the Great Indian Peninsula Railway and
the Madras Railway, both of them under the auspices of the East India Company,
which was still the colonial 'owner' of the sub-continent. Though he provided
advice based on local surveys in 1847, little happened until 1850, when a line
inland from Bombay (Mumbai) was approved, and he supplied cost estimates,
contacted potential contractors and supplied an engineer, James Berkley, from
the Great George Street offices. Consultation, facilitation and strategic advising
of this kind was typical of Robert's business activity in the 1850s. After several
years' delay because of financing problems, under a plan organised by Stephenson
and involving a British construction consortium, Brassey, Peto & Betts, the
Norwegian *Storting* approved the deal for the railway north from Christiania on
15 March 1851, with Stephenson to act as arbiter in the case of any dispute.[18] At
home, too, he remained in demand as an arbitrator in disputes between railway
companies. Such work was largely sedentary in its nature and could mostly be
done in London. The triumphant opening of the Great Exhibition, in its 'Crystal
Palace' on 1 May 1851, did not mark the end of Robert's association with it, as the
Royal Commission continued to supervise the event, and then its winding-up
and the investment of its substantial profits; his last attendance at one of its meetings was on 14 May 1858.[19]

In a paper given at the Institution of Civil Engineers on 20 May, reporting on
his visit to Egypt, Stephenson first publicly announced his opposition to the Suez
Canal project. As far as a canal was concerned, '... it was evident, that it would not
be practicable to keep open a level cut, or canal without any current, between the
two seas, and that the project was abandoned'.[20] But only he had abandoned it.
The scheme merely receded from public view for the time being.

In a pattern which would continue until his death, he led the model existence
of an engineer at the top of his profession, one of its elder statesmen rather than
an active practitioner. Both at his office and his home, he was in demand from a
wide range of callers, with or without letters of introduction, wanting advice or
help – 'literally persecuted' by speculators, inventors and inquirers: 'When casual
indisposition kept the engineer from Great George Street, and confined him to
his house, swarms of talkative, and for the most part profitless clients intruded on
the privacy of the man whose too pliant temper laid him open to their annoyance.'[21] At the office it was easier for John Sanderson to keep invaders at bay. But
at Gloucester Square there were also invited guests, and 'few private entertainments in London were more pleasant than his "Sunday lunches" at which many
chiefs of literature and science were in the habit of meeting'.[22]

Among Stephenson's close friends were Sir Roderick Murchison, the geologist; Dr John Percy, the metallurgist; John Graham Lough, the sculptor (a fellow
north-easterner from Shotley in Northumberland); the artists John Lucas and

Frederick Lee; Samuel Sharpe, an Egyptologist; Joseph Bonomi, Egyptologist and
brother of Ignatius Bonomi, the Newcastle architect; Baden Powell, scientist and
theologian, who Robert first met through the Royal Society and whose 'Broad
Church' views Stephenson would have found congenial; and Dr Mayo – either
Thomas Mayo, then president of the Royal College of Physicians, or his brother
Herbert, a physician and mesmerist. Brunel was a frequent guest, as well as engineers from the Great George Street office. Apart from the publisher George
Routledge, whose cheap 'Railway Library' tapped a lucrative new market, literary
men are notably absent – though Samuel Warren, an almost-forgotten novelist, successful at the time, was a guest on *Titania*[23] – as are politicians. The circle
closely reflected Stephenson's own interests, and its sedate, conservative character
was far removed from the storms and restless schemings of the younger generation of Royal Society scientists like Thomas Henry Huxley and John Tyndall,
determinedly hacking into the foundations of received doctrine and setting up
standards of investigation that would leave amateur geologising and mushroom-
hunting as the merest hobbies.

A reminder of his father's influence, going right back to 'Dial Cottage', could
be seen in the drawing rooms, '… so liberally stocked with works of curious con-
trivance, and philosophical toys, that they had almost the appearance of a museum.
Singularly constructed clocks, electric instruments, and improved microscopes,
by Smith and Beck, and Pillischer, were arranged on all sides.'[24] Display cases
held geological samples and mineral sections prepared for the microscope. On
the walls hung paintings and engravings bought by or presented to Stephenson.
They included a Landseer, *The Twins*, of two sheep dogs with a ewe and lambs;
a work by Francis Danby, *The Evening Gun – a Calm on the Shore of England*; and
*Tilbury Fort – Wind Against Tide*, by Clarkson Stanfield. Robert's favourite painter
was John Lucas, a celebrated portraitist of the time, and he owned several Lucas
works, including a full-length portrait of George Stephenson at Chat Moss, a
study of Robert himself at the Britannia Bridge and another of Robert seated,
with a drawing of the 'long-boiler' locomotive, which had been presented to
him by the shareholders of the London & North Western Railway in 1845. Also
by Lucas were *Killingworth Colliery* and *The Stepping Stones*, showing a girl crossing a stream with the Britannia Bridge in the background. Lucas' most famous
Stephenson paintings were group studies. His *Conference of Engineers at the Menai
Straits Prior to Floating a Tube of the Britannia Bridge*, presented to Stephenson by
a group of friends led by Samuel Peto, shows a gathering which, if true in spirit,
never happened in reality. Peto also commissioned (in 1851) the even more fictitious *Stephenson Family Group*, showing George in a plaid suit and miner's hat,
with his 'Geordie lamp', a bare-kneed Robert with a short clay pipe, an elderly
man who is probably 'Old Bob', a woman with a little girl, perhaps Robert's
mother and his sister, who died as a tiny baby and another young woman carrying a tub on her head and holding a metal jug, who may be Aunt Nelly. In

the background is Killingworth colliery, and a *Blücher*-type engine approaches, passing George Stephenson's cottage. Two dogs and a miner's pick complete the rustic-industrial idyll.

As an art collector,[25] Robert was of the kind dismissed by John Ruskin in 1838, who 'remain in reverence and admiration before certain amiable white lambs and water-lilies, whose artists shall be nameless. We see them, in the Royal Academy, passing by Wilkie, Turner and Callcott, with shrugs of doubt or scorn, to fix in gazing and enthusiastic crowds upon kettles-full of witches, and His Majesty's ships so-and-so lying to in a gale, etc., etc.'[26] But Ruskin's views on art and life had not penetrated the carapace of 1850s middlebrow opinion. Incidentally, though most surviving portraits and photographs of Robert Stephenson show him as clean-shaven, Jeaffreson noted that on occasions (perhaps to conceal the effects of his illness) he sported a moustache and beard, and in some photographs and engravings he is shown with a fringe of facial hair under his chin.

Stephenson was an excellent and considerate host: 'No description of his demeanour in the society of men would be complete which did not contain the word "jolly". He was the embodiment of joviality, without the faintest touch of boisterous awkwardness.' Furthermore, 'his courteous bearing to ladies possessed the style of ancient chivalry'.[27] Something about these expressions prompts one to remember that Jeaffreson was writing an 'authorised' biography, but there is every reason to believe that Robert was charming and engaging company, when he felt relaxed and well; just as many little asides in his letters show glints of characteristic humour. He felt most at his ease in his clubs, where he could smoke cigars and talk engineering and topics of the day with men whom he knew well, in an ambience of masculine comfort. As a Conservative, he was elected to the Carlton Club; as a man of learning, to the Athenaeum. He was also a member of the Geographical Society. Between these, his home, his office, the Royal Society, the House of Commons, the Institution of Civil Engineers and the RSA, his London life was structured. In all these contexts, his status was assured as one of the country's leading engineers – perhaps *the* leading engineer, though the Rennie brothers' friends might have disputed that. Isambard Kingdom Brunel was not seen as a rival for this position: Stephenson and the Rennies were more 'establishment' figures than Brunel, whose work continued to be strongly experimental and individual. During the 1850s, Stephenson and Brunel drew closer to each other and their friendship deepened from a mutual professional respect into a real liking and a shared correspondence in various topics that interested them both.

Harbours were expanding to cope with increased maritime trade, and the tidal range round Britain's coasts made it necessary to build docks in order to cope with ships of increasing size and draught. Complex issues of construction in relation to tidal flows, silt deposits, building costs and future extensions, combined with the financial anxieties of port authorities, made dock construction a rich field for consulting engineers, and in the second half of 1850, Robert Stephenson was commissioned to report on the plans for Grimsby Docks. In June 1851, with Admiral Sir Francis Beaufort (Hydrographer to the Admiralty and deviser of the wind-speed scale), he made a detailed inquiry into a scheme for completion of the partially constructed basin and docks at Birkenhead. Such requests testify to his expertise and pre-eminence, but in two important areas of engineering development in the 1850s he failed to take the side of progress, and in both, his judgement as a mechanical and a civil engineer is open to question. One related to locomotives; the other was the already ongoing saga of the Suez Canal project.

Even though in 1829 *Rocket* had shown that a locomotive could pull a train (admittedly lightweight) up a gradient previously supposed to be possible only for fixed rope haulage, both Robert and his father had kept a narrow view of the potential performance of locomotives on steep or long gradients. In this they were not alone, though a few engineers, with Joseph Locke as the leading British exemplar, were much more prepared to build up-and-over railways like the fell-breasting Lancaster–Carlisle line. Brunel pitched his steep South Devon gradients only in the expectation of atmospheric traction sucking the trains uphill. As railways spread, it became more difficult to avoid mountain barriers, and a challenge was presented in the Austrian Empire, where a line was under construction in 1850–51 to link Vienna, the Imperial capital, with its major seaport, Trieste, on the Adriatic Sea.

The lowest way through the mountains took the line over the Semmering Pass, at a height of 2,880ft (878m) with long gradients, in each direction, as steep as 1 in 40 (2.5 per cent). Suitable locomotives were needed and, in a replay of Rainhill, a competition was announced in May 1850, with a prize of 20,000 florins for the builder of the winning locomotive, which would have to draw 140 tons, exclusive of its tender, up the line at 8mph (12.5km/h). The Rainhill debate was also replayed, with a strong body of opinion declaring that locomotives could not work the line and that cable operation was the only real possibility. Here, too, more than prize money was at stake – a locomotive that could pull trains over the Semmering line would be of immense interest to many other countries.

The tests took place between 31 July and 16 September 1851. It is believed that three or four British builders (and some Americans) expressed interest, but two were eliminated because 'they had no engineering workshops of their own in Austria and could not therefore guarantee spare parts',[28] and the only one mentioned by name, Sharp Brothers of Manchester, was said to be too much engaged in the Exhibition of 1851 to take part. Two expatriate British engineers, John Cockerill at Seraing in Belgium and John Haswell at the State Railway Works in Vienna, built special locomotives for the purpose. Britain's premier locomotive builder did not compete. While direct comments by Robert Stephenson on the Semmering trials are not recorded, it is likely that he thought the task was beyond a steam locomotive. His own recent survey work on Swiss railways had provided for cable working of comparable, but less daunting routes. Though the plans for the Norwegian Trunk Railway also involved gradients of 1 in 40, they were not of Alpine length, and the entire line was only 42 miles (67km) long. But an engine with eight driving wheels, *Bavaria*, built in Munich by Josef Anton Maffei and his English draughtsman, Joseph Hall, pulled a 130-ton train up the gradients at 11.4mph (18km/h) and won the prize. Cockerill's and Haswell's engines also performed respectably.

Once again the steam locomotive had triumphed over scepticism, and this time the scepticism was shared by the man who had fought so hard for it twenty-two years earlier. It was evident that much development work remained to be done, but the result was a considerable boost to the fortunes of the German locomotive builders, who now set out to produce bigger locomotives with scientifically assessed tractive power. In Britain, locomotives remained small, mostly with a single driving axle. A promising market slipped away from the British builders. Even as the Great Exhibition was promoting Great Britain as the world's workshop, in some fields at least, other countries were showing greater inventiveness and imagination.

Late in July 1851, Robert Stephenson was engaged in an exercise which shows how completely he was now dissociated from George Hudson's activities: he was retained by the York, Newcastle & Berwick Railway to act as arbiter in a case against Hudson's leasing of the Newcastle & Carlisle and Maryport & Carlisle railways 'without any sanction or assent of the shareholders'. The YN & B wanted £3,768 11s 3d; Stephenson, having considered the matter, awarded them £793 7s 11d.[29] Compared to other claims against Hudson, this was hardly more than a flea bite.

At the end of the month he went to Norway again, either with, or joining up with, George Bidder, to see work start on the Norwegian Trunk Railway. On 28 July they viewed a total eclipse of the sun in southern Norway, a 'solemn spectacle', as described by Bidder,[30] then after a few days spent following the route of the line up to Eidsvold and travelling up Lake Mjøsa to Lillehammer, they were back in Christiania for the ceremony on 8 August. Actual construction was subcontracted to the English firm of Earle and Merrett. Here, too, Bidder was engineer-in-chief, but Robert was consulting engineer and took a personal interest in the work. By now, George Bidder was doing a great deal that Stephenson would have done a few years earlier, but he was something of a workaholic, and was also earning a lot of money – an important consideration to a man with a large and growing family.

The Stadholder, or King's Representative, cut the first sod, to the accompaniment of a specially composed song from the united singing clubs of the town. At a dinner in the evening, Stephenson made a speech, complimenting John Crowe, the British consul, as 'father of the enterprise'. The role of officials like Crowe, with a remit often virtually ambassadorial rather than merely consular, in the export of British technology and the overseas investment of British capital, was often a crucial one. Stephenson and Bidder then headed south to Denmark with Samuel Morton Peto, who was proprietor of the East Anglian port of Lowestoft and keen to expand its activity, with an eye on the Danish trade as well as on possible contracts for railway construction there.[31] Stephenson's role in Denmark was limited to drafting a pattern of strategic rail routes, and Bidder undertook the role of engineer-in-chief when construction actually began. That summer Robert had sent Michael Borthwick back to Egypt to carry out a full survey of the Cairo–Alexandria Railway, but instead of the direct line running west of the Nile Delta, as desired by the British, whose interest was in the direct sea-to-sea link, the viceroy insisted that the railway should cross the delta region to serve the town of Tanta; a route that would require crossing two major branches of the Nile.

While the continental locomotive builders were fettling up their designs, in the second half of August 1851 Robert Stephenson was at the Royal Yacht Squadron's Cowes regatta on *Titania*. The event was of special interest this year because a big new yacht, *America*, from the New York Yacht Club, had sailed across the Atlantic by invitation to compete. There was a link with the Great Exhibition, which by title was for the manufactures of 'all nations', and *America* was intended to show the qualities of US boat-building. On 22 August she was entered for the Royal Squadron Cup race – a circumnavigation of the Isle of Wight, along with eighteen other yachts, including *Titania*. Only fifteen raced, and *Titania* was not one of them, although she took up a starting position. The reason for Stephenson's non-participation has not survived.[32] *America* won, with the British *Aurora* in second place. At the Needles, *America* was so far ahead that no competitor could be seen; the story survives that Queen Victoria, on the royal yacht, inquired at that point: 'Who is first?' to which the answer was '*America*, ma'am.' 'And who is second?' 'There is no second, ma'am.' After his victory, hungry for further action, and prize money, John Stevens, *America*'s skipper and Commodore of the NYYC, posted a notice challenging any vessel in the Royal Yacht Squadron to a further race, for a prize stake of anything up to £10,000. Having seen *America*'s prowess, the paladins of British yachting prudently sat on their hands. Robert Stephenson, no doubt regretting his failure to take part in the race on the 22nd, was the only one to answer the challenge, though the stake was a sportsman's £100 rather than a gambler's thousands. The two-ship contest took place on 28 August: 20 miles out to the Nab Light and back again. *America* was the winner by a wide margin. Speaking at a celebratory dinner on his return to New York, Stevens recalled the event:

In the race with the *Titania*, I suspect – although I do not know the fact – that too much of her ballast was taken out. It gave her an advantage in going before the wind but told very much against her in returning. There was a steady breeze and a good sea running, and she fell so rapidly to leeward as to be hull down and nearly out of sight. We beat her, according to the Secretary's report, three or four minutes going down, and some forty-eight or fifty minutes in returning, on a wind.[33]

Though it has been stated that *Titania*'s race was the inception of the later America's Cup international races,[34] this is not the case: the 'Cup' race was the one in which she failed to take part. At that time, trophies were kept permanently by each year's winner, and Stevens bore the cup home in triumph; in due course it became the America's Cup, the most prized yachting trophy of all. But Robert Stephenson has the credit of taking up a challenge that all others declined. *America*'s design, based on New York harbour pilot boats, was a good deal broader in the beam than that of British racing yachts. *Titania*'s builder, John Scott Russell, later criticised Britain's 'antiquated theories of yacht measurement by which the keel served as the measure of tonnage instead of the waterline'. He had been 'forced to cut two large slices off the *Titania* on each side of the waterline'.[35] This accounted for the 'wetness' Stephenson had experienced on her trials, and reduced her speed. The RYS rule that penalised broad-beamed yachts was abandoned, to the benefit of British yachting.

Continuing to cut back on his involvement with business affairs, Robert Stephenson sold his holdings in the Clay Cross mines and mineral works on 17 October 1851. Three other directors, Sandars, Hudson and Claxton, also sold out, leaving Walmsley, Peto and Jackson as owners. Hudson, now being pursued through the Chancery Court by the vengeful York & North Midland Railway Company, was badly in need of money and had sold off most of his other property. Robert was rich: even if he took on no new commitments, he enjoyed a steady income from the locomotive works and his role as consulting engineer to the London & North Western (with Joseph Locke), the York, Newcastle & Berwick, and the Chester & Holyhead Railways.

In 1851–52, not years of great industrial expansion, he seems to have been very lightly employed, though he acted as arbitrator in a number of disputes between railway companies. Michael Bailey states: 'The last years of Stephenson's life were devoted to a volume of work that few men of lesser constitution could have withstood',[36] but there seems little ground for this assertion. These were years of winding down. Robert himself wrote from Newcastle to Smiles on 4 October 1855: '... I have decided (and I have acted on the decision for the last two years) to withdraw myself entirely from all new professional engagements. This I have done chiefly on account of my health not being very good.'[37] As with his father, intention and practice in this respect did not wholly correspond, but Bidder, only three years younger, was far busier and led a much more strenuous life.

In the case of many ongoing projects originated at Great George Street, Robert's role was that of engineer-in-chief, which, while it required him to be up to date with what was going on, to offer comments and advice, to help with contracts and contacts, and to nominate new or replacement staff, did not need his physical presence and direct participation in the details of construction. He did not visit the Great Indian Peninsula Railway works, where there was a capable team under James Berkley; although he (with Bidder) advised the Swedish government in 1853 on the creation of a railway system in the south of the country, it was done through reports and surveys sent back to London by his emissary, William Lloyd, and without a personal visit. The overseas projects in which he took the greatest personal interest were the Norwegian Trunk Railway and the Alexandria–Cairo line, and both offered the prospect of holiday as well as work.

Construction of the first public railway to be laid on the African continent began on 9 February 1852. Here, too, Stephenson had an effective team, in two divisions under the generalship of Michael Borthwick. Its main engineering features, wholly Stephensonian, were designed by his cousin George Robert and Charles Wild,[38] though Robert himself would certainly have been involved in the conceptual work. Three long bridges, with central opening spans, were of tubular girders, but with the rails on top rather than inside, and until the last one over the Rosetta branch of the Nile was completed, a giant train ferry was used. This vessel was designed by George Robert and manufactured at the Forth Street Works before being sent, in parts, to Egypt for assembly.

Undoubtedly Robert was busy, conscientious in what he took on and always aware of his overall responsibilities, but he was setting his own pace and choosing what to do. After the Britannia Bridge he undertook no project which stretched the limits of technology or human resource. The contrast with Brunel, heaven-storming to the end though equally beset by ill-health, is plain. But Stephenson had a range of businesses to mind. While it is, of course, the fate of the senior executive to end up reading documents in the office rather than solving challenges in the field, he chose to search for commercial opportunities rather than personal engineering challenges. Involvement with the affairs of the Exhibition Commission, and the Institutions of Civil and Mechanical Engineers, and the RSA, also made demands on his time. And he was still a Member of Parliament.

Following the fire that had razed the old House of Commons in 1834, the present building was inaugurated in 1852. Immediately members began to complain about the ventilation and lighting of the debating chamber, and Stephenson and Locke were drafted on to a Select Committee to work with Sir Charles Barry, the architect, and D.B. Reid, the engineer, to resolve the problems. With members' comfort at stake, the task was accomplished with un-parliamentary speed, by May. Stephenson revealed his feelings in a letter to Brunel on 7 April 1852: the Chamber's technical problems 'have saddled Locke and myself with the management of Barry & Reid during the Easter recess and to superintend the alterations proposed to be made in the H. C. to improve if possible the present insufferable state. I feel as Serjt. Murphy says "like a cat in hell without claws" but as we have undertaken it I am determined to stick to it.'[39] As already noted, Stephenson was liable to be beset by importunate callers wanting his professional advice, services or support, and found it difficult to turn them away. His brother-in-law John Sanderson was better at this, and with a touch of Victorian humour, Jeaffreson gives a little dialogue between Sanderson and the inventor of a new railway brake (a preoccupation of the time):

Inventor: I have a proposal for a new break, about which I should like to ask
Mr Stephenson's opinion.

Mr Sanderson: Indeed, sir. And what can you do with the break?

Inventor: I can stop a train instantly.

Mr Sanderson (with an expression of horror in his countenance): Good heavens, sir! If you did that, you'd kill all the passengers!

Inventor (suddenly modifying his statement): But, sir, – I can stop a train gradually.

Mr Sanderson (bringing the interview to an end): So can anybody else.[40]

Such unwanted callers were one reason why Robert Stephenson enjoyed being
on his yacht; in a letter of 25 May 1852 he wrote to his old acquaintance C.R.
Moorsom (who, despite a railway-based career, from 1830 had risen up the naval
status ladder to the rank of Rear Admiral): '*I find I can get no peace on land* … I find
it no easy matter to get rid of a multitude of questions which follow on a tolerably
long professional life … *Ships have no knockers, happily.*'[41] In addition, a large yacht
gave him the perfect support structure, with its combination of a disciplined crew
and chosen guests, in which he could both relax, be looked after and be in control.
It must have come as a terrible jolt when George Robert Stephenson came to
Gloucester Square one evening early in May 1852, interrupting a dinner party, to
blurt out that *Titania* had been destroyed by fire. Also a keen sailor, he occasionally
borrowed the yacht, and having sent an order ahead for fires to be lit on board,
he had arrived to find that an overheated flue had set the vessel alight. Robert's
reaction was to console his distraught cousin[42] and to commission a new yacht
from John Scott Russell. The letter to Admiral Moorsom, already quoted from,
notes that he was 'again cogitating with Scott Russell over the lines of another
yacht'. But for a year he had to forgo the relief of sailing trips. It may be at this
time that he stayed with Joseph Locke – one of the few men outside the family
circle with whom he was on first-name terms, 'Joe' and 'Robert' going back to the
teenage years before they adopted the formalities of grown men – and his party
at Cranstoun's Hotel at Moffat, in the Scottish Borders, the location of Locke's
annual retreat for the shooting season. Sharing his father's detestation of game-
bagging, Stephenson 'enjoyed the conviviality of the life but excused himself from
the shooting expeditions and wandered off on geological excursions'.[43]

One of his rare parliamentary speeches was made on 25 March 1852, when
he rose to address the House on the subject of the London (Watford) Spring
Water Company, which was seeking a Bill to enable it to draw water from the
chalk near Watford and pipe it to London. This was the latest of the series of
attempts, from 1840 on, to tap the vast reserves of water held in the chalk deposits north-west of London. Stephenson's remarks were measured: 'He would not
pledge himself to the allegations of the promoters of this Bill; he merely rose to
assure the House that it possessed merits which ought to be discussed before a
Committee.'[44] Like previous efforts to muscle into the lucrative business of water

The *John Bowes* arrives at the Collier Dock of the East & West India Dock Railway, August 1852. (*Illustrated London News*)

supply to the capital, the Bill was rejected. Robert's political leader, Lord Derby, became Prime Minister in February 1852; he was a Protectionist, but as the head of a brief minority government, was not in a position to carry through any sort of controversial legislation, and Parliament was dissolved four months later.

The impending election overshadowed a significant event on Tyneside on 30 June – the launch of what has been described as the first bulk carrier, the iron-hulled screw steamer *John Bowes* from Palmer's Shipyard in Jarrow. This three-masted ship, with its 60ft-long cargo hatch, its water ballast that could be mechanically pumped in and out, and its ability to carry 650 tons of coal to London and be back again to reload within five days and six hours, made the entire fleet of collier brigs instantly obsolescent, and was hailed by the *Newcastle Journal* as: '... the commencement of a new era in the history of shipbuilding and as the precursor of a fresh order of things imperatively called for by the altered circumstances of the times.'[45] A two-cylinder engine from Robert Stephenson & Co. powered the vessel, but Robert's involvement went beyond that: 'Robert Stephenson believed that *he was, in some measure, in conjunction with Mr Palmer, in first bring* [sic] *into notice the system of screw colliers*.'[46]

Stephenson knew more than a little about iron ships, and his involvement, unnoticed by previous biographers, is attested by an exchange from the General Election of 1859 at Whitby, when, having reiterated his support for the shipping industry, he was interrupted by a voice calling '*John Bowes*'. His reply was sharp:

With reference to that barque of my own, the *John Bowes* has done you no harm. The mind of man has always been contriving something to save his hand, and the more the mind has been given to it the better it has been for man. He had devoted all his life to the investigation of such like matters, and would say to them: 'Do not be afraid of the *John Bowes* injuring your interests; let it excite you to make improvements.'[47]

The ship had cost £10,000 to build and was owned by the General Iron Screw Collier Company, founded by Charles Mark Palmer in April 1852 with a capital of £250,000 – no Whitby brig-owner could compete with that. Stephenson was certainly not the sole owner as his answer suggests, but it is likely he was a part-owner, and was notably unapologetic about it despite the hostility of sail-driven coal-carriers. There is no surviving contemporary record of his involvement, though his friend Nicholas Wood was a vice-chairman at the dinner for 200 given at the shipyard after the launch. Palmer, a self-promoter in the George Stephenson mode, always claimed the entire credit for *John Bowes*, but Stephenson would not have courted unpopularity with his electors without good reason.[48]

At the General Election of July 1852, Stephenson stood for re-election at Whitby. This time he was not unopposed. Some former supporters had been dissatisfied with his performance as an MP. He had rarely spoken in the House, and then – apart from his defence of the Great Exhibition – only on matters to do with engineering. Instead of a robust defender of the Navigation Laws and of Protectionist Conservatism, the Whitbians found they had elected the Member for Great George Street, where the Institution of Civil Engineers was housed. Nor did Stephenson have a good voting record. The Liberals put up the Hon. Edmund Phipps to oppose him. G.M. Trevelyan characterised the politics of this period as '... of quiet Whig-Peelite rule ... when everything was safe, when nothing seemed to matter very much either at home or abroad, and when even to provoke a war with Russia involved only a limited liability'.[49] None of that stopped the election of 1852 in Whitby, and many other places, from being a fine old English faction-fight, complete with hired thugs, open bribery, threats and heavy pressures on individual electors, and a quota of broken heads. Polling was not yet by secret ballot and the men eligible to vote had to proclaim their choice in public. The Whig *Yorkshireman* and the Tory *Yorkshire Gazette* reported the campaign in partisan style; the latter describing Phipps as a 'Whig-Radical' and his supporters as a 'rabble'.

At the hustings on 8 July, a show of hands seemed to favour Phipps, and Stephenson's supporters demanded a poll, which was fixed for the following day. Rival gangs of supporters, meanwhile, went wild in the streets. 'There was a dreadful fight in the yard of the Angel Hotel, and stones flew in all directions'[50] and the town was in uproar until late in the evening. The Liberals accused the Conservatives of bringing in paid thugs by special train; the Conservatives accused the Liberals of dragging down their pink flags and bunting with boat-hooks. On the 9th, Stephenson had a majority of 109, a handsome margin with an electorate of only around 400.

He resumed his seat as a silent backbencher under a Tory/Liberal coalition government headed by the elderly Lord Aberdeen, whose Free Trade sympathies were not congenial to Stephenson. His friend Hudson was also back, struggling with debts and litigation, but supported by the still-loyal citizens of Sunderland. Hudson, like many others, had abandoned his commitment to Protection and was now a vigorous Free Trader, but Stephenson remained one of the minority, still clinging to what was now a forlorn and even slightly laughable cause. Corn prices had not collapsed with the repeal of the Corn Laws, and the country's industrial lead ensured that Free Trade went along with increased prosperity. Protection was 'the one solitary subject about which in his last years he would in discussion lose his temper'.[51] He could be very forceful: listening to Robert argue at a dinner party, Thomas Sopwith reflected: 'He comes to any discussion with great powers of argument ... and states his views at times with a bluntness and hardness of decision which contrasts very strongly indeed with the graceful gentleness – the winning ease and perfect suavity which are the essence of his true character.'[52] Since Stephenson never spoke publicly on the subject, his reasons for this opinion, or article of faith, are opaque. He had little in common with the backwoodsmen of Protection, most of whom were landowners, and he had good friends on the opposite benches, not only Locke, but Walmsley, Peto and others whom he had come to know through business and his clubs; he was after all, a most 'clubbable' man.

In August he made another trip to Norway, combining inspection of the advancing railway works with a visit to the Steinsbye Estate, a big house and farm with a few miles of wild country, which Bidder had just bought and which 'Robert who is in high glee and health will persist in calling Mitcham Stadt'.[53] Mitcham Hall was the Bidders' home. Bidder's letters to his wife now regularly refer to Stephenson as 'Robert', but when Stephenson invited them both for a special evening in November that year, he wrote to 'My dear Bidder':

> I have not for years past been in my native land on my birthday, next Tuesday Mr & Mrs Sanderson intend to commemorate it by a quiet party at 34 – If you and Mrs Bidder will join the party I need hardly say, that it will afford me great pleasure as I shall regard it as a testimony of a long & satisfactory private as well as professional friendship between two who were at Edinburgh together and who were afterwards thrown together by accident in a course of Engineering which can scarcely be said to have a parallel.
>
> Yours sincerely, Robt. Stephenson
>
> P.S. Of course I expect you and your wife will stay overnight – Dinner will be on table at ½ past 6.[54]

# 20

# HEAD OF
# THE PROFESSION

Late in 1852, or very early in 1853, Robert Stephenson had a visit from his former assistant on the Chester & Holyhead Railway, Alexander Ross, newly back from Canada, where he had become engineer-in-chief to the recently incorporated Grand Trunk Railway. In order to complete its route from Portland, Maine, into the Canadian interior, this company faced a major task which had often been considered before, but never attempted – the bridging of the St Lawrence River at Montreal. Here the stream was more than a mile and a half wide, and its bridge would be the longest in the world. A Canadian engineer, Thomas C. Keefer, had already proposed a tubular bridge in 1851, and Ross took up the idea and was authorised to consult with his old employer on all aspects of the project. Stephenson took on the job with enthusiasm. He had not seen the St Lawrence since 1828, but using Ross' notes, and aided by copies of previous surveys, they drew up a design within a few weeks. The river, though wide and fast-running, was relatively shallow, its bed a shifting jumble of boulders. In winter it froze up, and the time of greatest hazard to any bridge or adjacent structure was the thaw, when the current moved piled-up sheets of ice downstream in what the locals called 'the shove' – these could spill right over the banks and wreck unprotected buildings. Any design for a bridge had to take account of the possibility of piled ice and of water-borne boulders, and Stephenson's plan allowed for twenty-four massive piers, their lower faces sloping in the direction of flow, and reaching a height of 60ft above summer water level, well above the maximum height of 'the shove'. The tubular spans were of 242ft (73.75m), except for that over the navigable span, which was 330ft (100.5m) – all shorter than the main spans of the Britannia Bridge. At an estimated cost of £1.4 million the bridge was a national project in the colony, and the plan was sent to the Hon. John Ross, Speaker of the Canadian House of Assembly, by Stephenson on 18 March, and presented to the Board of Railway commissioners at Quebec by Alexander Ross on his return to Canada on 6 June.

Despite his desire to draw back from business activities, it was difficult when old friends were involved. R.S. Illingworth had kept up his South American connections and corresponded with Stephenson about developing machinery for use in gold extraction, but the order went elsewhere. Stephenson wrote that he did not regret losing it, and evidently had doubts about his friend's business associates:

> ... for I am persuaded that however perfect our machinery might have been, it would have been always going wrong and giving me very great annoyance, for I know full well the power of an adverse party at a distance from home to do mischief – Some people are apparently sent into the world for doing mischief and my only regret (which I assure you is most sincere and full of apprehension) is that you should in any way be mixed up with that party to which I need not allude specifically – Do not make yourself at all uneasy, for R.S. & Co. have 18 months work before them.[1]

Another link with his South American years, Thomas Richardson, died on 27 April at the age of 82. His holding in the locomotive works was transferred to Edward Pease's son, Joseph, and the elder Pease recorded in his diary a partners' meeting with Robert Stephenson in Newcastle on 16 May: 'We were most pleasantly met by R.S. who appeared to have a very sincere satisfaction in having his (T.R.'s) shares transferred into Joseph's name, so after my decease my three dear sons will stand possessed of two fifths of that concern.'[2] Things had plainly changed since Robert's grouching at 'Quaker impudence'.

Even before official notification from Canada confirmed approval of his bridge design, Stephenson had left England for Montreal in mid-July, arriving at Portland, Maine, from where a special train was laid on to take him and his companion, Samuel Bidder, northwards. Robert's first ride on an American train, as recalled by Bidder, was an exhilarating one. The Atlantic & St Lawrence line, of 5ft 6in (1,676mm) gauge, was newly opened and:

> ... we found the embankments looking more like the hollows between the crests of the waves in the Atlantic, than anything else I can compare them to. Mr Stephenson and myself stood on the platform of the cars, and had to hold on to the rails by both hands, as hard as possible, to prevent ourselves from being thrown off ... I shall never forget the expression of Mr Stephenson's countenance during the journey. On our arrival at the boundary line, which divides Canada from the States, Mr Stephenson jumped off, and examined the engine, and said it was the most dangerous ride he had ever had in his life, but that he was not sorry he had had an opportunity of proving and witnessing what this extraordinary machine was capable of performing.[3]

It was an opportunity for Stephenson to examine the kind of locomotive design now standard in America, already on a scale much larger than a British engine: 'An engine built on the principle of those now used in England would not have kept on the rails a hundred yards, and yet this Bogie engine took us a distance of 58 miles, over such a road as I have described, in perfect safety.'[4] For Robert Stephenson, this experience of North American railroading and of the qualities of a locomotive when fitted with an articulated set of leading wheels must have been something of an eye-opener, but it did not prompt him to apply the bogie to locomotives built at Forth Street. Admittedly, he would have expected his engines to run on a much more substantial, less undulant and curvy track-bed.

On 19 August 1853, the day on which approval of the bridge design was announced, Stephenson was entertained to a banquet in Montreal. In a long speech of thanks, he warned the Canadians against the unregulated spread of railways such as had taken place in England: 'It was the interest of the whole country that was involved, not that of individuals, nor of particular localities … What was to be gained by ruinous competition?'[5] On the subject of the St Lawrence bridge, he reassured his audience about the ice factor, and hoped he might come to Canada again to see the breaking up of the ice for himself, but, '… there were facts attending the breaking up of ice, which were susceptible of mathematical demonstration, and the pressure of the shove was also capable of being valued. It was therefore as easy as the addition of two and two to estimate the pressure necessary to retain the piers in their places.'[6]

The bridge contract confirmed Stephenson as engineer-in-chief and Ross as co-engineer on site. Building was to be done by the consortium of English contractors already engaged on the Grand Trunk line: William Jackson of Birkenhead, and Peto, Brassey and Betts of London. The completed bridge had to be approved by Robert Stephenson or, in the event of his death, refusal or inability, by 'an eminent civil engineer, to be appointed by the president for the time being of the Institution of Civil Engineers in England'.[7] Construction was to be a lengthy process, hampered by the short building season. The ironwork, 9,044 tons in total, was fabricated at Birkenhead, test-assembled, then each piece was numbered and shipped to Canada. No floating of the tubes was required – they were built out along temporary scaffolding as the piers were completed. Stephenson did not, in the end, return to Canada and never saw the finished bridge, which was eventually opened on 19 December 1859, two months after his death. Even at the time, comments on the structure were respectful rather than awestruck. It was very long – yes – and there had been difficulties of climate to be contended with in its building, but the general public was used to engineering heroics by now, and Stephenson himself had taught them how bold and dramatic a big modern bridge could be. Though the Canadian press hailed it as the 'Eighth Wonder of the Modern World',[8] the Victoria Bridge, as it was named, did not arouse great interest elsewhere, despite its record-breaking length.

Its repetitive series of piers and iron spans did not show the blend of inspired vision and full use of the qualities of the material which had made the Britannia Bridge both breathtaking to look at and novel in construction. The tubular-girder technology was already becoming obsolete. Compared to Brunel's new bridge going up over the Tamar at the same time, the Victoria Bridge was a prosaic structure, described by one contemporary writer as 'a gigantic centipede creeping across the flood'.[9] Even in his supplementary notes to Jeaffreson's biography, Professor William Pole could find no complimentary comment for the design, and merely notes, 'the bridge remains a lasting monument, not only of the engineering knowledge and skill which designed it, but of the energy and perseverance of those who had to carry the design into execution'.[10]

Less than a month before his departure for Canada, Stephenson's new yacht had been launched at Millwall on 21 June. Also named *Titania* (the original boat was sold, rebuilt and renamed *Themis*), it was substantially larger, 90ft (27m) long and of 184 tons burthen; although 'Somewhat deficient in speed, it had every other good quality'.[11] Like the first *Titania*, it was iron-hulled and designed according to Russell's 'wave-form' theory. It may seem surprising that Stephenson did not install a steam engine in this larger vessel, but coal bunkers and an engine room would have used up a great deal of space and he wanted accommodation for his guests and for storage: the new ship was intended for long-distance cruises.[12]

Stephenson might have been looking forward to his first voyage after the fitting-out, but on his return in September, he was buffeted twice in quick succession. In August William Hutchinson had died in Newcastle, and the Forth Street Works lost its leading figure on the mechanical side. Much more of a personal blow was the death of John Sanderson in September, by which Robert lost one of his closest friends and business aides, and the stalwart defender of his privacy. Robert travelled up to Newcastle in October to meet his partners, and William Weallens was appointed in Hutchinson's place as general manager. In the later 1850s there were only three partners: Robert Stephenson and Joseph Pease Jr, each with 42 per cent of the shares, and Weallens with 16 per cent. There were problems to be dealt with at the Swinburne Glassworks, where George Hudson, the largest shareholder, was mortgaging some of his shares in a business which already had large liabilities. In order to help Hudson, Stephenson advanced a large sum to the company.[13] He went on to stay a few days with Edward Pease, who noted in his diary on 23 October:

My friend, Robert Stephenson, the engineer, to spend two or three days with me – a man of most highly gifted and talented power of mind, of benevolent, liberal, kindly, just, generous dispositions, in company most interesting. My dear sons John and Henry dined with me. At tea at my son Joseph's, a considerable and interesting company. At home to sup, and after it some social interesting subject occupied us to near eleven ... The evening pleasantly spent nearly alone, expressing to Robert Stephenson my anxious desire that smoking and taking wine might be carefully limited. Oh my soul, be upon the watch.[14]

Pease's generous comments on Robert do not suggest a man in deep grief, yet Jeaffreson records that so affected was he by Sanderson's death that he '... could not for months endure the solitude of Gloucester Square. Closing the house, he took apartments in "Thomas' Hotel", Berkeley Square, and did not return to his residence until twelve months of mourning had expired.'[15] It may be churlish, but not irrelevant, to ponder on whether his own convenience played a part in this. The Sandersons had effectively run his household, but the mid-Victorian moral code would not have allowed Mary Sanderson, as a widow, to have acted as a surrogate chatelaine for him even if either of them had wished it to happen. In fact, 1854 was to be a peripatetic year even for Stephenson, and he was not in London for long at any one time.

In March, Britain and France declared war on Russia, and the Crimean War lurched into disjointed action. Stephenson and Bidder were preoccupied at this time by the affairs of the Norwegian Trunk Railway, where the contractors had pulled out only six months before the completion date, but there was also an important event for Robert in April, when a statue of his father, sponsored by the Institution of Mechanical Engineers and sculpted by Edward Bailey, was unveiled at Euston. A fuller perception of George's achievement was steadily emerging, and *The Times* felt the event was too low-key for proper commemoration of 'a man who stands more nearly and intimately associated with the spirit of the century than we are yet willing to recognise'.[16]

Late in June he and Bidder had a stormy crossing of the North Sea in Bidder's yacht *May Fly* – Bidder noted quaintly that 'even Robert felt promiscuous'[17] – to arrange matters on the Norwegian railway, with Peto, Brassey and Betts taking over the work. After only a few days, he was back in the north-east of England. Also in his company at this time was Samuel Smiles, who after a four-year gap had reappeared in the summer of 1854, being conveniently based in Newcastle. Now he took up the biography of George in earnest, beginning his researches and interviews, including a session with Edward Pease. Soon Stephenson seems to have crossed the North Sea again, the Netherlands Land Enclosure Company having been successful in 'inducing' him to advise Bidder 'with other eminent engineers' on the works needed to enclose a vast extent of flooded polder-land in Holland.[18] He and Bidder submitted a joint report in August.

On the 25th of that month he crossed yet again, from Dundee to Denmark, to meet up with Bidder and travel on to Christiania for the opening ceremony of the railway on 1 September. He was in Newcastle on 6 October, the day of the horrendous fire and explosion in a Gateshead warehouse, which caused enormous damage on both sides of the river and the deaths of over fifty people. His own factory suffered only slight damage, but '... it was a heartrending sight to see scores of poor families wandering about the streets having lost every thing but the clothes on their backs'.[19] A few days later he was tramping the heather in the Trossachs with Isambard Brunel, examining the proposed route of Loch Katrine water to Glasgow which they had been asked to advise on, and he wrote in the same letter: 'For the last 10 days I have been roaming in the Scotch Highlands & Lakes and the weather having proved propitious I have upon the whole enjoyed myself with Brunel as my companion.' This was one of a number of city water-works projects on which he and Brunel acted together as consultant engineers.

Diplomatic rivalry between French and British interests in Egypt was made more delicate by the two countries' alliance in the Crimean War, but a decisive swing to the Gallic side took place with the assassination, in July 1854, of Abbas Pasha and the installation of his cousin Saïd Pasha as his successor. Saïd had been educated in France and was a Francophile; he also believed that Murray, the British consul-general, had intrigued against him with Abbas.[20] The new British consul-general, Frederick Bruce, found that his French colleague already had his foot in the door. Clear evidence of the changed state of things was not long in coming. From his home in France, a former vice-consul in Egypt, Ferdinand de Lesseps, wrote to congratulate the new viceroy on his accession. Saïd was a good friend of de Lesseps from the days when the French diplomat had been kind to him as a fat and lonely adolescent.[21] Invited to Egypt, de Lesseps, now a private citizen, arrived in Alexandria on 7 November, and was welcomed into the court circle. Since his first posting there in 1832, he had been fascinated by the idea of a canal across the isthmus, and now, despite the railway construction, which had reached the Nile on 4 July 1854, the prospect of a canal became once again a live topic in Cairo.

By 30 November de Lesseps had obtained a firman from the new viceroy, giving him the exclusive right to the formation of a Universal Canal Company. Also in Cairo that winter was Robert Stephenson, as before, on a visit that combined business and recreation. With the Alexandria–Cairo Railway well advanced, it was desirable to move ahead with plans for the second stage, from Cairo to Suez; the revived Canal Scheme made it all the more important to establish the complete sea-to-sea rail link. British diplomacy in Constantinople began to grind into a familiar anti-Canal mode. Frederick Bruce made it clear to Saïd that Britain would not be pleased by the granting of the Canal concession, but the viceroy, though polite, was unmoved. Stephenson was received in audience, but no progress was made on the question of the Cairo–Suez Railway. Writing to Lord Clarendon, the Foreign Secretary, Bruce said: 'I beg to refer your Lordship to Mr Stephenson who is fully master of the whole subject.'[22] Accompanied for part of the time by a Cairo resident, Frederick Ayrton, Stephenson went on a trip into the isthmus which he later described as a search for 'specimens', but was also a quest for further objections to the Canal proposal.[23] Though both men were in Cairo, there is no evidence that Stephenson and de Lesseps met. Later, the Frenchman would claim to have seen the extent of Stephenson's investigation, his carriage tracks ending only three leagues out of Cairo.[24] By mid-December, Stephenson was back in London,[25] and spent most of January in France, 'in search of quietude without I am sorry to say much benefit'.[26]

General dissatisfaction and indignation at the way in which Britain was conducting the Crimean campaign came to a head in Parliament in January 1855. Robert Stephenson, though he shared the general view about military and civilian incompetence in the field, blamed the commanders on the spot rather than the government. He voted for the government on 29 January, but his was the minority view; Aberdeen resigned and Palmerston became Prime Minister. Writing to Moorsom from Gloucester Square on 1 February 1855, Stephenson described Britain's level of support for its army as 'distressing, execrable, and contemptible ... If I had been in stronger health I think I should have been off in the Titania, and I am not sure that I shall hesitate much longer. I long to be at Balaklava to get the stores away to the camp. I believe I could be useful there as long as I felt myself beyond the range of the artillery. When nigh them I am persuaded I should run away and disgrace my country.'[27] Early this year he was elected as president of the 'Lit & Phil' in Newcastle, whose finances he rescued by paying off half an accumulated debt of £6,200. A message read to the committee on 14 February offered: 'That if the Members and Friends of the Society would, before the holding of this Anniversary Meeting, pay off one-half of the Society's Debt, Mr. Stephenson would discharge the remainder, upon condition that the annual subscription shall be reduced to one guinea.'[28] The balance was duly raised. Robert's request for the halving of the members' annual subscription from 2 guineas to 1 was a curious action for someone who professed to believe that working men did not need education.

A great regatta was held at Lowestoft in August, and both *Titania* and *May Fly* were among many yachts taking part. September found Stephenson in Paris for almost three weeks, visiting the *Exposition Universelle*.[29] A new Stephenson locomotive for the Lyons–Mediterranean Railway was on view, one of only two British locomotives among the twenty-two exhibited. He was surprised by the quality of the engineering section, remarking of the French: 'It is probable that knowing their inferiority in this particular branch, they have made a very unusual effort.'[30] Numerous British experts were on the various juries, but Robert played no official part in the Exposition. Two distinctions came his way: a Great Gold Medal of Honour, awarded by the Council of Presidents and Vice-Presidents of the Exposition, 'for the invention and introduction of tubular plate-iron bridges'; and appointment as a Chevalier of the Legion of Honour, along with some twenty other Britons of distinction in science and commercial arts, including Brunel and William Fairbairn.[31] The medal would perhaps have pleased him more, since Fairbairn (who was on the mechanical engineering jury panel), was awarded a First Class Silver Medal, along with Hodgkinson, 'for their co-operation in the experiments'[32] – pre-eminence was thus confirmed.

Back in England, he spent some time in Newcastle, writing from there on 4 October to Smiles that he had decided 'to withdraw entirely from all new professional engagements ... I intend leaving England for a cruise to southern climes in about three weeks.'[33] This was a relatively short cruise, to Madeira,[34] since in late December and early January he was working with Bidder on his presidential address to the Institution of Civil Engineers, which he gave on 8 January 1856, on the subject of the development and present state of the railway system.

Stephenson's presidency of the ICE has generally been seen as another bloom in his garland of honours, as of course it was, but the position was not purely honorific. As a senior figure in this body (and as president of the then provincially-based Institution of Mechanical Engineers between 1849 and 1853) he was also one of a small group responsible for the organisation and future of the engineering profession. In this respect they can be accused of lacking foresight. The lessons British engineers had taught themselves earlier in the century were now world currency, and in Europe and the United States they were being developed through an infrastructure of technical colleges and universities. In England, argument about the need for formal technical education had been going on for twenty years.[35] King's College, London, had set up a course in civil and mechanical engineering in 1838, and University College, London, established a chair in civil engineering in 1841, with C.B.Vignoles as first Professor. Trinity College, Dublin, followed in 1842. But the prime way into the profession was a five-year pupillage in a working practice, which cost anything between £500 and £1,000. In 1849 Stephenson had been a member of an ICE sub-committee considering university courses, which concluded that, '... it was not yet possible to lay down any clear requirements for those intending to enter the profession by this route'.[36] Perhaps the success of his own career, and that of such fellow engineers as Locke, unduly influenced his attitude. With other leaders of the profession, he failed to perceive the need for a new generation of educated engineers, just at a time when the scientists (the word only became current in the 1840s) were battling to establish their own academic disciplines. Stephenson was well placed to assess what was happening elsewhere, and (as at the Paris Exposition of 1855) occasionally commented on it, but he was as myopic as most of his peers about the implications, and some of his comments reveal a complacent, even arrogant attitude to 'foreign' competition and achievement. William Fairbairn was more perceptive:'The French and Germans are ahead of us in theoretical knowledge of the principles of the higher branches of industrial art; and I think this arises from the greater facilities afforded by the institutions of these countries for instruction in chemical and mechanical science.'[37] As public esteem for the university-educated scientist rose, Robert Stephenson, amiably, and quite unintentionally, presided over the apogee of the workshop-trained engineer's prestige and the beginnings of its decline.

Personal interest in continuing progress is evident from his involvement with the *John Bowes* and with electric telegraphy. Some long-distance undersea cables had been laid, but were not reliable, and in the winter of 1856/57, Stephenson was considering this issue, writing to Brunel: 'I do not like the Atlantic Telegraph … I am persuaded the matter is not ripe enough for you & me to declare ourselves opening in reference to this undertaking which I am confident has not yet been sufficiently considered.'[38] In the summer of 1859 Stephenson, who had chaired the Electric Telegraph Company for a period in 1858, and Bidder were both appointed to a government-organised committee on submarine cables, whose first meetings took place at his home.[39]

April 1856 was a month of short cruises in *Titania*, including a visit to the Solent from the 21st to the 25th with Bidder to see the great naval review marking the end of the Crimean War, before lending the vessel to the astronomer Charles Piazzi Smyth for a scientific visit to Tenerife between June and September. In these years Robert Stephenson's favourite London pleasures were 'the periodical dinners of learned societies, or of coteries composed of certain members of learned societies',[40] and on 26 April he was elected to a most select body, the Royal Society Club, formed of Fellows of the Royal Society, who dined from time to time at the Freemasons' Tavern. Proposed by Sir Charles Wheatstone and seconded by Sir Roderick Murchison, Stephenson dined on this occasion with a group that also included George Rennie and Dr Peter Roget, medical scholar and compiler of the eponymous *Thesaurus*. He was always one of the last to leave such dinners, keen for 'just another cigar and a little more talk' – and retiring at midnight to a friend's house, or another convenient club, for 'a little more talk and just another cigar'.[41]

In the course of this year, the ICE connection proved useful in filling the vacancy caused by Starbuck's death. Charles Manby, FRS, Secretary of the Institution, aged 52, moved next door to the Stephenson offices to become European agent for the locomotive company, and Robert Stephenson's personal representative. He retained an ICE link as its honorary secretary. Fluent in French, Manby was involved in the ongoing Suez Canal proposal as a member and joint secretary of the 'International Scientific Commission for the Canalisation of the Suez Isthmus', formed by de Lesseps in Paris in 1855. Not an engineer himself, this was de Lesseps' way of ensuring an international and unified approach to construction. France and Great Britain each provided four members; Austria two (including Negrelli), with F. W. Conrad from the Netherlands, a Spaniard and an Italian. The other Britons were J.M. Rendel and John Robinson McClean, both civil engineers, and Captain Harris of the East India Company. In view of his new employer's anti-Canal stance, Manby must have experienced some conflict of loyalties, but he did not give up his seat at de Lesseps' table.

Since the opening of the Norwegian Grand Trunk Railway there had been a stream of complaints from Norway about poor-quality work. Things came to a head in September 1856 with a critical report made by a Norwegian military engineer, which was presented to Stephenson as arbitrator. Having always taken a direct interest in the project, with visits in 1851, 1852 and 1854, and having gone to some trouble to arrange for its financing and to keep the costs within the budget of £450,000, Stephenson took the matter personally and his reaction lacked the detachment and balance for which he was famous. He considered the report 'despicable'.[42] His comment to the Norwegians that they would not have had a railway at all but for him was irrelevant to the matter at issue. Possibly ill-health affected his temper, as he adjourned the proceeding for 'a few days, for I am suffering at present from illness, and I do not think I could apply my mind to it so suddenly'.[43] The engineer who had drawn up the list of defects with-drew from the case, but Stephenson's eventual ruling did list certain shortcomings which should be put right, and he awarded £10,925 against the contractors.[44]

Samuel Smiles had completed his biography of George Stephenson, and one evening he went round to Gloucester Square to read some of it to Robert. Thomas Sopwith was also there. Smiles settled down to the not uncongenial task: 'I read on and on, and when I looked up, Sopwith was drowsy, and Robert Stephenson was profoundly asleep! ... my audience had dined, and dined well. When I stopped, Stephenson suddenly looked up, and said, "Oh! I hear you very well. Go on, if you please."'[45] Perhaps this incident prompted Smiles' mildly critical comment in later editions about Stephenson's drinking. Despite similar comments from another abstainer, Edward Pease, Robert seems to have enjoyed good wine, and claimed to consider it medicinal,[46] but was not a heavy drinker.

When well, he had all his old humour and empathy, as a letter to the Bidders' younger daughter shows:

> Dear Bertha,
> I have received with much pleasure, the beautiful little shirt purse you have been so very kind as to knit for me.
>
> It is exactly the kind of purse I wanted, for several young ladies have made me a similar present, but thet are generally so long and covered with such heavy metallic ornaments that I find it very inconvenient to wear them ... I have now got nearly a drawer full of them – Yours however is so neat, elegant & compact that you may rest assured that it will not share the same fate, but be worn in remembrance of your romping with me. Yours sincerely Rob Stephenson.[47]

Just after that, Robert left for a leisurely journey through France with two com-panions, Sopwith and the artist Frederick Lee, RA, spending several days in Paris where they dined with Paulin Talabot and met some other Suez veterans who were not part of the de Lesseps team, as well as meeting Locke and Brassey. Paris was clearly a favourite place for Robert, who always stayed at the Hotel de Castille in the rue de Richelieu, on the left bank of the Seine. On this occasion they visited the Sainte Chapelle and the Père Lachaise cemetery, and enjoyed an evening at the Cirque Napoléon. By 20 November they had reached Nîmes, and Stephenson made his third visit to the Pont du Gard. In the hotel notebook, he wrote: '... although I continue to admire the magnificence of the structure, I cannot admit that the Romans approached the Moderns in the science and art of Bridge building.'[48]

At Marseille they watched *Titania* enter the port, and on 2 December set off from Toulon for Malta and Egypt. Apart from fully-stocked larders and a wine-cellar, the yacht was well prepared for long voyages; Sopwith noted that she was equipped with 'a variety of swords, cutlasses – Rifles – blunderbusses and pistols – which with four large pieces of cannon (six-pounders) and two brass bulwark guns afford suitable means of defence if occasion should arise to put them to use'.[49] For the amusement and edification of passengers, there was a library with a microscope and other scientific instruments. One feels Stephenson cruised the seas almost like Jules Verne's Captain Nemo, though without the submersible capacities of *Nautilus*.

Landfall at Alexandria was made on 13 December, when some political realities set in. Writing from there to Edwin Clark on 22 December, Stephenson was not optimistic about having an audience with the viceroy, who was in 'Abyssinia' (in fact The Sudan, and with de Lesseps among his entourage): 'I shall probably take a run over to Constantinople', but he was cheerful about his own health, which 'is quite a different thing here. I am quite in good spirits, without an atom of hypochon-driac feeling, and actually recovering my flesh.'[50] The Canal question had caused a rift between the Ottoman government, which was heavily under British influence, and the viceroy; hence the suggestion of a visit to Constantinople, though it does not seem that Stephenson made the trip. With the Alexandria–Cairo Railway now complete, he was keen to conclude arrangements for Cairo–Suez. But the project was eventually entrusted to a French engineer, Mouchelet. On his way home, on 19 February 1857, Stephenson wrote to George Bidder from Algiers, enthusiastic about his voyage: '... I am persuaded traveling is better than physic.'[51]

A General Election was held in April 1857, and Stephenson, with no intention of giving up his seat, spent a few days at Whitby at the end of March in order to be re-adopted as the Conservative candidate. By now he was firmly established and was returned unopposed. Palmerston, with a large Liberal majority, remained as Prime Minister. George Hudson, clinging to public life by his fingertips, was also returned as the second member for Sunderland.

There had been a Meteorological Society in London since 1823, and the British Meteorological Society was established in 1850. Stephenson is not on the list of founding members, but joined the society and was elected president in May 1857, 'at a time when the society was seeking a meeting-place'.[52] From 1858 the weath-ermen used the theatre of the Institution of Civil Engineers for their meetings. Further distinction came when the University of Oxford, reacting with com-mendable promptitude to the role played by engineers in the country's affairs, awarded an honorary doctorate of Civil Law to Robert Stephenson on 25 June 1857. Isambard Kingdom Brunel received the same honour, and among the other honorands were David Livingstone and Sir Colin Campbell of the 'thin red line' at Balaclava.

Shortly after that, the Suez Canal issue arose in the House of Commons, when a motion critical of the 'jealous hostility on our part to the project' was debated on 7 and 17 July. If Robert Stephenson and Lord Palmerston differed on many issues, they were at one on this. The Prime Minister, on 7 July, took leave to affirm 'on pretty good authority, that this plan cannot be accomplished, except at an expense which would preclude its being a remunerative undertaking'; he called it a 'bubble scheme' and virtually suggested it to be a confidence trick on investors. In his speech, Stephenson confined himself entirely to the technical aspect, reiterating the discoveries of the 1847 surveys, repeating his view that a railway was a preferable solution and insisting that a canal would be 'an abortive scheme, ruinous to its contractors'.[53]

Stephenson's references to his travels, 'partly on foot, over the country to which the project applied' and to his 'investigation' of the subject in 1847 with Talabot and Negrelli led later writers (and perhaps led his hearers) to believe he had been in Egypt at that time. The motion was defeated, but Stephenson's expression of general support for Palmerston's point of view infuriated de Lesseps, who took it to include the scarcely-veiled suggestion of fraud. He came over to London and wrote to Stephenson citing the offensive passage: '*Je vous demande, Monsieur, une explication écrite à ce sujet, soit par vous même soit par un de vos amis.*'[54] The letter asked Stephenson to name two seconds, in effect a challenge to a duel; Stephenson made a conciliatory reply, saying he concurred with Palmerston only in believing the canal not to be commercially viable and that nothing was further from his intentions than to make '… a single remark that could be considered as having any personal allusion to yourself'.[55] Formal peace was restored, but de Lesseps intimated that there would be a response from his Canal Commission on the engineering and commercial aspects. Evidently feeling vulnerable, Robert wrote on 2 August to his friend in Cairo, Frederick Ayrton, asking for help:

I am anxious to have someone to confirm the fact that I went over the country twice, for unfortunately both Borthwick and Swinburne who did accompany me are now no more.

The Commission of Engineers who last went over the country are I understand to write a reply to my remarks in the H. C. which will annihilate all my personal and professional veracity –

I have begun already to ship my guns but I shall wait to see what comes forth before I fire. if these remarks are simply professional I do not think I shall fire at all, if personal I must discharge in self-defence.[56]

In the second paragraph, 'aim to' should be understood before 'annihilate'.
Stephenson wrote in the same letter:

> I had long kept silent on the matter because I did not, as a R.wayman, wish to
> push any opinions I might hold against the Canal proposed by Lesseps, but his
> recent proceedings in England by getting up public meetings … where resolu-
> tions were passed calculated, as I thought, to mislead, I felt silence on my part
> indicated acquiescence.

The International Scientific Commission published a reply by its Italian member,
Pietro Paleocapa, on 4 August, pointing out that Stephenson had misrepresented
Negrelli, who had always believed a canal to be feasible, and asking how a railway
could conceivably cope with the volume of freight traffic currently going via the
Cape of Good Hope, which was of the order of 3 million tons a year and growing
steadily, quite apart from the costs and problems of double transhipping.[57] Stephenson
did not feel it necessary to respond. Paleocapa failed to understand that the British
were perfectly happy for the 3 million tons to continue to take the long way round.

This year Robert Stephenson also consented to become godfather to the
newborn son of his friend Baden Powell[58]; the infant, Robert Stephenson Smyth
Baden-Powell, would grow up to be the hero of Mafeking and founder of the
Scouting movement. In the autumn, with Charles Manby and Matthew Bigge,[59]
Stephenson once again returned to the scene of his boyhood. At Killingworth,
they knocked on the door of Dial Cottage and were admitted by the occupants.
In the main room there was still a piece of furniture made by George Stephenson:
a combined writing desk and bookcase set against the wall. Remembering that
his father had built in a 'secret' drawer, he tried the mechanism, and to the surprise
of the residents it flew open. But this last vestige of the magician's chamber was
disappointingly empty.[60] During this visit Robert remarked to Thomas Sopwith
that '… his father always considered Ovingham to have been his birth place &
that soon afterwards his father removed to the house which is commonly reputed
to be his birth place'.[61] This knock at Wylam's fame has not been noticed before.
Perhaps the family was living with Mabel's parents before moving the few miles
to High Street House.

Mabel's grandson, in 1857, had *Titania* waiting at the Sunderland quay. With
George Bidder and a number of other friends, Robert set off on another nostalgia trip, following the eastern coast all the way up to Inverness and the northern
end of the Caledonian Canal. Near the Bell Rock a violent storm almost carried
away the ship's boat, and Robert, in the saloon, was thrown off his feet, cutting his
head on a bronze lamp fitting and bruising himself. However, his guest William
Kell, who reported the incident in a letter, added that he 'recovered ere we got
through the Caledonian Canal, and we were hearty as crickets'.[62] In 1823 he had
made a happy geological walking tour along the Great Glen; now his yacht was

towed through it by a steam tug, but with numerous stops for land excursions. After calls at Belfast and the Isle of Man, they made for Holyhead, and from there to the Britannia Bridge. With three others he climbed on to the top of one of the tubes and 'smoked a cigar in silent contemplation'.[63]

Back in the north-east in late November, he inspected reconstruction work on the pioneering iron-arched bridge over the Wear at Sunderland. As engineer-in-chief of the project, he had to justify a cost over-run of £10,000 and respond to local criticisms of the new structure; his appearance at a public meeting demonstrated complete mastery of design and site details and of his audience. Crisp, no-nonsense replies dealt effectively with financial grumblers and would-be engineers, and at the end he was warmly applauded.[64]

*Titania* did not make a winter cruise in 1857/58 and her owner remained in England. His final day as president of the Institution of Civil Engineers was on 15 December, when he handed over to Joseph Locke. On the following day he attended a meeting of the 'Geological Club' in Covent Garden and walked the 3 miles home with Thomas Sopwith.[65] In the course of November and December he made several visits to Russell's shipyard on the Isle of Dogs, where Brunel's vast steamship *Great Eastern* was obstinately resisting all efforts to launch it, sideways-on, into the Thames. One evening:

> ... a note was put into his hands from his friend Brunel ... to ask Stephenson to come down to Blackwall early next morning and give him the benefit of his judgment. Shortly after six next morning Stephenson was in Scott Russell's building-yard, and he remained there until dusk. About midday ... the baulk of timber on which he stood canted up, and he fell up to his middle in the Thames mud. He was dressed as usual, without great-coat (though the day was bitter cold), and with only thin boots on his feet. He was urged to leave the yard, and change his dress, or at least dry himself; but with his usual disregard of health, he replied, 'Oh, never mind me – I'm quite used to this sort of thing;' and he went paddling about in the mud, smoking his cigar, until almost dark ... the result of this exposure was an attack of inflammation of the lungs, which kept him to his bed for a fortnight.[66]

A letter to Brunel confirms that the inflammation was 'a little obstinate',[67] but a few days later he was recovered enough to go to Newcastle.

# 21

# THE LAST
# JOURNEYS

A man who often mused on death, who paid visits to the churchyard where his wife had lain for fifteen years,[1] and who had lost two close friends in more recent years, might well share the growing interest of the times in spiritualism and the 'beyond', but Stephenson was not impressed. While he was in Newcastle, friends took him to a seance, and he wrote afterwards to Brunel to describe it:

Dear Brunel,

My Spirit rapping visit was unsatisfactory indeed I regarded it as a perfect failure altho I was induced from the apparent sincerity of the parties to promise that I would make another visit to another house, as they held out the prospect of my seeing two or three Ladies lift a table by their finger ends. They were very respectable people and doubtless believed all they said, but I stand amazed at their credulity, in short their credulity in my mind beats the spirits hollow. I could sooner believe in spirits than I could conceive such frustration of mind. The chief 'medium' as they call it was a tolerably good looking girl who talked sensibly enough, but all the party knew she was suffering from epilepsy and therefore had a diseased brain. During the evening she was thrown into hysterics because I ventured to say that I thought one of the knocks came from a part of the table where she was sitting. Notwithstanding this knowledge of the girl's condition, the whole company implicitly believed her when she informed them that she constantly conversed with spirits, some of them very recently removed from this state of existence. She frequently sees spirits hovering in the air and she also sees halos or luminous atmospheres around the bodies of her friends ... This will give you some notion of the kind of evening I had, and I must add one more anecdote. When we were about to leave the table, the master of the house proposed to us to have another bottle of wine. The most of the company declined, but the master said he would consult the spirits and did so in our presence in the most solemn manner -- by looking earnestly down upon the table and by proposing himself in these terms – My dear spirit must we have any more wine? Whereupon three distinct knocks were heard under the table, which is always construed to mean yes. Whereupon we had another bottle of wine and I ventured to say I thought the party was getting more spirituous than spiritual. You must treat this as a private communication as I must meet the parties once or twice again in the year.

Yours sincerely, Robert Stephenson[2]

This intriguing letter gives an insight into Stephenson's social life, his humour, his stout scepticism of the fashionable occult, his tact vis-à-vis his Newcastle hosts and his consideration in trying to amuse Brunel who was in a state of extreme tension because of the problems of launching the *Great Eastern*. He found nostalgia for the past more difficult to resist than mysticism.

Back in London by the end of the month, he caught another chill and was unable to witness the launching of the giant ship, but wrote with glad congratulations: 'I slept like a top after I received your personal message – I felt desperately anxious all day, but my physic would not permit me to venture so far away as Millwall.'[3] In the spring his health improved considerably, and social life resumed, including a dinner for an American couple, Mr and Mrs Carr from Philadelphia: 'The evening passed off very agreeably – Mr Stephenson was in high spirits at dinner and kept up a most amusing round of conversational anecdote and repartee, not unmixed with matters of solid information on various points. We had some good piano and vocal music in the drawing room.'[4] Miss Gooch, daughter of Thomas Longridge Gooch, was at the piano playing 'The Last Rose of Summer' and other Moore-style pieces. That spring Stephenson gave a ball at Gloucester Square for around 190 people, 'mostly young' according to Sopwith.[5] Sopwith, Bidder, Brunel and many other friends had children in their late teens or early twenties. There was 'sumptuous fare' and dancing went on until 3 o'clock in the morning. Those who did not want to dance could play with the microscopes and stereoscopes.

Inside and out of the British Parliament, a growing number of influential people were questioning the official attitude towards a Suez Canal. Another debate was held in the Commons on 1 June 1858, on a motion in favour of the Canal from J.A. Roebuck, with William Ewart Gladstone among the parliamentary heavyweights who supported him. Opposing it, Stephenson re-marshalled his arguments in his usual measured manner, pointing out his personal knowledge of the terrain, and condemning the scheme as technically impracticable and commercially undesirable, since the railway would be more certain, more expeditious and more economical. His comments on its feasibility are considerably toned down from earlier remarks, 'because engineers whose opinions he respected had been to the spot since, and had declared the thing to be possible'. Again, his speech managed to convey the suggestion that he had been in Egypt in 1847, without actually saying so, referring to: '... the three gentlemen, one from

Austria, another from Paris, and himself from England – who first investigated the subject in 1847. They examined the physical features of the country, and deliberated over the matter in the most cautious manner ...'[6]

Later in the debate Palmerston (temporarily out of office) reiterated his own opposition: 'My firm opinion, founded upon the engineering and geographical reasons which have been so ably stated by my hon. friend the member for Whitby is that ... as a remunerative commercial enterprise it is, as I said before, in reality nothing more than a mere bubble.'[7] Sixty-two members supported Roebuck's motion and 290 opposed. After the debate Stephenson wrote to Tom Gooch: 'I have been pitching into my dear friend Lesseps again about the Suez Canal.' He had tried to persuade Roebuck to withdraw his motion, 'as I knew any discussion on it could only engender bad feeling on the part of the French. He was perfectly resolute ... I therefore had no alternative but to repeat what I had formerly said, and to stop, as far as I could, the English people from spending money on an abortive scheme.'[8] He was right about the French reaction: a few days later de Lesseps stormed into the office at Great George Street where he found Manby, who wrote: 'I have had Lesseps here bothering me & foaming at the mouth about the Cheif and the part taken by him relative to the Canal.'[9]

An open letter to Stephenson from Negrelli was published in Vienna by the *Oesterreichische Zeitung* on 18 June. More damaging than Paleocapa, it criticised him for not having replied to Paleocapa, noted that he had neglected his obligations to the original Suez study group, and 'sent nobody into Egypt', and when he had finally gone himself, he '... went into Egypt without consulting his colleagues ... with the intention of entering into a negotiation with the Government on the subject of executing a railway from Alexandria to Suez ...' Casting aspersions on Stephenson's knowledge of hydrography, Negrelli went on to expound his view of how a combination of the tides at each end and 'the motion of the lakes' would ensure the canal waters had movement.[10]

Stephenson sent a carefully composed, though strongly worded, reply to Negrelli's broadside. Expressing pained surprise that his continental colleagues continued to be so hot and bothered about his opposition to the canal project, he suggested that his position was, and always had been, clear. Giving his own brief version of events around 1846–47, he avoided the issue of whether he had sent a survey team, and stated he had only thought a canal possible when it seemed there was a considerable difference in sea level between the two ends. He pointed out that he had made his views public in the ICE Proceedings of May 1851, and referred to Talabot's report and its conclusion that there was '*une difficulté insurmontable*' for any canal to the east side of the Nile Delta. This, he suggested, had closed the matter for him; he had personally paid his share of the expenses ('one third of about £4500') and all correspondence on the subject ceased. Without mentioning the railway project at all, he relates how in the autumn of 1850, '... I sought health and recreation in a yacht voyage to the Mediterranean. Arrived at Alexandria, I determined to make a personal investigation of the district in which four years previously I had been so deeply interested … all that I saw and ascertained on this expedition confirmed and ascertained my convictions as to the Suez Canal project, and the finality of M. Talabot's report.'

Negrelli had somewhat rashly repeated a private claim of de Lesseps,[11] that Stephenson had never set foot on the Isthmus of Suez, and Robert was able to counter this with details of his two visits in 1850 and 1854, and expressed indignation at the slur on his honour. He explained that in the burning of the first *Titania*, he had lost specimens which he had gathered on the isthmus in 1850, and 'in the winter of 1854, being at Cairo, I felt a desire to replace these specimens', and so again visited the isthmus 'with Mr Ayrton, an English gentleman resident in Cairo'. In order to illustrate his own point about tidal influence, the Austrian had also suggested that the Thames was tidal as far as Windsor, an error which Stephenson pointed out with a retaliatory touch of contempt.

Robert was lucky to have these easily refutable, though inessential, points to attack, since otherwise he was fighting from a rather vulnerable position. Two issues were notably avoided: firstly, he had not told his colleagues in the *Société d'Etudes* that he was negotiating for a railway contract; secondly, he had personally examined only a very short section of the proposed canal route. A response to Stephenson's reply came in the *Oesterreichische Zeitung* of 26 September 1858; Negrelli regarded Robert's letter as insulting. His own report on Suez, in the Imperial Archives, had been available for consultation if not actually published (it seems strange that copies were not sent to his colleagues in the *Société d'Etudes*). He noted that Stephenson had not attempted to argue against the conclusions of the International Commission, which included calculation of the tidal ebbs and flows. It was the final shot in a dialogue between two sick men; only a week later, Negrelli died, aged 59. But Conrad kept up the defence, pointing out that all the Dutch canals were without current, but nevertheless perfectly navigable.[12]

In August Stephenson was in Newcastle, where a meeting of the Institution of Mechanical Engineers was held on the 25th. Nicholas Wood gave a paper on the improvements in the mining industry over recent years, and a dinner for 160 people followed.[13] With a considerable amount of his capital still tied up in the Swinburne Glassworks, Stephenson kept an eye on its activities and accounts, and was alarmed enough during his visit to arrange for its books to be examined by an outside accountant, at which point it emerged that the company was heavily indebted, not only to Robert himself, but owed large sums to other interests. Throughout his involvement, the business was run by Walter Swinburne, a notable glass technician who seems to have been less successful as a businessman. Some re-aligning of the industry took place that year and Swinburne became head of a national syndicate of plate-glass makers. Perhaps encouraged by this, Stephenson made a further loan to the company of £25,918 6s 4d in September.[14] Hudson was in financial exile in France and Nicholas Wood had mortgaged his own share to

Stephenson for £3,000 ten years earlier.[15] Having already made a loan five years before, substantial security must have been available for the prudent Robert to lend such an enormous extra sum, even though his personal fortune by now was vast.

At the end of the summer he was feeling rather low and dispirited. Thomas Sopwith records a long conversation with him on 11 September in London, which prompted Sopwith to muse on the *losses* in Robert Stephenson's life; remembering how they had met in the middle of the troubles of the Stanhope & Tyne Railroad, and now there was this difficulty with the plate-glass company – all 'of a character to try even a stout heart – how much more a susceptible one – Yet it is such conditions that truly display the character of Robert Stephenson, and how gladly did I promise to spend the evening with him … I found him sitting in the garden, in the gloom of far advanced twilight …' The two friends went inside to divert themselves with the microscopes and stereoscopes.[16]

The phase of depression seems to have passed, and only a few days later he took the young Bidders and Sopwith's daughters to see the celebrated German illusionist 'Professor Trickell', by whose talents he was deeply impressed. Variety and music hall were very much his favourite form of entertainment, and he does not seem to have been interested in drama or opera. Further cheer came from an immensely convivial evening at the Falcon Dining Rooms near the Crystal Palace, with a group of engineering friends, in honour of George Phipps. Recalling a connection that went back to the building of *Rocket*, Robert made a speech praising Phipps as 'one of the ablest and worthiest coadjutors he had had in his early career'.[17] Later in the month he was in Leeds, for the meeting of the British Association and another grand dinner in the municipal splendours of the city's new Town Hall.

In October Stephenson visited Sunderland to inspect the continuing reconstruction of the Wearmouth Bridge. Hudson was still the local MP, but remained in refuge in France from his creditors. His affairs had steadily worsened. With the expansion of railway-linked coal docks in the north-east, all competing for business, the Sunderland Dock Company was doing badly, and he was no longer the local hero. To cap it all, he was in agonies from gout. In reply to an address from the workforce, Robert's speech celebrated the virtues of the skilled working man and his essential part in modern life:

> There are no members of society for whom I have a greater respect than for industrious and intelligent workmen. It is to them that the engineer is indebted for the full and efficient realisations of his conceptions … The advance of mechanical science, and *its application to useful purposes*, must always go hand in hand with the skill and also with *the comfort of the working classes* … skilled labour is the great fulcrum upon which *all our social progress* depends …[18]

These remarks show the extent and the limitations of Robert Stephenson's understanding of the society his work had done so much to change. His basic attitude

would always be conditioned by his own and his father's careers. Engineers, professional leaders, would emerge through some kind of natural endogeny – forming a
small cadre to guide the legions of 'industrious and intelligent workmen', whose
living conditions would be improved, but who would remain second- or third-
class citizens. This, for Stephenson, constituted social progress. Four months later, at a
dinner of the Royal Society Club, in discussing the education of working men (there
had been a Working Men's College in London since 1854, established by Christian
Socialists), he said: 'It is all nonsense Lord John [Russell] preaching and preaching
education to the working classes. What the artisan wants is special education for his
own particular specialty, and the more he leaves everything else alone the better.'[19]
There was no hint of social progress in this vision of industrial helotry. He was certainly no radical; he was not a democrat, but in his own way he was a good utilitarian.

His winter cruise in 1858–59 on *Titania* was to Egypt, the voyage lasting from
14 October to 3 November, and was marked by a terrific storm between Algiers
and Malta which threatened to sink the yacht. However, in writing to Tom
Gooch on 5 December 1858 from Alexandria, Stephenson describes it as 'upon
the whole, remarkably fine, rather too much so, as the wind was generally light,
with occasional calms of two or three days' duration'.[20] Among the party were
the Bidders and their elder daughter Lizzy, and in the same letter to Gooch, he
says that, '… a portion of our party will most likely go up to Thebes; but having
once been there I do not mean to go again. I shall stay quietly at Cairo and
enjoy a daily drive into the desert, which I have always found most invigorating.
By the last post I had a message from Brunel, inviting me to dine with him on
Christmas-day. This I shall endeavour to do, though at some inconvenience.'[21]

The French engineer Mouchelet had just completed the Cairo–Suez Railway,
which opened on 7 December, and Stephenson was an early passenger. Writing to
Bidder, he described the chosen route as '… a huge Engineering blunder'. Meeting
Mouchelet at Suez, he tried to avert direct criticism: '… I deeply regretted so
much ability had been spent on so disgraceful an example of R.way Engineering,
alluding of course to the designing of the line which I knew was the Viceroy's …'[22]
On the Alexandria–Cairo line, the Nile bridge at Kafr el-Zaiyat was still being
built, and some work was involved for the engineer-in-chief. On the subject of a
disputatious employee, he wrote to Bidder on 22 December: 'I have already tried
persuasion, I intend to apply very strong language as a last resource – my natural
impulse would be a kick, but recollecting the abundance of peace, which your last
two or three letters have so forcibly inculcated, I do not intend to adopt this last
natural impulse of my nature but adopt that which has been so constantly brought
before me by that apostle of peace GPB since the year 1835 …'[23]

Brunel's party, which included his wife and his younger son Henry, then 17,
who recorded the trip in both English and French, arrived on the 20th. With no
professional commitments in Egypt, and no involvement in the Canal question,
Brunel was in a relaxed mode, and they rode about on donkeys until Brunel

became tired and took refuge in a carriage. Stephenson accompanied them to see a display of dervish dancing, but (after earlier experience) not to the baths, where 'a man who looked like John the Baptist came in and rubbed my father's skin off while another old man came and did the same to me'.[24] The Christmas dinner duly took place, with no apparent inconvenience beyond the short trip from Shepheard's Hotel to the Orient, where the Brunels were ensconced. Henry described it as '*une soirée très agreeable*'.[25] On the 28th the Brunels went off up the Nile, and Stephenson resumed his desert drives.

Commander Bedford Pim, RN, who had been a guest on the last Egyptian cruise, gave a talk to the Geographical Society on 11 April 1859, his subject 'The Isthmus of Suez, With Specific Reference to the Proposed Canal'. Fund-raising and other preliminaries meant that de Lesseps had not yet commenced construction; Pim set out to demolish the, as yet, non-existent canal. Pim was a navigator and explorer, not an engineer, and it was Stephenson's arguments that he presented, to 'show cause how the cutting a canal across the Isthmus of Suez is a hopeless undertaking'.[26] Robert Stephenson was in the audience as Pim affirmed that: 'Mr Robert Stephenson has made a personal inspection of the ground, and thoroughly mastered all the scientific and technical details of the project. The result of these admirable arrangements was a complete exhaustive survey of the locality.' The International Commission of Engineers had by a majority favoured an open cut, without locks. In the discussion that followed, George Rennie supported this, giving his opinion that there were no insoluble problems. Daniel Lange, the Canal Company's British representative, pointed out that numerous engineers 'of equal eminence' to Stephenson, including Conrad, an expert in waterway construction, considered the scheme perfectly viable. Stephenson rose to repeat his objections: '… the physical difficulties are as I believe insurmountable.'[27] He considered that Port Saïd would be inaccessible from the sea because of silting from the Nile, and 'with respect to the Canal itself, now that it is proved there is no difference of level, it would really remain a stagnant ditch, and must ever remain so. Whoever has travelled over that district, and seen the moving sand, must see that it would be necessary to dredge, not only the harbour, but the canal itself.'[28] In those last remarks, there is a hint of shifting ground on his own part. There was nothing technically problematic about dredging. This was Stephenson's last public word on Suez, but in any case, opponents of the Canal could now talk their heads off and it would make no difference – Lange announced that the Canal Company had got its funds together, and construction was about to begin. Just two weeks later, Ferdinand de Lesseps drove his spade into the sand to mark the beginning of work on the Canal.

Canal historians have put Stephenson's opposition down simply to his involvement with railways,[29] and therefore as cynical if not hypocritical. For the biographer, the crucial question is whether his anti-Canal stance was genuinely based on engineering, geographical and economic judgements. A critical moment was surely in the spring of 1847, when Bidder's anticipated Egyptian trip failed to materialise. When he began to make enquiries about the Canal project and found that British policy was solidly against it, but was favourable to an Alexandria–Suez Railway, Stephenson was immediately in a tricky situation. Quite apart from the fact that his venerated father had once planned to build that railway, it could be his, and Britain's project, while the Canal, if it came about, was obviously going to be a French-led enterprise.

Following on from this, a case against him can be set out. He never sent out a survey team. From his own testimony, he had considered the Canal impossible since the sea levels were confirmed in 1847,[30] but he made this judgement without examining either the route or survey reports, and said nothing about it to his colleagues when they met in Vienna at the end of 1847. Nor did he inform them of his involvement in promoting the railway. When he went to Egypt at the end of 1850 it was in order to secure the railway project, and he did not announce his opposition to the Canal until May 1851, by which time the railway contract was assured. He made great use of Talabot's report. but this came out three years after his own decision to oppose the Canal. He earned £56,000 from the railway works,[31] of which he was engineer-in-chief, and his company ran the line for the Egyptian government after it was opened. The economic element in his case was not developed until after de Lesseps reignited the debate at the end of 1854. His Commons speeches in 1857 and 1858 are framed in a way that suggests he had been to Egypt and made intensive studies in 1847, when in fact he had done neither. In his letter to Ayrton of 2 August 1857, he wrote that in Parliament he 'could not go into anything like a technical exposition of the whole question as that would have led me into unintelligible detail',[32] but he never set out a full technical argument. His reply to Negrelli's criticism is evasive and obfuscatory. No other senior British engineer backed his view: Rendel, Manby, McClean and George Rennie believed it was possible to build the canal.

No wonder a mixture of discomfiture and truculence is discernible in Stephenson's letters on the subject of the Suez Canal. Two further questions also arise – was he influenced by anti-French prejudice, or putting up a front for the British government's policy? From the Foreign Office papers, a consistent, deep and largely unworthy suspicion of French motives on the Canal question is evident: some people from Palmerston down found it impossible to believe that visionary idealism drove the French planners and not some deep-laid anti-British conspiracy. The notion of a general benefit to mankind, which de Lesseps certainly had, could not penetrate their mindset. Stephenson, however, was a Francophile, and men like Talabot and his colleague Dédéon remained his warm friends. Whether or not they ever met face to face, he did not like de Lesseps, 'my dear friend'. But he never expressed a paranoid view about a French conspiracy. He was undoubtedly a sturdy patriot – was he then screening British policy with any engineering reason that came to hand? – acting in effect as a government stooge? No one who has examined his career could believe that he would ever be Lord Palmerston's poodle. Stephenson was his own man, and showed a strain of unshakable obstinacy in minority or unfashionable political views.

The fairest and most reasonable interpretation of his behaviour is that he accepted the place on the *Société d'Etudes* in good faith, at the invitation of an old friend. Soon he discovered that a canal was opposed by the British government, who saw it as adverse to British interests, whereas a railway would be of value to Britain. At that point he stalled, and did not send a survey team. It probably seemed a waste of money and effort – most people assumed that a canal could never happen as long as Great Britain opposed the idea. Conversely, with British backing, a railway was very likely to go ahead. For Stephenson this was a business opportunity not to be missed. He could claim that a railway line was quite a separate affair from a canal feasibility study, and that normal business confidentiality made it impossible for him to mention it. But he waited until he had a reason other than opportunism to drop the Canal: this came with the sea-level discovery, and it would be unfair to doubt his sincerity merely because it was also convenient. Then the political tumult of 1848 in Europe and Egypt helped to push the Canal question into the background. The accession of Saïd and the appearance of de Lesseps created a new situation – the direct-route canal was back on the agenda, but Stephenson's French friends were not involved. De Lesseps, neither an engineer nor a businessman, but a self-professed humanitarian and world-improver, was in Stephenson's eyes an adventurer ready to pour other people's money into an impossible cause, and, having made himself the British expert on the isthmus, he felt it a duty to oppose him. Without shifting from the technical point on which he first chose to plant his case – the lack of difference between the sea levels – he extended his argument into economic feasibility as well.

Did Stephenson's hostility to the Canal make any difference? British government policy was established long before he became involved, but he spread a welcome drape of engineering respectability over its naked imperialism. His speeches, widely reported, played an important part in discouraging British shipping and commercial interests from investment in the Canal. Despite de Lesseps' barnstorming tour of port cities to promote the scheme, only 2 per cent of the shares in the original flotation of Canal Company stock were taken up by the richest trading nation on earth.[33]

More than any other controversy in his life, Stephenson's opposition to the Suez Canal displays the capacity of this complex and driven man to paint himself into a corner. If – as Negrelli came close to doing – the slender basis of his original anti-Canal judgement, and his inferences that he had been over the route, were exposed, he would have been in a difficult position. He could not recant his opposition without appearing either cynical or foolish, and he could not properly justify it without displaying its meagre foundations. As in the matter of the trussed compound girder, he clung rigidly to the only option that would not put him in the wrong, and maintained it in such a way that it could be contested only by challenging an entire long and distinguished career. Any serious assessment of Robert Stephenson must accept that he was capable of being devious in his own interest and defence. In the unique circumstances of his early life, and the strains of his adult career, this aspect of his character finds its context: he was never a cold paragon of excellence, but someone with a full share of conflicting, sometimes irreconcilable, demands and pressures, confined in a strongly emotional, but also self-repressing nature and always driven by the urge to success. He was 'The Chief' and had to be right. In the Suez affair, he can be most readily accused of a failure of imagination. Ultimately there was room for a railway and a canal. De Lesseps' vision of lines of steamships – mostly British – using the canal was right; Stephenson's prophecy of a stagnant ditch was wrong. One wonders if it ever occurred to him that, in this respect, he was occupying the same stance as those London engineers who had once scoffed at his own father's plan to build a railway over Chat Moss.

Despite worsened health, Robert presented himself for a fourth time to the Whitby electorate at the General Election of 29 April 1859. Sir Henry Hoare, banker and ship-owner, had come up from London to offer himself as a Liberal candidate, but, discouraged by the evident strength of support for Stephenson, went home again.[34] 'The prospect of a fight threw my "Liver" out of order', wrote Stephenson,[35] who was too unwell to conduct a public campaign, but was again returned unopposed, and made a barely audible speech of thanks. During his stay in Whitby, a delegation of temperance campaigners came to see him. Agreeing that 'habits of intemperance must be generally deplored', he admitted that he '… was not himself a teetotaller. Unfortunately his health would not admit of it', but he promised that 'they might rely on his best consideration of so important a subject'.[36] Referring to a popular view that to give the workers more money was merely to intensify their bad habits, he disagreed 'that high wages and crime went together. He believed that poverty and crime were concomitant. Good wages, sufficient to supply the comforts as well as the requirements of life, were necessary to produce independence and self-respect; and to raise the working class, they must be taught to respect themselves'[37] – once again a Liberal pulse is seen beating within the Conservative carapace.

The Bidders and Lizzy had come up to Whitby to join Robert and spend a few days in the north-east. This included a tour of the Swinburne Glassworks in Gateshead (still the subject of anxious discussion with Wood; they were hoping to sell it off), as well as the now traditional pilgrimages to Killingworth and Wylam.[38] Back in London, as a Conservative MP he was in opposition, facing Palmerston and a Liberal government with a large majority. His last remarks to the House of Commons, non-party political like all his contributions to parliamentary debates, were made on 11 August, in a debate on the cleansing of the Serpentine Lake in Hyde Park.

Two days later he signed his will, naming Charles Parker, George Robert Stephenson and George Bidder as executors. George Robert was to acquire the interests in the locomotive works and the Snibston collieries, the lease on Gloucester Square and its contents, together with half of the Great George Street house and contents, and the sum of £50,000. The other half of Great George Street was left to Bidder, with £10,000. Parker also received £10,000, and the cousins Robert and James Stephenson got £5,000 each. Ten female cousins each received £1,000. Nine people received £2,000, among them Edwin Clark, Thomas Harrison, George Phipps and William Weallens. A Newcastle friend, George Vaughan (a witness of George's will) was left £5,000. Miss Emily Lister received £4,000 and her married sisters £1,500 each – their father had been one of the engineering team; and £5,500 was left to the children of Edward Starbuck. Having thus looked after both his 'families', Stephenson left £10,000 to the Newcastle Infirmary, £7,000 to the 'Lit & Phil' and £2,000 to the North of England Mining Institute. The Institution of Civil Engineers received £2,000, as did the Society for the Propagation of Christian Knowledge and the Society for Providing Additional Curates in Populous Places. His housekeeper, Margaret Tomlinson, received £100 a year for life, and any servant who had been with him for more than twelve months got £20. The residue was to be divided among the executors.[39] Making his will was not an anticipation of imminent death: he was also planning a winter trip to Egypt, with Sir Joseph Paxton among his guests.[40]

September 1859 marked the end of a five-year period in which the building contractors had managed the Norwegian Trunk Railway, which now ran its own affairs. To celebrate, a dinner was arranged in Christiania in Stephenson's honour,[41] and on 15 August *Titania* and *May Fly* set sail across the North Sea. Arrived in Norway, the guests had a ride on the railway and visited Bidder's properties before the dinner on 3 September. Wearing the Cross of the Order of St Olaf, Stephenson, again ill, listened to a speech in his honour and though almost overcome with faintness and nausea, rose to reply. In his thanks, he disclaimed credit and mentioned instead Crowe and Bidder before wishing prosperity and happiness to Christiania. They returned to the yachts, but by the next day it was plain that Stephenson was seriously ill, and they left for England. On the 5th, Mr Perry, who was a retired doctor, transferred from Bidder's vessel to *Titania* just before heavy weather set in. Stephenson was suffering from a severe attack of jaundice on top of his already advanced nephritis, but when they landed on the 13th at Lowestoft, he felt well enough to walk, supported by two friends, to the railway station. Back in London, Dr Frederick Bird was immediately sent for.

Brunel, who had collapsed soon after the *Great Eastern* made its first trials, died on 15 September. Stephenson was too ill to attend the funeral on the 20th.[42] Knowledge of his grave illness had spread with all the speed that trains and the electric telegraph could bring. An anxious letter came from Talabot in Paris, suggesting that if he could travel, a course of Vichy treatment might work wonders.[43] Though at first he seemed to be regaining strength, after a few days there was a relapse, and his condition was evidently worsening. Members of the 'family' watched over him 24 hours a day, most notably Mrs Bidder, but he and they knew his death was fast approaching, and his main concern was for them rather than himself.[44] The end came just before midday on 12 October 1859, a few days before his 56th birthday. Even on that day, George Bidder kept to the engineer's inveterate habit of noting the weather first in his diary. The entry reads: 'Very wet. R. Stephenson died at 12am.'[45] The doctor's note on the death certificate said: 'obstinate congestion of the liver followed by dropsy of the whole system' – after so many years of inappropriate dosages and mounting mercury levels, his body, at last, simply could no longer function in its own defence.

Robert had made no provision for his own burial, and on the 13th Bidder went to make enquiries at Kensal Green Cemetery.[46] But some of Stephenson's friends, and others in high places, reflected on the significance of this death. 'To the enquiry where Robert Stephenson should be buried, there was only one answer. Without a dissentient voice, public opinion demanded the last and highest honour due to great Englishmen – sepulchre in Westminster Abbey.'[47] No engineer other than Telford, who died in 1834, had been honoured in this way; for all Robert Stephenson's great achievements and distinction of character, two additional reasons contributed to the feeling. One was the shock of his death so soon after Brunel's – the country had lost its two most famous engineers within a month. Brunel had been laid in the family tomb at Kensal Green – could Stephenson also be interred like any private man? The other reason is referred to obliquely by Jeaffreson as 'that powerful biography which made George Stephenson loved by thousands'.[48] Published in 1857, Smiles' *Life of George Stephenson* had become an immediate and ongoing bestseller, and with it, the nation realised that one of the most extraordinary men that England ever produced, whose influence on national life had been of incalculably vast impact, had been buried, without pomp and circumstance, in a small provincial town. The second Stephenson, in many ways more self-effacing than his father, but so much sharing and extending his father's achievement, could not escape national honours. The final accolade was bestowed by Queen Victoria, in response to a request made by Manby through the Duke of Cambridge, that the procession be allowed to cross the royal domain of Hyde Park:

… in acknowledgement of the high position he occupied, and the world-wide reputation he won for himself as an Engineer, his funeral, though strictly speaking private, as being conducted by his friends, partakes of the character of a public ceremony; and being anxious, moreover, to show that she fully shares with the public in lamenting the loss which the country has sustained by his death – she cannot hesitate for a moment in giving her entire sanction …[49]

On 21 October, ships' flags on Tyne, Wear, Tees and Thames flew at half-mast. At eleven in the morning, Robert Stephenson's coffin was carried into the Abbey by

Joseph Locke, Sir Roderick Murchison, George Carr Glyn, Samuel Beale MP, John Chapman and the Marquess of Chandos,[50] amidst a congregation of some 3,000 people, and laid next to Telford's. Of the prominent figures he had worked with, the only notable absentees were William Fairbairn, who would not come (though his younger brother Peter did); George Hudson, who could not; and Joseph Paxton, who was in Spain. From Bilbao he wrote to *The Times*, proposing that George Stephenson's remains be reburied at Westminster and a joint monument erected.[51] Two weeks later, it fell to Joseph Locke to salute the memory of his two friends in his presidential address to the Institution of Civil Engineers on 8 November. His own grief was apparent in the tribute he paid to Robert, who had been '... the friend of my youth, the companion of my ripening years, a competitor in the race of life; and he was as generous a competitor as he was firm and faithful as a friend'.[52]

Just thirty years after the locomotive trials at Rainhill, it was almost a cliché to comment on the seismic social shifts of those three decades: 'We elderly people have lived in that pre-railroad world, which has passed into limbo and vanished from under us. I tell you it was firm under our feet once, and not long ago. They have raised those railroad embankments up, and shut off the old world that was behind them. Climb up that bank on which the irons are laid and look to the other side – it is gone.'[53] Railways had brought industrial technology into the lives of everyone, and a new order of things was irreversibly under way. Robert Stephenson himself publicly contemplated these fundamental changes in his ICE presidential address of January 1856, speaking up for what he called 'the moral results of the railway system' – not just in terms of swift travel for the masses, encouragement of industry and expansion of the national wealth, and in its provision of direct and indirect employment, but in its effect on agriculture and fisheries, with consequent improvement in people's diet, and in making cheap coal available throughout the country: 'there is no contribution to the social comfort of society equal to warmth.' Acknowledging his own part in making it happen, he added: '... whatever may have been done, and however extensive may have been my own connection with railway development, all I know and all I have done, is primarily due to the Parent whose memory I cherish and revere.'[54] The lives of the Stephensons, father and son, had been useful to their country and to the world. For a joint epitaph, one could find worse than these lines written (not for them) by William Wordsworth in 1815, when George Stephenson was in a fever of constructive energy, and Robert was his eager and fast-learning pupil:

High is our calling, Friend! – Creative Art ...
Demands the service of a mind and heart
Though sensitive, yet in their weakest part
Heroically fashioned – to infuse
Faith in the whispers of the lonely Muse
While the whole world seems adverse to desert.[55]

# NOTES AND REFERENCES

## Archive Sources

B&W     Boulton & Watt Archives
BRIS     Bristol University Library, Brunel Papers
DRO     Durham Records Office
ICE     Institution of Civil Engineers
IMechE     Institution of Mechanical Engineers
Lilly     Lilly Collection, University of Indiana Library
LRO     Liverpool Records Office
TNA     The National Archives, Kew
NRM     National Railway Museum, York
NRO     Northumberland Records Office
ROB     Robinson Library, University of Newcastle
ScM     Science Museum Library
StA     St Andrews University Library, Forbes Papers
TWAS     Tyne and Wear Archives Service

References to 'Smiles' are to the 1873 edition of *The Story of the Life of George Stephenson*, except those indicated as 'Smiles, One', which are to the first edition of 1857.

References to Nicholas Wood's *Treatise on Railways* are to the 3rd edition (1838) except where noted as from the first edition (1825).

Edwin Clark (*Tubular Bridges*) is noted as Clark, Edwin; and E.F. Clark (*George Parker Bidder*) is noted as Clark, E.

*ODNB*: Oxford Dictionary of National Biography, 2004.

## Introduction

1. Smiles, *Autobiography*, 163.
2. See his letters to Smiles in NRM, 2006-7495.
3. Smiles, *Autobiography*, 254.

## 1. A Working Boy

1. Smiles, 14.
2. The north-eastern pits at least no longer employed women and girls underground at this time. In some coalfields this was common practice until 1842.
3. Halévy, *History of the English People*, II, 87.
4. Holmes, *Coal Mines of Durham and Northumberland*, 125.
5. Austin, *Autobiography*, 8.

6. Domestic details in this chapter are from Smiles, Chapter 1.

7. Wood, *Address*, 4.

8. Smiles, 18.

9. *Ibid.*, 21.

10. Rolt, *George and Robert Stephenson*, 14.

11. Also, 'I am reminded of a singular feature in him, viz., the absence of the use of writing and figures, which he seldom employed' (Summerside, *Anecdotes*, 62).

12. Smiles, 24.

13. Personal communications to the author from Professor Maggie Snowling of the University of York, England, and Professor P.G. Aaron of Indiana State University, who confirm typical dyslexic features in Stephenson's spelling and grammar.

14. Smiles, 21.

15. Wood, *Address*, 6.

16. Smiles, 26.

17. *Ibid.*, 29–30.

18. Grundy, *Pictures of the Past*, 124.

19. Jeaffreson, I, 5.

20. Smiles, 35.

21. *Ibid.*, 34.

22. *Ibid.*, 35.

23. Pole, *Life of Fairbairn*, 80.

24. Smiles places Robert's birthday on 16 October, having changed it, apparently at Robert's word, from 16 December. But in 1852, Robert invited the Bidders to a birthday dinner on 16 November (ScM: GPB 5/12/5a). George Parker Bidder notes this date as his friend's birthday in his diary for 1853, 1854 and 1857.

25. Jeaffreson, I, 8.

26. *Ibid.*, 9.

27. Jim Rees and Andy Guy (Bailey, *Early Railways 3*, 196) suggest this locomotive was built as a speculation in the hope that Blackett would buy it.

## 2. To Scotland, and Back Again

1. Skeat, *George Stephenson*, 14.

2. Smiles, 38.

3. This was Timothy Hackworth's father, foreman blacksmith at Wylam. The surviving records of Boulton & Watt have no details of the Montrose engines.

4. Low, *Industry in Montrose*, 6.

5. Smiles, 38.

6. *Ibid.*, 39.

7. Roper, R.S., 'Robert Stephenson, Senior, 1788–1837', in Lewis, *Early Railways 2*, 26ff. Smiles says that George returned to find Anne married to his brother, but if this is the case, he either spent more than a year in Montrose, or did not leave for Scotland until a year after his wife's death.

8. Wood, *Address*, 5.

9. Summerside, *op. cit.*, 24.

10. In 1824 Steel was killed by a boiler explosion on a steamboat at Lyon, France. One who escaped (seeing the safety-valve tied down) was Charles Manby, who in 1856 became a close business associate of Robert Stephenson. It was a small world for engineers.

11. IMechE: Stephenson, Presidential Address.

12. Hammond, *The Bleak Age*, 45.

13. Jeaffreson, I, 9.

14. *RT*, 6 July 1844.

15. Smiles, 52.

16. Sir Arthur Salter, 'John Stuart Mill', in Massingham, *The Great Victorians*, II, 310ff.

17. Smiles, One, 21.

18. Jeaffreson, I, 20.

19. *Ibid.*, 21.

20. Smiles, 43.
21. IMechE: Stephenson, Presidential Address.
22. *Ibid.*
23. Smiles, 50. Adrian Jarvis, 'The Story of the Story of the Life of George Stephenson', in Smith, *Perceptions of Great Engineers*, 38f, brands George as a drunk on very flimsy evidence, chiefly the comments of Edward Pease, an evangelical teetotaller. George was in fact notable for keeping a clear head, as a more objective witness testifies (Grundy, 129).

## 3. Steam and Firedamp

1. Wood, *Address*, 9.
2. *Ibid.*
3. Summerside, *op. cit.*, 30.
4. Addyman and Haworth, *Robert Stephenson*, 33.
5. Sykes, 'Local Records of Remarkable Events', quoted in Dendy Marshall, *A History of Railway Locomotives Down to the Year 1831*, 44.
6. Dendy Marshall, *op. cit.*, 83; also Mountford, *The Private Railways of County Durham*, 6.
7. Guy (Guy & Rees, *Early Railways*, 119) considers the locomotive *was* called 'My Lord', maybe after Lord Strathmore; Mountford (*op. cit.*) agrees and adds 'there is some evidence that a second geared engine, probably known as *Blücher* was developed in the same year'. There is no compelling evidence for this. The Strathmore suggestion is discredited by George's own double error at the opening of the Newcastle & Darlington Railway, 18 June 1844, when he said his first locomotive was built in 1812, with Lord Ravensworth's money – hence 'My Lord'.
8. Wood, *Treatise*, 288.
9. Summerside, *op. cit.*, 9.
10. Wood, *Treatise*, 1st edn, 142.
11. See Guy, 'North Eastern Locomotive Pioneers 1805 to 1827: A Reassessment', in Guy & Rees, *op. cit.*, 136f.
12. Guy (*op. cit.*, 137) suggests that the idea of the chain drive was taken by Stephenson from a proposal made by William Tindall and John Bottomley, of Scarborough, in a letter to the Society of Arts of 4 June 1814. They may have been passing on an idea of Chapman's.
13. Smiles, 92.
14. Holmes, *op. cit.*, 93.
15. *Ibid.*, 154.
16. *Ibid.*, 107.
17. *Ibid.*, 147.
18. *TNS*, 75.2, James, 'How Big is a Hole?' 180.
19. *Ibid.*, 183f.
20. Quoted in Smiles, One, 516.
21. Wood, *Address*, 13.
22. Summerside, *op. cit.*, 11.
23. *TNS*, 70, 1998–99. Watson, 'The Invention of the Miner's Safety Lamp: A Reappraisal', 137.
24. *TNS*, James, *op. cit.*, 217.
25. *TNS*, Watson, *op. cit.*, 137.
26. Parsons, *A History of the Institution of Mechanical Engineers*, 249.
27. Smiles, 98.
28. *PM*, XLVI, 459.
29. *PM*, XLIX, 204ff.
30. *Ibid.*
31. Holmes, *op. cit.*, 188.
32. *Ibid.*, 187.
33. A description of the Geordie lamp was published in the London *Star* on 18 December, in an article which also has details of Davy's gauze lamp. It is not unlikely that Alexander Tilloch, editor of the *Star* as well as of *PM*, had given Davy the details prior to publication (they had been published in Newcastle on 9 December).
34. *TNS*, 75.2, James, *op. cit.*, 190.

35. NRO: SANT/BEQ/18/11/13.

36. Quoted in *PM*, XLVII, 50.

37. Letter to William Locke, quoted in Webster, *Joseph Locke*, 29–30.

38. Rolt, *op. cit.*, 15; author's italics.

39. John Ferguson FRS was the prime scientific populariser of his day. He gave public lectures on mechanics, physics and astronomy, and published them in book form. *Mechanics* was first published in 1761 and several editions followed. The Stephensons also had his *Astronomy*.

40. A converse and equally hypothetical question might also be asked: if George Stephenson had remained an engine-minder, what would his son's career have been?

41. Jeaffreson, I, 34.

42. *Ibid.*

43 Smiles, 55.

44. Roper, *The Other Stephensons*, Family Tree 2.

45. Lit & Phil Bicentenary Lectures, 1993, 72, 196; quoted in Addyman & Haworth, *op. cit.*, 8.

46. Carlyle, *Signs of the Times*, in *Miscellanies*, ii, 263ff.

47. Jeaffreson, I, 42.

48. Smiles, 56.

49. Jeaffreson, I, 45.

## 4. Emergence of an Engineer

1. These were vertical cylinders opening into the boiler, using the 'elasticity' of the steam in the boiler to absorb shocks from the track and to keep the engine's weight evenly distributed: in fact, the first shock absorbers.

2. Smiles, 60.

3. George Stephenson, not without reason, has been accused of ignoring men whose work he profited from, so it should be noted that in 1846 he contributed £5 towards a testimonial for Dr Clanny, and wrote: 'I believe Dr Clanny was the first person who made the attempt to construct a lamp which should burn in an inflammable atmosphere without exploding' (Smiles, One, 133). Another early safety lamp experimenter was Willie Woolhave, the South Shields lifeboat pioneer; it has been suggested that Clanny got the bellows idea from him (see Snell, *op.cit.*, 2).

4. *TNS*, 75.2, James, *op. cit.*, 210. The whole safety lamps controversy is magisterially analysed and the account given here is indebted to it. James' final judgement is that: 'His reaction … does suggest a degree of uncertainty on Davy's part about his own claim to the invention.'

5. *PM*, XLVII, 314. John Buddle wrote on 22 March 1816 to inform George (IMechE: IMS135).

6. Quoted in Thompson, *Thunder Underground*, 113.

7. *NC*, 19 October 1816; *Durham City Advertiser*, 19 October.

8. Stephenson, *A Description*, 3f.

9. *Ibid.*, 5.

10. *PM*, February 1817, XLIX, 152, 204ff.

11. TWAS: DX 1242/1, letter from Davy to Hodgson, 6 November 1817. Even the current *ODNB* entry on Davy refers only to the fact that there were 'priority disputes', adding that Stephenson's claims were 'pressed sometimes scurrilously by his partisans' without mentioning Davy's unscrupulous comments.

12. Quoted in Rolt, *op. cit.*, 31f.

13. *PM*, LI, 65.

14. Smiles, One, 125.

15. Quoted in Davies, *George Stephenson*, 28.

16. Mountford, *op. cit.*, 9ff; see also Guy, *op. cit.*, in Guy & Rees, 117–144. Stephenson was also building stationary engines; a letter to Joseph Cabry of 2 March 1820 notes: 'I have got orders for two more Engines for Newbottle Colliery: one on the Waggon way and the other down the Pit' (IMechE: IMS133/1).

17. Tomlinson, *The North Eastern Railway*, 16.

18. Warren (*op. cit.*, 111) quotes a letter from Robert to the Rev. John Bruce, referring to his own 'taste for mathematical pursuits' and adding how necessary it is to adapt mathematical principles to practical circumstances: 'To an Engineer mathematical truths are like sharp

instruments, they require to be handled with care and circumspection – in short they are only truths so long as certain conditions obtain' (Letter printed in *Northern Daily Express*, 15 October 1859).

19. 'Report on Edinburgh Railway', 1818, quoted in Dendy Marshall, *op. cit.*, 29.

20. An alternative tradition suggested he came from a farming family near Jedburgh in the Borders (Roper, *op. cit.*, 3).

21. Paul Reynolds, 'George Stephenson's 1819 Llansamlet Locomotive' in Lewis, *Early Railways 2*, 165ff.

22. IMechE: IMS133/1, 21 December 1819.

23. *Ibid.*

24. Andy Guy, 'North Eastern Locomotive Pioneers: A Reassessment', in Guy and Rees, *op. cit.*, 121, quoting Blackett Papers in NRO.

25. Colin Mountford, 'The Hetton Railway', in Bailey, *Early Railways 3*, 78.

26. Quoted in *Mechanics' Magazine*, XLIX (1848), 500.

27. There is no first-hand record of whether Stephenson or Pease initiated the historic meeting. Nicholas Wood's *Address* makes clear that an appointment was made. Edward Pease's edited *Diaries* quote Francis Mewburn, the S & D's solicitor, as saying that the visitors came 'at the behest of Pease' (97). Tomlinson (*op. cit.*, 72) adds that John Dixon was Pease's emissary.

28. Smiles, 126.

29. Smiles, One, 186; also Summerside, *op. cit.*, 30. Stephenson had a lifelong belief in 'craniology'. Twenty years later, he told John Hart 'quite coolly that *I* could never succeed as he had done, for I had "too small a chin"' (Grundy, 126).

30. Quoted in Warren, *op. cit.*, 44.

31. Letter to Francis Mewburn, 5 January 1822, Newcastle Public Library, quoted in Tomlinson, *op. cit.*, 29.

32. Jeaffreson, I, 54.

33. *Ibid.*

34. Smiles, One, 194.

35. Quoted in Warren, *op. cit.*, 45. Stevenson and William James were among others concerned about the development of a reliable track.

36. Smiles, 134; though Rolt notes that wrought iron weighed less than half of cast iron of equivalent strength (*op. cit.*, 74).

37. Tomlinson, *op. cit.*, 77.


## 5. Music at Midnight

1. Smiles, 116–117.

2. The Carron Company built a three-arched iron railway bridge over the Carron in 1810 (illustrated in Dendy Marshall, *A History of British Railways*, 123).

3. LRO, 385 JAM 1/5/3.

4. Jeaffreson, I, 55.

5. Scott, Letter to Lord Montagu quoted in Lockhart, *Life of Sir Walter Scott*, III.

6. Jeaffreson, I, 58.

7. Clark, E., *George Parker Bidder*, 14–15.

8. Quoted in Webster, *Joseph Locke*, 29–30.

9. Quoted by George Robert Stephenson from a document of 1823 in Duncan, *The Stephenson Centenary*, 81.

10. Quoted in Skeat, *op. cit.*, 68.

11. Warren, *op. cit.*, 47.

12. Clause 5 of the articles of incorporation, reproduced in Warren, *op. cit.*, 53ff.

13. *Ibid.*, Clause 6.

14. The first purchase deed, in the names of George Stephenson and Michael Longridge, is dated 1 August 1823 (TWAS: DT/SC/305/3).

15. Tomlinson, *op. cit.*, 93.

16. Smiles, 136.

17. Bathurst, *The Lighthouse Stevensons*, 114.

18. *TNS*, vol. 50, 1978. Bailey, M.R., 'Robert Stephenson & Co., 1823–1829', 105ff.

19. Jeaffreson, I, 62.
20. Quoted in E.M.S.P., *op. cit.*, 55. Macnair, *William James*, says 'In the Stephenson family there was little true warmth and affection between son and father' (68), but there seem no grounds for this comment.
21. Privately-owned letter, quoted in Bailey, *Robert Stephenson*, 14.
22. ICE, 1824 STELG.
23. Jeaffreson, I, 73.
24. *Ibid.*, 70.
25. *Ibid.*, 68.
26. Smiles, 94.
27. See Spence Galbraith, 'William Hardcastle of Newcastle upon Tyne', in *Archaeologia Aeliana*, 5th series, Vol. xxvii, 161. In 1836 Empson published *Narratives of South America, Illustrating Manners, Customs and Scenery.*
28. Smiles, 281.
29. Jeaffreson, I, 77.
30. DRO: D/Ho/C/63/12.
31. NRM, MSL 8/1.

## 6. 'Did Any Ignorance Ever Arrive at Such a Pitch?'

1. Falk, *The Bridgwater Millions*, 122.
2. Quoted in E.M.S.P., *op. cit.*, 59.
3. IMechE: 130/1, letter to Longridge, 22 May 1824: Two emissaries have come 'very anxious of me taking the Birmingham intended line under my care ... we must not let such a valuable concern slip.'
4. IMechE: IMS130/2/3/4: letters to Longridge, 7 and 9 June, 11 July 1824.
5. Tomlinson, *op. cit.*, 96.
6. Kostal, *Law and English Railway Capitalism*, 12.
7. E.M.S.P., *op. cit.*, 66.
8. DRO: Pease-Stephenson Papers, D/HO/C/63.
9. DRO: Hodgkin Papers, D/140/C/63/10.
10. *TNS*, 70, Bailey, *op. cit.*, quoting Robert Stephenson & Co. Board Minutes.
11. Warren, *op. cit.*, 63.
12. Carlson, *The Liverpool and Manchester Railway Project*, 43.
13. Biddle, *The Railway Surveyors*, 59.
14. An agent of the Russian Tsar was studying locomotive design; George wrote to Longridge from Chester on 23 October 1824, 'I suppose I shall have the Russian gentleman with me next week' (IMechE: IMS130/6).
15. *Practical Mechanics' Journal*, Vol. III, 1850–51, 225.
16. DRO: Pease-Stephenson, U415J, Vol. III, Item 7.
17. Lewis, *The Cabry Family*, 9.
18. Rolt, *op. cit.*, 108.
19. Francis, *A History of the English Railway*, 110.
20. *Ibid.*, 103.
21. Smiles, 161.
22. Pendleton, *Our Railways*, I, 43.
23. Smiles, 165.
24. Thomas, *The Liverpool & Manchester Railway*, 27.
25. IMechE: IMS, 133/4.
26. Warren, *op. cit.*, 162.
27. Quoted in Rolt, *op. cit.*, 112.
28. Dendy Marshall, *History of Railway Locomotives*, 131.
29. Credit for this improvement has been claimed on behalf of Timothy Hackworth, and Ralph Dodds' teenage nephew Isaac (later a Stephenson apprentice) but there is no clear evidence.
30. Dendy Marshall, *op. cit.*, 123.
31. *NC*, 1 October 1825.

## 7. Testing Times

1. Jeaffreson, I, 77.
2. Empson and a party of miners went by sea via Cartagena.
3. Jeaffreson, I, 82.
4. Smiles, 215.
5. IMechE: IMS/164/1.
6. NRM, 1992-239, MSL 8/1-7.
7. *Ibid.*, MSL 8/5.
8. Jeaffreson, I, 89ff.
9. *Ibid.*, 91.
10. *Ibid.*, 155.
11. Quoted in Webster, *Joseph Locke*, 36.
12. NRM, 1992-239, MSL 8/7: 'The situation in which we are placed is nearly as bad as transportation.'
13. Kostal, *op. cit.*, 22.
14. Thomas, *op. cit.*, 33.
15. Carlson, *op. cit.*, 183ff.
16. *Ibid.*
17. Thomas, *op. cit.*, 37, 44, 50.
18. Wood, *Treatise*, 1st edn, 65.
19. Letter to Edward Riddle, in Vignoles, *Charles Blacker Vignoles*, 31f.
20. Quoted in Carlson, *op. cit.*, 189.
21. Thomas, *op. cit.*, 47.
22. Carlson, *op. cit.*, 189.
23. *TNS*, 52, 1980-81, 'George Stephenson – A Commemorative Symposium on the 200th Anniversary of his Birth', 171f.
24. Biddle, *op. cit.*, 110.
25. *TNS*, Vol. 50, 1978. Bailey, M.R., *op. cit.*, 122.
26. Jeaffreson, I, 97.
27. *Ibid.*, 101.
28. NRM, 1992-239, MSL8/7.
29. Jeaffreson, I, 92.
30. Quoted in Skeat, *op. cit.*, 246. The letter is discussed by Jack Simmons in *JTH*, Vol. 2, Sept. 1971, 108-115.
31. Macnair, *op. cit.*, 92.
32. Quoted in Smiles, *Lives of the Engineers*, iii, 250.
33. Jeaffreson, I, 103.
34. *Ibid.*, 97.
35. Trevithick, *Life of Richard Trevithick*, II, 274.
36. Letter from Fairbairn to Watkin, 27 November 1864; in Trevithick, *op. cit.*, II, 273.
37. *Ibid.*, 273-274.
38. Trevithick, *op. cit.*, I, 173.
39. Rolt (*op. cit.*, 43) says 'Between 1805 and 1808 the great Cornishman visited Newcastle on several occasions', but gives no source for this, and Dickinson and Titley note that in 1808 John Buddle and another leading Tyneside coal viewer, William Stobart, 'did not know Mr Trevithick' (*Richard Trevithick*, 101). On 3 June 1815 in the *Newcastle Courant*, Thomas Waters & Co., builders of Christopher Blackett's second locomotive in 1812-13, advertise themselves as 'manufacturers of Trevithick engines', and 'solely appointed by the inventor to make them for Home consumption or Exportation', taking over from 'the old Agent, John Whinfield, appointed in 1803'. It is possible that Trevithick visited this company without the news of his presence reaching the public prints. Incidentally, the *Oxford English Dictionary* does not record the word 'nurse' in the sense to hold caressingly, in the arms or on the lap, before 1850: too late for Trevithick, but not for Hall.
40. See Warren, *op. cit.*, 261f; Lecount, *A Practical Treatise on Railways*, 376; Skeat, *op. cit.*, 137.
41. Illingworth later returned to England to live in Hastings. He remained a friend of Stephenson's and they corresponded about horses, mining machinery and railway shares, among other things.

42. Jeaffreson, I, 106.
43. *Ibid.* Jeaffreson notes (I, vi) that Empson, 'shortly before his death ... contributed to my store of materials a most interesting collection of letters and documents; consisting of Robert Stephenson's early journals, and of nearly all the letters which he either received *from* or had written *to* friends and relations, between the termination of his life on Killingworth Moor and his return from South America'. One wonders how Empson came to have them.
44. Jeaffreson, I, 110.
45. *Ibid.*
46. *Ibid.*, 110f.

## 8. 'Worthy of a Conflict'

1. Rolt, *op. cit.*, 155.
2. Thomas, *op. cit.*, 50.
3. *Mr Telford's Report*, 1.
4. Thomas, *op. cit.*, 52f.
5. *Ibid.*, 15.
6. *Mr Telford's Report*, passim.
7. Thomas, *op. cit.*, 54.
8. Telford's entry in *ODNB* states that following his inspection he was: '... instrumental in persuading the company to abandon the idea of fixed engines and inclined planes in favour of a level line suitable for locomotive haulage.' This is quite untrue: the idea of fixed engines was abandoned only after the Rainhill trials, and the line was not levelled or adjusted after Telford's visit. Apart from the time and extra cost this would have required, it was not necessary, since *Rocket* and its successors could climb the gradients with a train.
9. Quoted in Carlson, *op. cit.*, 207.
10. Parry, *Whistler's Father*, quoted in Ellis, *British Railway History*, 26.
11. Jeaffreson, I, 119f.
12. Payen, *La Machine Locomotive*, 40.
13. Letter to Longridge from London, 1 January 1828, quoted in Warren, *op. cit.*, 72.
14. Jeaffreson, I, 121.
15. Warren, *op. cit.*, 149.
16. Webster, *op. cit.*, 57.
17. Jeaffreson, I, 133.
18. *Ibid.*
19. *Ibid.*, 131.
20. *Ibid.*, 135.
21. Bailey, *Robert Stephenson*, 168.
22. Jeaffreson, I, 123f.
23. *Ibid.*, 137.
24. *Ibid.*
25. *Ibid.*, 136.
26. It has always been accepted in both France and Britain that though Seguin was first to design a tubular boiler for locomotive use, the Booth-Stephenson model was independent of his. His tubes were brass, theirs were copper, and not least *Rocket* had a separate firebox and relied on blast from the exhaust to draw heat through the tubes. If they had modelled their work on Seguin's, *Rocket* would have had a fan to drive the heat through the tubes. Seguin and Stephenson did not discuss tubular boilers, and in the course of a later visit to England in 1834, Seguin said: 'I was ignorant of the fact that Mr Stephenson's first engine, which carried off the prize ... was with tubes' (See *TNS*, 7, 1926, 97ff; Achard, F., and Seguin, L., 'Marc Seguin and the Tubular Boiler').
27. *TNS*, Vol. 50, 1978. Bailey, M.R., 'Robert Stephenson & Co., 1823–1829', 127.
28. Quoted in Warren, *op. cit.*, 177.
29. R.L. Hills, 'The Development of Machine Tools in the Early Railway Era' in Bailey, *Early Railways 3*, 255, notes however that the Stephenson Works 'lagged behind Manchester workshops with their range of machine tools'.
30. Warren, *op. cit.*, 177.

31. Whitting, H.S., *Alfred Booth: Some Memories, Letters and Other Family Records*, 21–22.
32. NA Rail 1008/88/15. Robert Stephenson to Booth, 31 August 1829.
33. Warren, *op. cit.*, 181, says this was an invention of George Stephenson's. If so, he did not patent it.
34. Payen, *op. cit.*, 61.

## 9. Rainhill and Afterwards

1. *Mechanic's Magazine*, quoted in Warren, *op. cit.*, 188f.
2. Webster, *op. cit.*, 54.
3. Warren, *op. cit.*, 191.
4. Dendy Marshall, *op. cit.*, 147, reproduces the drawings.
5. *Mechanics' Magazine*, Vol. XII, 146; 24 October 1829, quoted in Warren, 195.
6. Quoted in Rolt, *op. cit.*, 174f.
7. Booth became secretary of the Grand Junction Railway, which, given George Stephenson's behaviour towards that company, might account for a coolness between them.
8. Quoted in Thomas, *op. cit.*, 58.
9. Smiles, 187.
10. *Ibid.*, 220.
11. *Ibid.*, 264.
12. Walmsley, *Sir Joshua Walmsley*, 67.
13. Letter of 17 December 1829, quoted in Jeaffreson, I, 152.
14. Kemble, Frances, *Memoirs of a Young Girl*, Vol. 2, 158ff.
15. Walker, *An Accurate Description*, 9.
16. John Foster the younger was at that time Liverpool's leading architect, with a preference for Greek models.
17. Walker, *op. cit.*, 39.
18. *Ibid.*, 45.
19. Walmsley, *op. cit.*, 72.

## 10. The Utilitarian Spirit

1. Bailey, *Robert Stephenson*, 171.
2. ICE, 5830 STELG.
3. Jeaffreson, I, 168.
4. Quoted in Skeat, *op. cit.*, 138.
5. TWAS: DT/SC/317.
6. Walmsley, *op. cit.*, 75.
7. Cole & Postgate, *The Common People*, 246.
8. Walmsley, *op. cit.*, 68.
9. Roper, *op. cit.*, 3.
10. Warren, *op. cit.*, 76.
11. See Bailey, *Robert Stephenson*, 175f.
12. Quoted in Bailey, *op. cit.*, 35.
13. Jeaffreson, I, 178.
14. *Ibid.*
15. Letter of 15 November 1832, quoted in Bailey, *Robert Stephenson*, 33.
16. *Edinburgh Review*, No. CXI, 131.
17. Bailey, *Robert Stephenson*, 36.
18. Cole & Postgate, *op. cit.*, 257.
19. Webster, *Britain's First Trunk Line*, 37f.
20. Rolt, *op. cit.*, 223.
21. Addyman & Haworth, *op. cit.*, 60.

## 11. London and Birmingham

1. Mike Chrimes, 'London & Birmingham', in Bailey, *Robert Stephenson*, 43.
2. Jeaffreson, I, 182.

3. Quoted in Warren, *op. cit.*, 312.

4. Summerside, *op. cit.*, 64.

5. Smiles, 235.

6. *Ibid.*, 261.

7. Devey, *Joseph Locke*, 102.

8. Webster, *Joseph Locke*, 47.

9. Clark, E., *op. cit.*, 29, 82.

10. Webster *op. cit.*, 49.

11. Quoted Rolt, *op. cit.*, 232.

12. NA: RAIL 384/2.

13. Conder, *Personal Recollections*, 14–15.

14. *Ibid.*, 15.

15. Smiles, 247.

16. Jeaffreson, I, 203.

17. Rolt, *op. cit.*, 242.

18. From 'Second Report to the London, Westminster & Metropolitan Water Company', 141; quoted by Chrimes in Bailey, *Robert Stephenson*, 256.

19. Clark, E., *op. cit.*, 36.

20. BRIS: DM1306/2, from the L & B engineer's office.

21. BRIS: DM1306/7, letter from Stephenson to Brunel, 11 March 1835.

22. Surprisingly, this line was designed at first for horse traction, but the reason was to save the cost of making bridges and other works strong enough to support locomotives.

23. Tomlinson, *op. cit.*, 267.

24. Clark, E., *op. cit.*, 32.

25. *Ibid.*

26. L & B London Committee minutes, quoted in Bailey, *Robert Stephenson*, 69.

27. Smiles, 266.

28. Dambly, *Vapeur en Belgique*, 74.

29. Quoted in Skeat, *op. cit.*, 161.

30. NA: RAIL 384/67.

31. Quoted in Warren, *op. cit.*, 87f.

32. BRIS: Letter from Stephenson to Brunel, 4 August 1836.

33. Bailey, *Robert Stephenson*, 60.

34. Clark, E., *op. cit.*, 40.

35. BRIS: DM1306/16.

36. Quoted in Rolt, *op. cit.*, 244.

37. Quoted in *ibid.*, 246.

## 12. Lives of Engineers

1. Samuel Sidney, *Rides on Railways*, 1851, quoted in Gloag, *Victorian Taste*, 8.

2. The roundhouse was '… designed, so it seems in collaboration between Robert Stephenson and R.B. Dockray, the Company's resident engineer' (Simmons, *The Railway in England and Wales*, 164).

3. Pugin, *Contrasts*, 1.

4. Gloag, *op. cit.*, 3.

5. Brunel, letter to Benjamin Hawes, 27 March 1831, quoted in Buchanan, *Brunel*, 48.

6. Smiles, 280.

7. *RT*, Supplement, 16 June 1838.

8. Clark, E., *op. cit.*, 43.

9. Thomson, *History of Tapton House*, 31.

10. George wrote a cheerful letter from there to Mrs Longridge on 7 August, excusing himself for not having replied to one of hers, and promising a visit: 'I hope you will have one or two good Crabs for me … but I hope you wont be crabby' (IMechE: IMS130/7).

11. Jeaffreson, I, 234.

12. *Ibid.*, 231.

13. Webster, *op. cit.*, 199.

14. Jeaffreson, I, 237. Nevertheless, Robert used his crest, a mailed fist clasping a scroll, with the motto *Fides in Arcanis*, as a seal on correspondence, and it was engraved, without the motto, on the plate used on his yacht *Titania* (Clark, E., *op. cit.*, 368). Incidentally, the motto 'Faith in Secret Things' suits this very private public man.

15. 29 August 1838, quoted in Jeaffreson, I, 232.

16. Jeaffreson, I, 243.

17. The testimonial is preserved (IMechE: IMS188). Edward F. Starbuck and William Fairbairn were among the subscribers.

18. Smiles, 263.

19. Quoted in Rolt, *op. cit.*, 277.

20. Lambert, *op. cit.*, 48.

21. Quoted in Rolt, *op. cit.*, 258.

22. *RT*, 17 July 1841.

23. IMechE: IMS168/2.

24. Smiles, 287.

25. Stephenson, 'Report on the Atmospheric Railway System, 1844', quoted in Bailey, *Robert Stephenson*, 195f.

26. *London Post Office Directory*, 1841.

27. Quoted in Warren, *op. cit.*, 94.

28. Jeaffreson, I, 248.

29. Jeaffreson, I, 249.

30. *Ibid.*

31. *Ibid.*, 108. His membership certificate (IMechE) in English, French and Spanish, is clearly intended as an introduction should help be needed.

32. Lambert, *op. cit.*, 45, 59. In 1862 Joseph Devey wrote, referring to another of George's announcements, of his imminent retirement: 'He had been gifted with immense force. All that force had been expended, and expended most usefully. Withal, it was spent … Unfortunately, he was introduced to George Hudson; and still more unfortunately Faust was not inattentive to Mephistopheles …' (*Joseph Locke*, 109). Nowadays we might be less certain about this characterisation. The two were made for each other. John Hart gives a splendid description of Hudsonian hospitality on a visit George and he made to York around 1841 (Grundy, *op. cit.*, 129).

33. Storey's departure seems to have been prompted by a dispute with the board over costings. Later he claimed that the construction costs were deliberately underestimated in order to get the Company's Bill passed (see *R.T.*, September–October 1842).

34. *HRM*, 13–20 March 1841, 261.

35. *Ibid.*

36. Kirby, *Origins of Railway Enterprise*, 126.

37. *HRM*, 15 May 1841, quoting from the *Yorkshire Gazette*.

38. Lambert, *op. cit.*, 61; no source given.

39. *Ibid.*, 65.

40. *HRM*, 12 March 1842.

41. Rolt, *op. cit.*, 272–273.

42. Bailey, *Robert Stephenson*, 107.

43. Lambert, *op. cit.*, 65.

44. *Ibid.*, 66.

45. *Ibid.*, 67. Lambert's hostility to Robert Stephenson is at no point given supporting evidence.

46. Grundy, *op. cit.*, 118.

47. Smiles, *Autobiography*, 189.

48. Rolt, *op. cit.*, 274.

49. *RT*, 7 August 1841.

50. Richardson, *Thomas Sopwith*, 177.

51. Owen, *The Leicestershire & North Derbyshire Coalfield*, 198.

52. *RT*, 27 February 1841.

53. Smiles, 276.

54. *Ibid.*

55. Quoted in Warren, *op. cit.*, 95.

56. Richardson, *op. cit.*, 185.

57. BRIS: DM1758/9, Letter from Newcastle, 12 October 1841.

58. Warren, *op. cit.*, 346.

59. Robert Stephenson, 'Report to the Directors of the Norfolk Railway', 21 January 1846, quoted in Warren, *op. cit.*, 350.


## 13. 'I am George Stephenson'

1. Grundy, *op. cit.*, 125, 127. Summerside (*op. cit.*, 57) says George never bought a railway ticket. But he would have had a Free Pass for any line on which he cared to travel.

2. Grundy, *op, cit.*, 123.

3. *Ibid.*, 135.

4. Smiles, 281. Summerside (*op. cit.*, 60) says George 'was not at all an adept at public speaking' and dreaded the prospect.

5. Grundy, *op. cit.*, 132.

6. NA: RAIL 1008/91.

7. Jeaffreson, I, 255.

8. *Ibid.*

9. Quoted in Warren, *op. cit.*, 96.

10. IMechE: IMS/189/1. Robert wrote to Starbuck on 25 October 1842, 'Mr J. Pease I have no doubt has been instrumental in this.'

11. Jeaffreson, I, 260.

12. Grundy, *op. cit.*, 135.

13. Smiles, 309.

14. Lambert, *op. cit.*, 69.

15. Parliamentary Papers, 1852–53, xxxviii, 123; quoted in Simmons, *The Railway in England and Wales*, 26.

16. *RT*, 26 August 1843, Part II.

17. *RT*, 18 November 1843.

18. Lambert, *op. cit.*, 90.

19. TWAS: DT.STB 131/1.

20. *Ibid.*

21. *Ibid.*, letter from Robert Stephenson to Conrad, 18 January 1844.

22. Richardson, *op. cit.*, 212.

23. Lambert, *op. cit.*, 109.

24. Rolt, *op. cit.*, 276.

25. Smiles, 308.

26. Letter from Tapton House, 30 November 1843, quoted in Parsons, *History of the Institution of Mechanical Engineers*, 251.

27. Letter to *Derbyshire Courier*, 27 April 1844, quoted in Thomson, *op. cit.*, 42.

28. *Ibid.*, 40.

29. Owen, *op. cit.*, 205.

30. Parsons, *op. cit.*, 124.

31. Northumberland Railway: Lord Howick's Circular, November 1844, quoted in Bailey, *Robert Stephenson*, 109.

32. 'To the Landowners of the County of Northumberland', November 1844, quoted in *ibid.*, 110.

33. Lambert, *op. cit.*, 142.

34. Francis, *op. cit.*, II, 212.

35. Smiles, 309.

36. BRIS: DM1306/24; letter of 21 March 1845 from Great George Street.

37. Gooch, *Diaries*, 13, 56.

38. In 1843 Pasley had circulated all leading engineers with the question: 'Supposing that there were no railways in England … what would you, with your present experience, recommend as being the best uniform gauge to be adopted for the whole of them …?' George Stephenson replied that 4ft 8.5in 'was most economical in construction, not only as regarded the engines and carriages, but more particularly of the railway itself', adding that he had once considered 5ft 2in to be preferable, 'but having discussed the matter fully with his son, he had again

. changed his opinion'. Most engineers opted for something between 5ft and 5.5ft (quoted in Francis, *op. cit.*, II, 64).

39. *Report of the Gauge Commissioners*, 1846, quoted in Warren, *op. cit.*, 377.
40. Warren, *op. cit.*, 383.
41. *Railway Chronicle*, December 1845, quoted in Warren, *op. cit.*, 383.
42. Ellis, *op. cit.*, 106. A more complete and balanced account is given in Warren, *op. cit.*, Chapter XXVIII. Bidder's biographer, E.F. Clark, ignores Ellis' accusation, though he quotes Ellis in other contexts.
43. George Hudson duly obliged in 1846.

## 14. Big Bridges

1. Clark, E., *op. cit.*, 56.
2. Francis, *op. cit.*, II, 166.
3. Simmons, *op. cit.*, 54.
4. Jeaffreson, I, 280. The Lords Spiritual are not exempted.
5. *Ibid.*, I, 287.
6. Smiles, 165.
7. *Ibid.*, 292.
8. Lambert, *op. cit.*, 166; also Francis, *op. cit.*, II, 150.
9. Clark, E., *op. cit.*, 58.
10. Francis, *op. cit.*, II, 155.
11. In 1838 Robert had given support to James' widow in an unsuccessful effort to have a memorial set up, and a letter to Richard Creed at the London & Birmingham Railway shows him again supporting a proposal for a subscription to James' family in November 1844 (NA, RAIL1008/90). George was regarded by the James family as the wrecker; see *The Two James' and the Two Stephensons*, by James' daughter Ellen ('E.M.S.P.'), 106. Some writers have speculated that George had been jealous of Robert's close relationship with James in 1821–23; see Macnair, *op. cit.*, 119ff.
12. Grundy, *op. cit.*, 149ff.
13. *Ibid.*, 134. 'George Stephenson's Scrapbook', a thick volume of newspaper cuttings now in the ICE archives, is more likely the work of Betty, compiled with an eye on George's activities and interests. As well as many items of railway and industrial news, it includes reports of murders, along with anecdotes and miscellanea, and a great many sentimental verse works of the 'Dying Girl's Lament' type. Most of the items are from 1830–37, but some are from the 1840s. There are only one or two from after her death.
14. Clark, E., *op. cit.*, 53.
15. *Ibid.*, 64.
16. *Ibid.*, 85.
17. Smiles, 351.
18. Walmsley, *op. cit.*, 148.
19. Ernouf, *Paulin Talabot*, Chapter VI.
20. Walmsley, *op. cit.*, 153.
21. *Ibid.*, 155.
22. IMechE: IMS130/8.
23. Walmsley, *op. cit.*, 156.
24. Thomas, *op. cit.*, 225.
25. It was suggested that Hudson had authorised publication of a list of subscribers before they had even been approached. Lambert ascribes the story to the gossipy William Bridges Adams, adding: 'George Stephenson, whose name was included in the list, denounced the trick and threatened to insert his refusal in the papers. But he was dissuaded by a host of directors, who warned him that his actions would cause a calamitous fall in the price of railway shares' (*op. cit.*, 144).
26. *York Courant*, 10 July 1845.
27. Quoted in Lambert, *op. cit.*, 198.
28. Smiles, 291.
29. Jeaffreson, II, 130.

30. Clark, Edwin, *The Britannia and Conway Tubular Bridges*, Introduction.

31. Jeaffreson, II, 304.

32. Foreword to Clark, Edwin, *op. cit.*, I, 31f.

33. Pole, *Life of Sir William Fairbairn*, 198.

34. *TNS*, 52, 1980–81. Rennison, 'The High Level Bridge, Newcastle', 180ff.

35. Tomlinson, *op. cit.*, 294.

36. NA: RAIL1008/90.

37. Quoted in Warren, *op. cit.*, 99.

38. Warren, *op. cit.*, 101.

39. Howe, a pattern-maker, was responsible with a Forth Street gentleman apprentice, William Williams, for the development of the first really effective locomotive valve gear, in 1842. Never patented, it was very widely used and known as the 'Stephenson link' gear, though the Stephensons played no part in its design.

40. Warren, *op. cit.*, 391f.

41. Quoted in Davies, *op. cit.*, 296.

42. *RT*, 6 July 1844.

43. Kostal, *op. cit.*, 50.

44. Quoted in Tomlinson, *op. cit.*, 493.

## 15. Invitation to Egypt

1. Quoted in Pole, *op. cit.*, 203.

2. Jeaffreson, II, 88.

3. Quoted in James Sutherland, 'Iron Railway Bridges', in Bailey, *Robert Stephenson*, 323.

4. Rapley, *The Britannia Bridge*, 41.

5. *Ibid.*, 97.

6. Clark, Edwin, *op. cit.*, II, 496. Rapley, *op. cit.*, 47, considers many of Clark's claims to be suspect.

7. Fairbairn, *The Britannia and Conway Tubular Bridges*, 93.

8. *Ibid.*, 94.

9. ICE: EC1 b/3.

10. Clark, E., *op. cit.*, 65, 230ff.

11. *Ibid.*, 231.

12. *Ibid.*, 232.

13. Cranfield, *The Railways of Norway*, 5.

14. Clark, E., *op. cit.*, 232.

15. Quoted in Markham, *Paxton and the Bachelor Duke*, 174.

16. Colquhoun, *A Thing in Disguise*, 145.

17. IMechE: IMS131/1, letter of 8 February 1830.

18. *TNS*, 75.2, Helps, H., 'Luigi Negrelli, Engineer, 1799–1858'; ICE: Negrelli; also PRGS, Pim, 177ff.

19. Pole, *op. cit.*, 212.

20. Quoted in Fairbairn, *op. cit.*, 194.

21. Quoted in Rolt, *op. cit.*, 296.

22. Parsons, *op. cit.*, 11, 254.

23. Clark, E., *op. cit.*, 67.

24. NA: FO 97/411.

25. NA: FO97/411, letter sent on 13 March 1847 from Empress of Austria Hotel, Vienna.

26. ICE: 1857 PALCSI, Negrelli.

## 16. An Engineer at Bay

1. Rolt, *op. cit.*, 302.

2. Jeaffreson, II, 51.

3. Conder, *op. cit.*, 138.

4. *The Times*, 5 June 1847.

5. *Ibid.*

6. *Ibid.*, 18 June 1847.

7. *Ibid.*

8. Rolt, *op. cit.*, 303.

9. In his 1856 article on 'Iron Bridges' for the *Encyclopaedia Britannica*, Stephenson wrote of trussed girders: 'Provided the upper flange were sufficient to resist the thrust to which it is subject, it is evident that such girders are far less liable to accident than simple castings ... The determination of the strength of such girders is, however, a difficult task. A serious accident, moreover, which resulted from the failure of one of these girders on the Dee Bridge near Chester, has entirely put an end to their employment ... the evidence given on that occasion renders it highly probable that even in that case the fracture was occasioned by the train running off the line.' This was a perverse reading of the evidence and verdict. Francis Conder was surprised that: '... the practical eye of Mr Stephenson should have been for a moment blind ... if these [wrought-iron] tie-bars had no tension put on them, they were not only useless, but by their weight, and in consequence of the weakening of the girder by the holes bored for their attachment, were injurious to the strength of the structure. On the other hand, if tension were applied, either by keying up, or by deflexion of the girder, the direct effect would be to break the beam which the bars were supposed to strengthen ...' (Conder, *op. cit.*, 141).

10. Pole, *op. cit.*, 208–209.

11. Clark, Edwin, *op. cit.*, II, 812; Addyman & Haworth, *op. cit.*, 162. In October 1846 the C & H board had objected to Fairbairn's company sharing in the fabrication contract.

12. Quoted in Roper, *op. cit.*, 2. William Hardcastle had been the Stephensons' doctor in Newcastle and remained a close family friend. See Galbraith in *Archaelogia Aeliana*, *op. cit.*, 155ff.

13. Ann died in Pittsburgh in 1860. Her first husband had died and she had remarried; being known as Mrs Anna York (Roper, *op. cit.*, Family Tree 9).

14. Whitworth, *Long Live the King!* 58ff.

15. *The Yorkshireman*, 19 June 1847.

16. Smiles, 144.

17. Quoted in Whitworth, *op. cit.*, 61.

18. *Yorkshire Gazette*, 31 July 1847.

19. *Ibid.*

20. ICE Presidential Address, January 1856.

21. Ruskin, *Fors Clavigera*, in *Works*, Vol. 27, 167.

22. Clark, Edwin, *op. cit.*, II, 629.

23. BRIS: DM1306/27, letter from Great George Street of 6 August 1847.

24. ICE: ECLB/5.

25. Pole, *op. cit.*, 210.

26. The atmospheric traction system on the South Devon Railway was dismantled in September 1848; the last to survive was the Dublin & Kingstown, replaced by locomotives in 1854.

27. Letter to *The Times* from 'G.C.W.', 4 April 1903; quoted in *SLS Journal* March–April 2006, 67.

28. Hallberg, *op. cit.*, 98.

29. NA: FO 141/12.

30. ICE, PALCSI, Negrelli's open letter in the *Oesterreichische Zeitung*. Negrelli says January 1848.

31. Emerson, *Jounals*, Vol. X, 190.

32. *Ibid.*

33. *Ibid.*, 300.

34. Smiles, 342.

35. Walmsley, *op. cit.*, 145.

36. Smiles, 347.

37. NRM 1875–0003.

38. Walmsley, *op. cit.*, 157.

39. Thompson, *A History of Tapton House*, states that Robert attended the wedding (64).

40. Lilly, ILLING, undated.

41. Clark, Edwin, *op. cit.*, II, 645.

42. *Ibid.*, 812.

43. Sutherland, *op. cit.*, in Bailey, *Robert Stephenson*, 328.

44. Pole, *op. cit.*, 175.

45. Northumberland Record Office, quoted Davies, 269.

46. Smiles, *Autobiography*, 215.

47. Smiles, 145.

## 17. 'The Tubes Filled My Head'

1. Walmsley, *op. cit.*, 158.
2. Clark, E., *op. cit.*, 70.
3. Skeat states that George was buried 'with full masonic honours' (*op. cit.*, 252), but the account given in the *Chesterfield Gazette* on 19 August 1848 makes no mention of this. There is no evidence that George was a Freemason.
4. Pease, *Diaries*, 261, 17 August 1848.
5. Walmsley, *op. cit.*, 159. In fact, Stephenson is buried inside the church.
6. Held at the IMechE.
7. NA: Prob 11/2083.
8. Summerside, *op. cit.*, 78f.
9. Smiles, 376.
10. Emerson, *op. cit.*, Vol. X, 342.
11. NRM, MS1152/2.
12. Clark, E., *op. cit.*, 72.
13. Pease, *op. cit.*, 264.
14. Warren, *op. cit.*, 395.
15. StA: Forbes: Letter ref. 83, Robert Stephenson to Professor Forbes.
16. ICE: ECLB, 10.
17. *Ibid.*, 11.
18. Lambert, *op. cit.*, 249.
19. *The Times*, 10 April 1849.
20. Greville, *Diaries*, VI, 162.
21. Tomlinson, *op. cit.*, 499.
22. Bailey, *Robert Stephenson*, 133.
23. Lambert, *op. cit.*, 258.
24. *Ibid.*, 16. When Hudson returned to London from financial exile in 1868, he would reminisce in the Carlton Club (where he was re-elected as chairman of the smoking room) about George as 'the best of fellows and the best of friends' (Lambert, 298).
25. Letter to Captain Moorsom, 5 May 1849. NRM, MS1152/3.
26. Jeaffreson, II, 96.
27. Quoted in Bailey, *Robert Stephenson*, 331.
28. *ILN*, 30 June 1849.
29. Clark, Edwin, *op. cit.*, II, 684.
30. ICE: ECLB, 27.

## 18. Calomel

1. Smiles, *Autobiography*, 163.
2. ICE: ECLB/36. Sutherland, in Bailey, *Robert Stephenson*, 331, wrongly ascribes this letter to 29 December.
3. ICE: ECLB, 37.
4. *HRM*, 27 October 1849.
5. Pole, *op. cit.*, 212.
6. ICE: ECLB, 44.
7. *Carnarvon Herald*, 9 March 1850, quoted in Clark, Edwin, *op. cit.*, II, 706.
8. Jeaffreson, II, 108.
9. Dempsey, *Tubular and Other Iron Bridges*, 92.
10. Clark, Edwin, *op. cit.*, I, 35.
11. NRM, MS1504. Ure made the claim in a new edition of his *Dictionary of Arts, Manufactures & Mines*. Stephenson also had robust supporters: Francis Head in *Stokers and Pokers* (1850) recalled his statements on 5–6 May 1845 to the Commons Committee on the Chester & Holyhead Bill: 'My own opinion is, that a tube of wrought iron would possess sufficient strength and rigidity to support a railway train.' He thought the chains essential only for the construction phase (Head, 208).
12. See Rapley, *op. cit.*, 35, 99, 115.

13. IMechE: IMS/190, Letter to Thomas Brodrich, 6 May 1850.

14. RYS Archives, communication to author, 10 April 2007.

15. Paxton's recollection transcribed by Charles Dickens, in *Household Words*, 18 January 1851.

16. Stephenson, *The Great Exhibition*, 17.

17. Colquhoun, *op. cit.*, 168.

18. Half the glass for the structure was supplied by the R.W. Swinburne Works in South Shields, in which Robert Stephenson was a partner.

19. *Hansard*, 3rd Series, 112, 913f.

20. Clark, Edwin, *op. cit.*, II, 711.

21. *Gateshead Observer*, 3 August 1850.

22. *Ibid.*

23. Jeaffreson, II, 139.

24. *Punch*, 1850,Vol. 19, 113, quoted in Addyman & Haworth, *op. cit.*, 146.

25. Jeaffreson, II, 136.

26. *Ibid.*, 232.

27. Cameron, *Kidney Disease*, 103.

28. Sigmond, *Mercury*, 51.

29. *Ibid.*, 75.

30. BRIS: DM1306/11/20/43/4.

31. Grundy, *op. cit.*, 119.

32. *Encyclopaedia Britannica*, 11th edn, 'Bright's Disease'.

33. Smiles, 368.

34. Jeaffreson, II, 232.

## 19. *Titania* and *America*

1. Clark, E., *op. cit.*, 367.

2. *Ibid.*

3. *Ibid.*, 75.

4. Lewis, Brian, *op. cit.*, 75, 98.

5. Clark, E., *op. cit.*, 78.

6. Tomlinson, *op. cit.*, 502.

7. Jeaffreson, II, 173.

8. *Scientific American*,Vol.VI, Issue 16, 4 January 1851.

9. ICE, Swinburne, *Diary*, 11 November 1850. Swinburne did die as a cholera victim in December 1856.

10. *Ibid.*, 21 December.

11. *Ibid.*, 1–7 January 1851.

12. *Ibid.*, 16 January.

13. ICE: 1857 PALCSI; Stephenson.

14. Jeaffreson, II, 173.

15. Hallberg, *op. cit.*, 106.

16. *TNS*, 75.2, Help, *op. cit.*, 322.

17. IMechE: IMS189/2, letter of 9 February 1851.

18. Cranfield, *op. cit.*, 5.

19. Jeaffreson, II, 166.

20. ICE: 1857 PALCSI; Stephenson.

21. Jeaffreson, II, 168.

22. *Ibid.*, 159.

23. ScM: SMK: GPB 5/12/7d.

24. Jeaffreson, II, 160.

25. The *Burlington Magazine*,Vol. 137, February 1995, 108–118, in an article on Brunel's ambitious 'Shakespeare Room' at 17, Duke Street, also notes Stephenson and Samuel Morton Peto as engineers who bought paintings, though refers only to Jeaffreson's list for Stephenson's collection.

26. Ruskin, *The Poetry of Architecture*, 3.

27. Jeaffreson, II, 161.

28. *TNS*, 27, 1949–50, Haut, 'The Centenary of the Semmering Railway and its Locomotives', 19ff.
29. NA: RAIL772/82.
30. Letter to Mrs Bidder, 30 July 1851. ScM, GPB 5/1/2.
31. Clark, E., *op. cit.*, 254.
32. *ILN*, 30 August 1851.
33. *New York Times*, 3 October 1851. See also 'Eyewitness' letter to *The Times*, 2 September 1851.
34. Most recently in Julia Elton, 'Robert Stephenson in Society' in Bailey, *Robert Stephenson*, 236.
35. Scott Russell, *The Modern System of Naval Architecture*, I, 613, quoted in Emmerson, *John Scott Russell*, 71.
36. Bailey, *op. cit.*, 161.
37. NRM, 2006–7495.
38. Sutherland, 'Iron Railway Bridges', in Bailey, *Robert Stephenson*, 333.
39. Quoted by Julia Elton in Bailey, *Robert Stephenson*, 231. Murphy was Francis Stack Murphy, serjeant-at-law, and Liberal MP for Cork.
40. Jeaffreson, II, 169.
41. *Ibid.*, 71.
42. *Ibid.*, 170.
43. Webster, *op. cit.*, 194.
44. Hansard, 3rd Series, 120, quoted in Bailey, *Robert Stephenson*, 363.
45. *NJ*, 2 July 1852.
46. Clarke, J.F., *Building Ships on the North-East Coast*, 121; Clarke's italics.
47. *WG*, 30 April 1859.
48. See also TWAS: DF. CMP/1; Dougan, *The History of North-East Shipbuilding*, 41; Finch, *Coals from Newcastle*, 166–167. *ILN*, 17 July 1852. John Bowes himself was by 1854 the owner of Killingworth colliery, with Nicholas Wood as agent.
49. Trevelyan, *op. cit.*, 647.
50. *Yorkshire Gazette*, 10 July 1852.
51. Jeaffreson, II, 145.
52. ROB: Sopwith, *Diary*, 25 July 1858.
53. Letter to Mrs Bidder, 31 August 1852, ScM, GPB 5/2/1.
54. ScM, GPB 5/12/5a.

## 20. Head of the Profession

1. Lilly: ILL, 3 March 1853.
2. Pease, *Diaries*, 304.
3. Jeaffreson, II, 181.
4. *Ibid.*
5. *Ibid.*, 185.
6. *Ibid.*
7. *Ibid.*
8. McCord Museum, *The Victoria Bridge*, 14.
9. W.H. Withrow, quoted in *Toronto Globe & Mail*, 6 February 2006.
10. Jeaffreson, II, 229.
11. *Ibid.*, 172.
12. In normal cruising service, the yacht had a crew of sixteen, with 'a good cook, and a first-rate cellar ... and a capital library' (letter from a guest, quoted in Jeaffreson, II, 241ff). Corke skippered both *Titanias*.
13. Elton, *op. cit.*, in Bailey, *Robert Stephenson*, 217.
14. Pease, *Diaries*, 321f.
15. Jeaffreson, II, 186.
16. *The Times*, 11 April 1854.
17. ScM, GPB 5/2/4. Letter to Mrs Bidder, 29 June 1854.
18. Clark, E., *op. cit.*, 208.
19. Lilly: ILL, Letter to R.S. Illingworth from Edinburgh, 18 October 1854.
20. In a letter to Mme Delamalle, of 16 December 1854, de Lesseps quotes Saïd as saying: 'I don't know how this Mr Murray has the *cheek* [*le toupet*] to ask to see me' (Lesseps, *Souvenirs*, II, 71).

21. Karabell, *Parting the Desert*, 8.
22. NA: FO78/1156.
23. Lilly, MANBY, Letter to Frederick Ayrton, 2 August 1857.
24. Lesseps, *Souvenirs*, II, 663; letter to B. St-Hilaire, 18 August 1858.
25. StA: Forbes: Letter ref. 127 to Professor Forbes from Great George Street, dated 15 December 1854.
26. StA: Forbes: Letter ref. 35, from Gloucester Square, 27 February 1855.
27. Jeaffreson, II, 154.
28. Lit & Phil Annual Report, 1854–55.
29. Clark, E., *op. cit.*, 377f.
30. *Ibid.*
31. *The Times*, 16 November 1855.
32. PRS, Vol. 10, 1859–60, xxix–xxxiv.
33. NRM, 2006–7496; also mentioned in a letter to Brunel, BRIS: DM1306/48, 26 October 1855: 'I shall be leaving for my Winter's cruise about the end of next week.'
34. StA: Forbes: ref. 139, letter of 5 November 1855 to Professor Forbes.
35. Correspondents in *Herapath's Railway Magazine* as early as 1842 were urging the establishment of a civil engineering college while others wrote just as firmly to oppose them.
36. Armytage, *A Social History of Engineering*, 159; Watson, *The Civils*, 156.
37. *Report on the Paris Universal Exhibition*, 132. Some others agreed. A letter quoted in Pole, *op. cit.*, 373, from Lord Ashburton to Fairbairn deplores the ignorance of the masses through the lack of technical education, 'in which masses I include our peers, gentry, tradesmen and mechanics, as well as our manufacturers and operatives'.
38. BRIS: DM/1306/51; letter from Marseilles, 27 November 1856. The Electric Telegraph Company was not involved in the laying of the first Atlantic telegraph cable (1857), which did develop faults.
39. Clark, E., *op. cit.*, 275.
40. Jeaffreson, II, 230.
41. *Ibid.*, 231.
42. Quoted in Bailey, *Robert Stephenson*, 292.
43. Clark, E., *op. cit.*, 379.
44. Cranfield, *op. cit.*, 7.
45. Smiles, *Autobiography*, 217.
46. WG, 30 April 1859.
47. ScM: GPB, 5/12/8. Letter to Bertha Bidder, 6 November 1856.
48. ROB, Sopwith, *Diary*, 21 November 1856.
49. *Ibid.*, 21 December 1856.
50. Jeaffreson, II, 247.
51. Quoted in Clark, E., *op. cit.*, 95.
52. *Weather*, Vol 50, No. 2, February 1995; Pedgely, 'Pen Portrait of Robert Stephenson', remarks *en passant* that Fanny Stephenson was 'an accomplished portrait painter' but I have found no evidence for this.
53. Hansard, Series III, Vol. 146, 1706.
54. Lilly: MANBY, letter of 27 July 1857.
55. Lilly: MANBY, letter of 28 July 1857.
56. Lilly: MANBY, letter of 2 August 1857.
57. ICE: 1857PALC SI.
58. Addyman & Haworth (*op. cit.*, 150) note that 'Apparently Robert Stephenson had started a long-term affair' with Miss Henrietta Smyth, who married Baden-Powell in 1846, 'and it is very probable that the child was his'. But they themselves point out that there is no evidence other than mere gossip for this. It should be treated with extreme scepticism. Henrietta Powell bore ten children between 1847 and 1860 (ODNB, entry on Baden-Powell).
59. Bigge was a Northumbrian landowner, of Linden Hall, Morpeth. On 26 February 1824 his father, Charles William Bigge, was elected the first and life-long president of the Newcastle Mechanics' Institute, at the same time as George Stephenson became its chairman. The Bigges were a prominent Whig/Liberal family in Northumberland.
60. Jeaffreson, II, 239.

61. ROB: Sopwith, *Diary*, 13 August 1857.
62. Jeaffreson, II, 243. Kell was the first town clerk of Gateshead. He refers to Robert as 'My old friend and schoolfellow' (*Ibid.*, 242).
63. Quoted in Rolt, *op. cit.*, 330.
64. *Sunderland Herald*, 20 November 1857.
65. ROB: Sopwith, *Diary*, 16 December 1857.
66. Smiles, 367f.
67. BRIS: DM1306/11/20/43/6; letter of 8 January 1858.

## 21. The Last Journeys

1. Jeaffreson, II, 244.
2. BRIS: DM1306/56, letter of 17 January 1858.
3. BRIS: DM1306/11/20/43/9, letter of 1 February 1858.
4. ROB: Sopwith, *Diary*, 11 March 1858.
5. *Ibid.*, 11 April 1858.
6. Hansard, 3rd Series, Vol. 150, 1371.
7. *Ibid.*, 1382.
8. Jeaffreson, II, 155.
9. *Ibid.*
10. ICE, 1857PALCSI.
11. De Lesseps later wrote: 'I myself saw his carriage tracks, which did not get beyond a league from Suez. He did not make the most essential visit, to the Bitter Lakes, from Timsah to Pelusium, and the Mediterranean coast, since there were the only difficulties, at least the only ones which malevolence and ignorance indulged themselves by exaggerating' (Lesseps, *Souvenirs*, II, 663).
12. Conrad, 'The Isthmus of Suez Canal', in *The Engineer*, 8 October 1858, 278.
13. ROB: Sopwith, *Diary*, 25 August 1858.
14. Elton, *op. cit.*, in Bailey, *Robert Stephenson*, 217.
15. TWAS: DT/SC/319.
16. ROB: Sopwith, *Diary*, 10 and 11 September 1858.
17. *Ibid.*, 17 September 1858.
18. Quoted in Jeaffreson, II, 233f; author's italics.
19. Jeaffreson, II, 144.
20. Quoted in Jeaffreson, II, 245.
21. *Ibid.*
22. Quoted in Bailey, *Robert Stephenson*, 152.
23. Quoted in Clark, E., 329.
24. BRIS: English Notebook of Henry Brunel.
25. BRIS: French Notebook of Henry Brunel.
26. *PRGS*, Vol. 3, No 4, 11 April 1859. Pim, 188.
27. *Ibid.*, 203.
28. *Ibid.*, 213.
29. Karabell, *op. cit.*, 59.
30. Lilly: MANBY, draft of letter to de Lesseps, 28 July 1857.
31. A.R.E; also ScM, BIDD 27/1–3: The accounts for Forth Street show that in 1859 the Egyptian contract was worth £102,983. Coasts were £64,664, and the gross profit was £34,439.
32. Lilly: MANBY, letter to Ayrton of 2 August 1857.
33. *JTH*, Vol. 9, No 1, Spring 1978. Bradshaw, D.F., 'A Decade of British Opposition to the Suez Canal Project, 1854–64'. Then as now, misconceptions and misreadings of the future abounded; *The Times* on 27 November 1858 pronounced that: '… probably in all time to come, five-sixths of the heavy goods known to commerce will be transported in sailing ships.'
34. *WG*, 30 April 1859.
35. Lilly: ILL, letter to Illingworth, 1 May 1859.
36. *WG*, 30 April 1859.
37. *Ibid.*
38. Clark, E., *op. cit.*, 97.

39. TWAS: DT/SC/319.
40. *The Times*, 26 October 1859, letter from Paxton.
41. Jeaffreson and Rolt mistakenly say the public dinner was to celebrate the opening of the railway, which had been accomplished five years previously. See Clark, E., *op. cit.*, 383.
42. In its obituary of Brunel (19 September 1859), *The Times* noted: 'It is a remarkable circumstance that in the early part of his career he was brought into frequent conflict with Robert Stephenson, as Stephenson was with him, and that, nevertheless, their mutual regard and respect were never impaired. Brunel was ever ready to give his advice and assistance whenever Stephenson desired it, and the public will recollect how earnestly and cordially during the launch of the *Great Eastern* Stephenson gave his assistance and lent the weight of his authority to his now deceased friend. Such rivalry and such unbroken friendship as theirs are rare, and are honourable to both.'
43. Lilly: MANBY, letter to Charles Manby.
44. Jeaffreson, II, 261.
45. Clark, E., *op. cit.*, 97.
46. *Ibid.*, 384.
47. Jeaffreson, II, 263.
48. *Ibid.*, 262.
49. ScM, GPB 5/12/3.
50. Carr Glyn and Chandos were chairman and ex-chairman of the LNWR; Beale was chairman of the Midland; Chapman was a railway engineer who had been one of Stephenson's pupils. The *Illustrated London News* account (29 October 1859) gives George Rennie instead of Chapman.
51. *The Times*, 26 October 1859.
52. Webster, *op. cit.*, 196. Locke himself died only ten months later, aged 55.
53. W.M. Thackeray in 1860, quoted in Altick, *Victorian People and Ideas*, 75.
54. Smiles, One, 477ff.
55. Wordsworth, *Lines to Haydon* (1815).

# BIBLIOGRAPHY

1. Original Documents

Boulton & Watt Archives, Birmingham (B&W)
Brunel, Henry, English and French Notebooks, BRIS
Sopwith, Thomas, *Diaries*, ROB
Stephenson, George, Presidential Address to the Institution of Mechanical Engineers (1847): IMechE
Stephenson, George, Last Will and Testament (1839): IMechE
Stephenson, George, Last Will and Testament (1848): NA
Swinburne, Henry, Diary of a Cruise on the Nile, 1850–51: ICE


2. Pamphlets

Booth, Henry, *An Account of the Liverpool & Manchester Railway*, 2nd edn, Liverpool, 1831
*Mr Telford's Report to the Commissioner for the Loan of Exchequer Bills*, published by the Liverpool & Manchester Railway Co., with their responses. Liverpool, 1829
Stephenson, George, *A Description of the safety lamp, invented by George Stephenson, and now in use in the Killingworth Colliery*, London & Edinburgh, 1817
Wood, Nicholas, *Address on the late Eminent Engineers ...* Literary & Philosophical Institution, Newcastle, 1860


3. Journals, Magazines and Newspapers

*Archaeologia Aeliana*
*Burlington Magazine*
*Edinburgh Review*
*Gateshead Observer*
*Hansard*
*Herapath's Railway Magazine, Commercial Journal, and Scientific Review (HRM)*
*Illustrated London News (ILN)*
*Journal of Transport History (JTH)*
*New York Times*
*Newcastle Courant (NC)*
*Newcastle Journal (NJ)*
*North & South Shields Gazette*
*Philosophical Magazine and Journal (PM)*
*Proceedings of the Royal Geographical Society (PRGS)*
*Proceedings of the Royal Society (PRS)*
*Railway Times (RT)*

*Scientific American*
*Sea Breezes*
*Stephenson Locomotive Society Journal (SLS)*
*Sunderland Herald*
*Technology and Culture (TC)*
*The Times*
*Toronto Globe & Mail*
*Transactions of the Newcomen Society (TNS)*
*Weather*
*Whitby Gazette (WG)*
*York Courant*
*Yorkshire Gazette (YG)*


## 4. Books on the Stephensons

Addyman, John, and Haworth, Victoria, *Robert Stephenson: Railway Engineer*, Newcastle, 2005
Bailey, Michael R. (ed.), *Robert Stephenson: The Eminent Engineer*, Aldershot, 2003
Davies, Hunter, *George Stephenson: Father of Railways*, London, 1975
Jeaffreson, J.C., *The Life of Robert Stephenson, FRS*, 2 vols, London, 1864
Parsonage, W.R., *A Short Biography of George Stephenson*, London, 1937
Rolt, L.T.C., *George and Robert Stephenson: The Railway Revolution*, London, 1962
Roper, R.S., *The Other Stephensons*, Rochdale, 1990
Skeat, W.O., *George Stephenson: The Engineer and his Letters*, London, 1973
Smiles, Samuel, *The Story of the Life of George Stephenson*, London, 1857 (new edn, 1873)
Summerside, Thomas, *Anecdotes, Reminiscences & Conversations of and with the late George Stephenson, Father of Railways*, London, 1878


## 5. Other Books

*Alphabetical List of Patentees of Inventions*, London, 1854, reprinted New York, 1969
A.R.E., *Egyptian Railways Museum*, Cairo, no date
Allan, T. & G., *Allan's Tyneside Songs*, Newcastle, 1891
Altick, Richard D., *Victorian People and Ideas*, London, 1973
Armytage, W.H.G., *A Social History of Engineering*, London, 1961
Austin, Alfred, *Autobiography*, London, 1908
Bailey, Michael R. (ed.), *Early Railways 3*, Sudbury, 2006
Bathurst, Bella, *The Lighthouse Stevensons*, London, 1999
Baughan, Peter, *The Chester & Holyhead Railway*, Vol. I, Newton Abbot, 1972
Biddle, Gordon, *The Railway Surveyors*, London, 1990
Booth, Henry, *Henry Booth: Inventor*, Ilfracombe, 1980
Buchanan, R. Angus, *Brunel*, London, 2002
Burstall, A.F., *A History of Mechanical Engineering*, London, 1963
Burton, Anthony, *Richard Trevithick, Giant of Steam*, London, 2000
Cameron, S., *Kidney Disease: The Facts*, Oxford, 1981
Carlson, Robert F., *The Liverpool and Manchester Railway Project, 1821–1831*, Newton Abbot, 1980
*Catalogue Officiel, Exposition Universelle*, Paris, 1855
Clark, E.F., *George Parker Bidder: The Calculating Boy*, Bedford, 1983
Clark, Edwin, *The Britannia and Conway Tubular Bridges*, 3 vols, London, 1850
Clarke, J.F., *Building Ships on the North-East Coast, Part I, 1640–1894*, Whitley Bay, 1997
Cole, G.D.H. and Postgate, Raymond, *The Common People, 1746–1946*, 4th edn, London, 1949
Colquhoun, Kate, *A Thing in Disguise: The Visionary Life of Joseph Paxton*, London, 2003
Conder, F.R., *Personal Recollections of English Engineers*, reprinted as *The Men Who Built Railways*, ed. Jack Simmons, London, 1983
Cranfield, John, *The Railways of Norway*, London, 2000
Dambly, Phil, *Vapeur en Belgique*, Tome I, Bruxelles, 1989
Day, James, *A Practical Treatise on the Construction and Formation of Railways*, 2nd edn, London, 1839
Dempsey, G.D., *Tubular and Other Iron Bridges*, London, 1864, reprinted Bath, 1970

Dendy Marshall, C.F., *A Centenary History of the Liverpool & Manchester Railway*, London, 1930
———, *A History of British Railways Down to the Year 1830*, London, 1938
———, *A History of Railway Locomotives Down to the Year 1831*, London, 1953
Devey, Joseph, *The Life of Joseph Locke*, London, 1862
Dickinson, H.W. and Titley, A., *Richard Trevithick: The Engineer and the Man*, Cambridge, 1934
Dillon, M., *Some Account of the Works of Palmer's Shipbuilding & Iron Co. Ltd*, Newcastle, 1904
Dougan, David, *The History of North-East Shipbuilding*, London, 1968
Duncan, W. (ed.), *The Stephenson Centenary*, Newcastle, 1881
Durie, A.J., *The Scottish Linen Industry in the Eighteenth Century*, Edinburgh, 1979
'E.M.S.P.', *The Two James' and the Two Stephensons*, London, 1861
Ellis, C. Hamilton, *British Railway History, 1830–1871*, London, 1954
Emerson, Ralph Waldo, *Journals*, Vol. X, ed. Merton M. Sealts Jr, Cambridge, Mass., 1973
Emmerson, George S., *John Scott Russell*, London, 1971
*Encyclopaedia Britannica*, 8th edition, London, 1856
Ernouf, Baron, *Paulin Talabot, sa Vie et son Oeuvre*, Paris, 1886
Fairbairn, William, *The Britannia and Conway Tubular Bridges*, London, 1849
Falk, Bernard, *The Bridgwater Millions*, London, 1942
Farnie, D.A., *East and West of Suez: The Suez Canal in History*, Oxford, 1969
Finch, Roger, *Coals from Newcastle*, Lavenham, 1975
Francis, John, *A History of the English Railway*, London, 1851, reprinted Newton Abbot, 1967
Gloag, John, *Victorian Taste*, London, 1962
Gooch, Daniel, *Memoirs and Diary*, ed. R. Burdett-Wilson, Newton Abbot, 1972
Grinling, Charles H., *History of the Great Northern Railway*, London, 1898
Grundy, Francis, *Pictures of the Past*, London, 1879
Guy, Andy, and Rees, Jim (eds), *Early Railways*, London, 2001
Halévy, E., *A History of the English People in 1815*, Harmondsworth, 1937
Hallberg, C.W., *The Suez Canal: Its History and Diplomatic Importance*, New York, 1931
Hammond, J.L. and Barbara, *The Bleak Age*, new edn, Harmondsworth, 1947
Hartley, Harold, *Humphry Davy*, London, 1966
Head, F.B., *Stokers and Pokers*, 2nd edn, London, 1850
Holmes, J.H.H., *Coal Mines of Durham and Northumberland*, London, 1816
*Hommes et Choses du PLM*, Paris, 1911
Karabell, Z., *Parting the Desert: The Creation of the Suez Canal*, London, 2003
Kemble, Frances, *Memoirs of a Young Girl*, 3 vols, London, 1878
Kirby, M.W., *The Origins of Railway Enterprise: The Stockton & Darlington Railway, 1821–1863*,
     Cambridge, 1993
Kostal, R.W., *Law and English Railway Capitalism*, Oxford, 1994
Lamalle, Ulysse, *Histoire des Chemins de Fer Belges*, Bruxelles, 1943
Lambert, Richard S., *The Railway King*, London, 1934
Lecount, Peter, *A Practical Treatise on Railways*, London, 1839
Lesseps, Ferdinand de, *Souvenirs*, 2 vols, Paris, 1887 (translated into English as *Recollections of Forty
     Years*, London, 1887)
———, *Percement de l'Isthme de Suez: Rapport et Projet de la Commission Internationale*, Paris, 1856
Lewin, Henry Grote, *The Railway Mania and its Aftermath, 1845–52*, 1936, new edn, Newton Abbot,
     1968
Lewis, Brian, *The Cabry Family: Railway Engineers*, Mold, 1994
Lewis, M.J.T. (ed.), *Early Railways 2*, London, 2003
Lewis, P.R., *Disaster on the Dee*, Stroud, 2007
Lockhart, J.G., *Life of Sir Walter Scott*, 7 vols, London, 1837–38
Low, J.G., *Industry in Montrose*, Brechin, 1943
MacCarthy, Fiona, *All Things Bright and Beautiful: Design in Britain, 1830 to Today*, London, 1972
McCord Museum, *The Victoria Bridge*, Montreal, 2002
McGowan, Christopher, *The Rainhill Trials*, London, 2004
Macnair, Miles, *William James: The Man Who Discovered George Stephenson*, Oxford, 2008
Markham, Violet, *Paxton and the Bachelor Duke*, London, 1935
Marshall, W.P., *Description of the Patent Steam Locomotive of Messrs R. Stephenson & Co.*, London,
     1838

Massingham, H.J. and H., *The Great Victorians*, Vol. 2, London, 1938

Mitchell, David, *The History of Montrose*, Montrose, 1866

Mountford, Colin E., *The Private Railways of County Durham*, Melton Mowbray, 2004

Owen, Colin, *The Leicestershire and North Derbyshire Coalfield, 1200–1900*, Ashbourne, 1984

Parry, Albert, *Whistler's Father*, Indianapolis, 1939

Parsons, R.H., *A History of the Institution of Mechanical Engineers 1847–1947*, London, 1947

Payen, Jacques, *La Machine Locomotive en France*, Lyon, 1988

Pease, Sir Alfred E., *The Diaries of Edward Pease*, London, 1907

Pendleton, J., *Our Railways*, 2 vols, London, 1894

Pole, William, *The Life of Sir William Fairbairn, Bart.*, London, 1877, reprinted Newton Abbot, 1970

Popplewell, R., *A Gazetteer of the Railway Contractors and Engineers of East Anglia*, Southbourne, 1984

Pugin, A.W., *Contrasts Between the Architecture of the Fifteenth and the Nineteenth Centuries*, 2nd edn, London, 1842

Raine, Rev. James, *A Memoir of the Rev John Hodgson*, 2 vols, London, 1857

Rapley, John, *The Britannia and Other Tubular Bridges*, Stroud, 2003

*Report on the Paris Universal Exhibition*, London, 1856

Richardson, Benjamin W., *Thomas Sopwith*, London, 1891

Ruskin, John, *Works*, ed. E.T. Cook and A. Wedderburn, London, 1903–12

Sambre & Meuse Railway, *Grant from the Belgian Government, Report of Mr Stephenson, and General Statement*, London, 1845

Seguin, Marc, *Des Chemins de Fer*, Paris, 1839

Sidney, Samuel, *Gauge Evidence: The History and Prospects of the Railway System*, London, 1834

Sigmond, G.G., *Mercury, Blue Pill and Calomel: Their Uses and Abuses*, London, 1840

Simmons, Jack (ed.), *Rail 150*, London, 1975

——, *The Railway in England and Wales*, Vol. I, Leicester, 1978

Smiles, Samuel, *The Autobiography of Samuel Smiles, Ll.D.*, London, 1905

Smith, Denis (ed.), *Perceptions of Great Engineers: Fact and Fantasy*, Liverpool, 1994

Snell, S., *A Story of Railway Pioneers: Isaac Dodds and His Son*, London, 1921

Stephenson, Roberts, *The Great Exhibition: Its Palace and Its Principal Contents*, London, 1851

Thomas, R.H.G., *The Liverpool & Manchester Railway*, London, 1980

Thompson, Roy, *Thunder Underground: Northumberland Mine Disasters, 1815–65*, Ashbourne, 2004

Thomson, F.M.L. (ed.), *Horses in European Economic History: A Preliminary Canter*, Reading, 1983

Thomson, Len, *History of Tapton House*, Cranleigh, 2000

Tomlinson, W.W., *The North Eastern Railway*, new edn, Newton Abbot, 1967

Trevelyan, G.M., *History of England*, London, 1926

Trevithick, Francis, *The Life of Richard Trevithick*, 2 vols, London, 1872

*Tyneside Songs and Droleries, Readings and Temperance Songs*, reprinted Norwood, Pa., 1973

Vignoles, K.H., *Charles Blacker Vignoles: Romantic Engineer*, Cambridge, 1982

Walker, J.S., *An Accurate Description of the Liverpool & Manchester Rail-Way*, Liverpool, 3rd edn, 1831

Walmsley, H.M., *The Life of Sir Joshua Walmsley*, London, 1872

Warren, J.G.H., *A Century of Locomotive Building by Robert Stephenson & Co.*, Newcastle, 1923

Watson, Garth, *The Civils*, London, 1988

——, *The Smeatonians*, London, 1989

Webster, N.W., *Joseph Locke: Railway Revolutionary*, London, 1970

——, *Britain's First Trunk Line*, Bath, 1972

White, Andrew, *A History of Whitby*, Chichester, 1993

Whittle, G., *The Railways of Consett and North-East Durham*, Newton Abbot, 1971

Whitworth, Alan, *Long Live the King! George Hudson and Whitby*, Whitby, 2002

Wilson, Arnold T., *The Suez Canal*, Oxford, 1933

Wood, Christopher, *Victorian Painters*, Vol. I, Woodbridge, 1995

Wood, Nicholas, *Treatise on Railways*, 1st edn, London, 1825; 3rd edn, London, 1838

Young, Robert, *Timothy Hackworth and the Locomotive*, 1923, reprinted Shildon, 1975

# INDEX

Gooch, Daniel 184ff

Gooch, Thomas Longridge 103, 114, 119, 133, 151f, 160, 213, 231, 276, 279

'Grand Allies' 27, 33, 46, 64

Grand Junction Railway 125, 130f, 132, 134ff, 137, 156, 202

Grand Trunk Railway of Canada 262

Gray, Rev. Robert 38

Gray, Robert 17, 21, 227

Gray, Thomas 74

Great Exhibition 238f, 248, 252

Great Indian Peninsula Railway 248, 254

Great North of England Railway 160, 165ff, 183, 186, 229

Great Northern Railway 182, 190, 202

Great Western Railway 142, 150f, 160, 184ff

Great Yarmouth 173

Gregory, Ellen, *see* Stephenson, Ellen

Greville, Charles 230

Grey, Earl 181

Grundy, Francis 9

Gurney, Goldsworthy 102, 109

Hackworth, Timothy 36, 74, 85, 100, 102, 108, 111ff, 122, 158, 235, 292

Halévy, E. 13

Hall, Bruce Napier 93ff

Hardcastle, William 215, 301

Hardwick, Philip 150

Harrison, Thomas E. 123, 128, 227, 239f, 284

Hart, John 9, 168, 172, 175, 177, 291, 297

Hartley, Jesse 86f

Hawthorn, Robert 16, 20–1, 60, 63

Headlam, Dr 68

Hedley, William 36, 51

Henderson, Frances, *see* Stephenson, Fanny

Herring, Graham & Powles 68, 81, 83, 93

Hetton Colliery Railway 53, 61, 79

High Level Bridge 178, 184, 195, 199, 227f, 234, 237, 239f

High Street House 14

Hindmarsh, Elizabeth, *see* Stephenson, Betty

Hindmarsh, Thomas 54

Hodgkinson, Eaton 198f, 203f, 207, 211, 214, 223, 231, 237, 266

Hodgson, Rev. John 38, 41, 48, 51, 290

Holmes, J.H.H. 13, 40f

House of Commons 35, 149, 188, 215, 219, 255, 258, 275

Howe, William 200, 299

Howick, Lord 181ff, 202

Hudson, George 147, 160, 162, 163f, 166ff, 170, 175, 177, 181, 182f, 186, 189, 190f, 192, 193ff, 196, 199, 202, 207, 209, 215f, 219, 224, 227, 229ff, 234, 240, 252, 254, 259, 263, 277–8, 286, 297, 299, 302

Hulton, William 92

Huskisson, William 118ff, 227

Hutchinson, William 105, 170f, 199, 263

Hutton, Charles 42, 60

Illingworth, R.S. 82, 84, 96, 222, 261, 293

India 248

Institution of Civil Engineers 99, 121, 142, 177, 208f, 228, 240, 245–6, 248, 255, 262, 267, 270, 273, 276, 284, 286

Institution of Mechanical Engineers 208, 224, 240f, 255, 264, 267, 277

Invicta 115

Ireland 64, 128, 179, 191, 207

Italy 158, 169, 229

James, William 54f, 57–8, 62–3, 65, 70ff, 74, 78, 85, 103, 174, 189f, 242, 291, 299

James, William Henry 190

Jameson, Prof. Robert 59

Jarrow 38, 257

Jeaffreson, J. C. 9

Jessop, Josias 86

*John Bowes* 257f, 267, 304

Keefer, Thomas C. 260

Kell, William 272, 306

Kemble, Fanny 116

Kennedy, James 136

Kennedy, John 70, 110

Kenton & Coxlodge 35

Killingworth 21, 25, 32, 34, 36, 41, 47, 60, 70, 74, 101, 132, 156, 249f, 272, 283

Kilmarnock & Troon Railway 51, 89

Kilsby Tunnel 140ff, 148–9, 153–4

Lake District 234

Lambert, Richard 167f, 178

Lancashire Witch 102, 105, 113

Lancaster 165, 227, 251

Lange, Daniel 280

Lardner, Dionysius 129, 138

Lawrence, Charles 71, 125, 129

Lean, Charles 153

Lecount, Peter 137, 151

Lee, Frederick 249, 269

Lee, Hedworth 222, 227

Leeds 84, 174, 278

Leeds & Hull Railway 73

Leeman, George 240

Leicester 134f, 170

Leicester & Swannington Railway 105, 124, 128f, 169, 214

Leopold, King 191

Leopold Railway 196, 215, 229

Leslie, Professor John 59

Lesseps, Ferdinand de 265, 268, 270, 271f, 276f, 280ff, 306

Liddell, Hon. H.T. 180, 239

Liddell, Sir Thomas 27, 34, 36, 62

'Lit & Phil', Newcastle 39f, 43, 50, 125, 266, 284

Liverpool 43, 55, 68f, 96, 108, 117, 156, 180, 220

Liverpool & Birmingham Railway 71ff, 92, 127

Liverpool & Manchester Railway 54, 58, 64, 68, 70ff, 75ff, 85f, 99ff, 103, 115, 118ff, 121, 123, 125–6, 129, 132, 180, 195

'Liverpool Party' 123, 136–7, 147, 165, 179, 190f

Locke, Joseph 18, 61, 73, 78, 84, 85f, 88, 103, 106, 110, 119, 121, 128, 130, 132, 135, 137, 149, 157–8, 178–9, 183, 190, 202, 211, 213, 217, 231, 251, 254–6, 259, 267, 269, 286

Locke, William 13, 18, 60f

*Locomotion* 79f